# Relativistic Plasmas

**Relativistic Plasmas: Acceleration and Radiation** gives a consistent description of relativistic plasmas and focuses on recent advances in plasma accelerators, radiation sources and intense laser-plasma interactions. The book is a useful reference on plasma physics for postgraduate students and researchers in many different areas of study, including astrophysical plasmas, those working on the new areas of plasma accelerators and intense laser plasma phenomena, or new models to understand intense radiation processes.

**Key Features:**

- Discusses relativistic plasmas, plasma accelerators, radiation sources and laser-plasma interactions.

**José Tito Mendonça** is Professor Emeritus of the Instituto Superior Técnico (IST), University of Lisbon, and Docteur-ès-Sciences of the University of Paris-Orsay. He is a former Head of the Physics Department of IST. He launched the Euratom-IST Association, and worked at the Joint European Torus (JET) as scientific advisor. He created the Group of Lasers and Plasmas, and organised research groups in nuclear fusion, laser-plasma interactions and ultra-cold matter. He published five books and a few hundred papers. He made early contributions to laser-accelerators, plasma turbulence and spacetime optics.

# Relativistic Plasmas
## Acceleration and Radiation

José Tito Mendonça

CRC Press
Taylor & Francis Group
Boca Raton London New York

CRC Press is an imprint of the
Taylor & Francis Group, an **informa** business

Designed cover image: CRC Press

First edition published 2027
by CRC Press
2385 NW Executive Center Drive, Suite 320, Boca Raton FL 33431

and by CRC Press
4 Park Square, Milton Park, Abingdon, Oxon, OX14 4RN

*CRC Press is an imprint of Taylor & Francis Group, LLC*



ISBN: 978-1-032-84367-4 (hbk)
ISBN: 978-1-032-84735-1 (pbk)
ISBN: 978-1-003-51470-1 (ebk)

DOI: 10.1201/9781003514701

Typeset in CMR10 font
by KnowledgeWorks Global Ltd.

*Publisher's note:* This book has been prepared from camera-ready copy provided by the authors.

*To my family*
*and*
*To the bright side of human kind*

# Contents

# Preface

Long ago, I attended a plasma physics conference in Goa, India. My hotel, on the north bank of the Mandovi River, was a pleasant place surrounded by tropical forests, busy with Western tourists. It was a small meeting, only a few dozen people, but I remember enjoying the company of famous colleagues, some of whom were pioneers of relativistic plasma physics, the subject of the present book. One day, as I walked to the conference room, I was approached by a German tourist, who asked me if I was at the Plasma conference. After my positive response, she then asked, "Would you mind seeing my husband? He is not feeling well, and needs medical attention". She was, of course, disappointed when I told her that I was a physicist, not a physician, and that the plasma conference had nothing to do with blood or with medicine. Plasma is, indeed, a puzzling word, with erroneous medical connotations. That probably attracted me as a postgraduate student, when I had to choose between quantum optics and plasmas for my main area of research.

This book is a result of many years of work in areas related to plasma electrodynamics, with special emphasis on acceleration processes and radiation sources. Although many excellent books exist on plasma physics, I noticed that some new problems emerged in recent years, while some old topics still need a more complete description, in particular the aspects related with laser-plasma interactions in the relativistic intensity domain.

There is a new generation of students starting their work in the areas of laser fusion, astrophysics and novel accelerators, who could benefit from a more specialized introduction. Furthermore, postdocs and senior scientists still need some reference book to help their daily work, even if they have a good knowledge of this area of science. I therefore wrote this book hoping that it would be useful to a large audience. But, ultimately, I wrote this book in order to provide a global view of the subject, where basic theoretical methods, examples of application, and physical insight can be found.

This work gives a unified view of results and methods coming from many different sources, including books, papers, reports and personal notes, which are identified as much as possible. I would like to thank the involuntary contributions from collaborators and colleagues, including Robert Bingham, Gert Brodin, Kyriakos Hizanidis, António Serbeto, Hugo Terças, Jorge Vieira, and many others. I am particularly grateful to Bengt Eliasson for reading a part of the manuscript and for suggestions.

Finally, I would like to thank the editorial staff of CRC Press and, in particular, Carolina Antunes who established the first contacts and Danny Kielty who helped finalize this project.

Lisboa, October 2025

# 1

# *Introduction*

## 1.1 Historical Background

More than two thousand years ago, the Ancient Greeks used to identify four different constituents of matter, the *elements*: Earth, Water, Air and Fire. In the sophisticated language of modern-day science, we can similarly identify four different *states of matter*: Solid, Liquid, Gas and Plasma. By providing energy from an external source, or *heating* a piece of solid matter, we can fuse it, producing a phase transition into the liquid state. Similarly, by further increasing the temperature, we can produce a vapour and, with further heating, a plasma.

Apart from dark energy and dark matter, which are unknown, plasma is the most abundant state in the universe. The concept of a plasma as a physical medium was introduced by Langmuir in 1927. As told by one of his collaborators much latter, the word derives from a close resemblance of electrons in a gaseous discharge with the "way blood plasma carries around red and white corpuscules" [1].

Along the years, several excellent books have been published on the principles of plasma theory, such as [2, 3, 4, 5, 6], and many others. So, one wonders why a new book can be justified. My main motivation is that previous texts tend to ignore or minimize relativistic processes. Furthermore, the modern wave-kinetic theory, which has multiple applications in anomalous transport, plasma turbulence and quantum plasmas, is usually forgotten in general presentations of the field. These areas became relevant due to the recent advances on plasma accelerators, intense laser systems and laser fusion.

In this book, considerable emphasis will be given to new aspects of relativistic plasmas, and the associated processes leading to radiation, particle acceleration and scattering, along with the basic presentation of non-relativistic models. Finally, recent advances on plasma wave topology, mostly related with intense laser-plasma interactions, which include twisted plasmons and structured wakefields, are included. While covering these topics, I will try to emphasize the unity of the plasma theoretical approach, and its originality when compared with other branches of physics. Other modern topics, such as Rydberg plasmas (also known as ultracold neutral plasmas), where the view of a plasma as a very hot medium is challenged, have been discussed elsewhere (see, for instance, the third part of the book [7]).

DOI: 10.1201/9781003514701-1

Many examples of relativistic plasmas can be found, in laboratory experiments as well as in natural conditions, in space and astrophysical environments. Laboratory plasmas where relativistic plasma dynamics is relevant and, sometimes, determinant are free electron lasers, plasma accelerators, magnetically confined fusion devices (in particular, the excitation of runaway electron populations) and inertial fusion experiments (with emphasis on the fast ignition approach to laser fusion).

Astrophysical plasmas include galactic and micro-quasar jets, super-nova explosions, generation of cosmic rays, and gamma-ray bursts. In that respect, special attention has been given to electron-positron plasmas, which play an important role in neutron star environments. Quark-gluon plasmas can also be seen as extending the usual plasma electrodynamics onto the more complex plasma chromo-dynamics. Nonlinear quantum electro-dynamical (QED) problems, both in vacuum and in plasmas, have been explored in recent years, partly motived by the advent on ultra-intense laser systems.

Instead of a complete coverage of this area of knowledge, we have decided to mainly focus on problems that have been recently discussed in the literature, such as those related with the excitation of electron plasma waves and wakefields, the creation of quasi-static magnetic field structures and the plasma radiation processes. They have a direct relevance to plasma accelerators and to astrophysics. Our analysis of relativistic plasmas will be mainly based on classical physics, but quantum aspects of acceleration and radiation phenomena will also be considered. Complementary views on laser-plasma interactions can also be found in the recent literature [8, 9].

## 1.2 Plasma Parameters

Plasma is a gas of charged particles, such that it is globally neutral. If the plasma is made of electrons and $Z$ times ionized positive ions, this quasi-neutrality is defined by

$$n_e = Z n_i \simeq n_0 \,, \tag{1.1}$$

where $n_e$ is the electron density and $n_i$ is the ion density. The quantity $n_0$ is the mean density. For a plasma of electrons and protons, we have $Z = 1$. We should, however, note that in the near vicinity of every particle, there is a local deviation from neutrality. In order to understand this deviation, we consider a positive ion, located at some position $\mathbf{r} = 0$. In its vicinity, we expect a larger number of electrons than ions because electrons are attracted and ions are repelled by this *test charge*. If the electrons are in thermal equilibrium, they will satisfy a Boltzmann distribution in the local potential $V$, meaning that

$$n_e(\mathbf{r}) = n_0 \exp\left(-\frac{q_e V}{T_e}\right) , \tag{1.2}$$

where $q_e = -e$ is the electron charge. The electrostatic potential is determined by Poisson's equation, such that

$$\nabla^2 V = -\frac{e}{\epsilon_0}\left(1 - e^{eV/T_e}\right) . \tag{1.3}$$

The electron temperature $T_e$ is written in energy units. This equation can be rewritten in terms of a dimensionless potential $\Phi = eV/T_e$, as

$$\nabla^2 \Phi = \frac{1}{\lambda_D^2}\left(e^{\Phi} - 1\right) , \quad \lambda_D^2 = \frac{\epsilon_0 T_e}{e^2 n_0} , \tag{1.4}$$

where the quantity $\lambda_D$ is the *electron Debye length*. This equation can easily be solved if we assume that $\Phi \ll 1$, allowing us to use the expansion: $\exp \Phi \simeq 1 + \Phi$. For radial solutions, we can define

$$\nabla_\rho^2 \equiv \lambda_D^2 \nabla^2 = \frac{1}{\rho^2}\frac{d}{d\rho}\left(\rho^2 \frac{d}{d\rho}\right) , \tag{1.5}$$

where $\rho = r/\lambda_D$, and transform (1.4) into a simple dimensionless equation

$$\frac{d^2 U}{d\rho^2} = U , \quad U = \rho\Phi . \tag{1.6}$$

For physical solutions, we need to satisfy $U(\rho \to \infty) = 0$, or

$$U(\rho) = U_0 \exp(-\rho) . \tag{1.7}$$

This leads to an explicit solution for the electrostatic potential of the form

$$V(r) = \frac{Ze}{4\pi\epsilon_0 r} \exp\left(-\frac{r}{\lambda_D}\right) , \tag{1.8}$$

where the constant of integration $U_0$ was chosen in order to reduce it to the Coulomb potential in the immediate vicinity of the ion of charge $q_i = Ze$, as $V(r \to 0) \to q_i/4\pi\epsilon_0 r$. This means that the electron density around the ion increases with respect to the mean density, as

$$n_e(r) = n_0 \left[1 + \frac{Zr_0}{r} \exp\left(-\frac{r}{\lambda_D}\right)\right] , \tag{1.9}$$

where we have introduced the *Landau radius*, defined by

$$r_0 \equiv \frac{eV(r_0)}{T} = \frac{e^2}{4\pi\epsilon_0 T} . \tag{1.10}$$

This is another important plasma scale-length and is defined as the distance where the electrostatic potential energy between two electrons equals the thermal energy. Let us now turn to the plasma characteristic time-scale, which is determined by the *electron plasma frequency*, $\omega_p$. This can be defined, using the Debye length as a basic scale-length and the *electron thermal velocity* $v_{the} = \sqrt{T_e/m_e}$ as the characteristic velocity, such that

$$\omega_p^2 = \frac{v_{the}^2}{\lambda_D^2} = \frac{e^2 n_0}{\epsilon_0 m_e}. \tag{1.11}$$

Another important concept is the *plasma parameter*, $g$, defined as

$$g = \frac{1}{n_0 \lambda_D^3} = \frac{4\pi}{3N_D}, \tag{1.12}$$

where $N_D$ is the number of electrons inside a Debye sphere. In order to observe a collective behavior, a plasma with size $L$ and duration $T$ should satisfy the following inequalities

$$L \gg \lambda_D, \quad \omega_p T \gg 1, \quad g \ll 1. \tag{1.13}$$

This defines the *ideal plasma conditions.* In order to understand the meaning of this last condition, it is useful to introduce another characteristic distance, usually called the *Wigner-Seitz radius*, $r_{ws}$, defined as the mean distance between identical particles, electrons in this case, as

$$r_{ws} = \left(\frac{3}{4\pi n_0}\right)^{1/3} = \left(\frac{3}{4\pi} g\right)^{1/3} \lambda_D. \tag{1.14}$$

It is obvious that $g \ll 1$ implies the existence of many electrons inside the Debye sphere, $r_{ws}^3 \ll \lambda_D^3$. Another important question is related with the validity of the classical plasma description. Quantum effects become relevant when the de Broglie wavelength of a thermal electron, $\lambda_{the}$ becomes larger than the inter-particle distance $r_{ws}$, or

$$\lambda_{the} = \frac{h}{m_e v_{the}} = \frac{h}{\sqrt{T_e m_e}} \geq r_{ws}, \tag{1.15}$$

where $h$ is Planck's constant. From here we conclude that the quantum plasma regime occurs for high density and low temperature, such that $n^{1/3}\lambda_{the} \geq 1$. A more formal but nearly equivalent criterium for the classical-quantum regime transition is based on the definition of *Fermi temperature*, $T_F$, and uses the ratio $\chi = T_F/T_e$, between the Fermi and the plasma temperature, as

$$\chi \equiv \frac{T_F}{T_e} = \frac{\lambda_{the}^2}{2} \left(3\pi^2 n_0\right)^{2/3}. \tag{1.16}$$

The quantum plasma regime corresponds to the region where this new parameter is larger than one, $\chi \geq 1$. We should note that $\chi$ is nearly equal to $n_0^{2/3}\lambda_{the}^2$, and therefore it also implies that the thermal de Broglie wavelength is larger than the inter-particle distance.

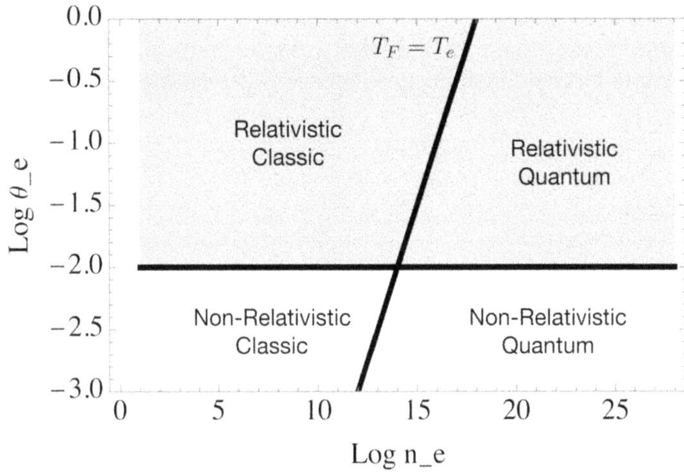


**FIGURE 1.1**
Plasma diagram, normalized temperature $Log_{10}\Theta_e$ versus $Log_{10}n_0$, where three main regimes are indicated: Classical non-relativistic, Relativistic, and Quantum. The line $\chi = 1$, or $T_F = T_e$, is also indicated.

Finally, the $\log n - \log T$ *plasma diagram* can be used to represent the different plasma regimes, classical and quantum. Although illuminating, this diagram is not so useful to identify all the areas covered by the present work. This is because, relativistic plasmas are characterized by a thermal velocity $v_{the}$ that is comparable to the velocity of light, $c$. But relativistic plasma effects can also occur in non-equilibrium situations involving relativistic particle beams, moving in a cold plasma background where $v_{the} \ll c$. Two new parameters are often used to characterize these two kinds of relativistic plasmas. They are

$$\theta_e = \frac{T_e}{m_e c^2}, \quad \gamma_a = \sqrt{1 + a^2}, \tag{1.17}$$

where $\theta_e$ is the normalized temperature, or equivalently, the ratio between thermal energy and the electron rest energy, and $\gamma_a$ is the electron relativistic factor in an intense laser field. The quantity $a$ is the normalized field amplitude acting on the electrons, to be defined later. Non-relativistic plasma physics is recovered in the limits $\theta_e \to 0$ and $a \to 0$, or $\gamma_a \to 1$. In Figure 1.1, we represent the plasma diagram in terms of the normalized temperature $\theta_e$ and indicate the main plasma regimes, classical, relativistic and quantum.

## 1.3 Plasma Creation

A plasma can be created in thermal equilibrium, just by increasing the temperature of the gas. This is described by the famous *Saha equation*. But they can also be created by ionization processes associated with energetic particle beams interacting with a background medium, or by applied electric fields. These non-equilibrium processes lead to what is called a *gaseous discharge*.

A plasma can contain a fraction of neutral particles, atoms or molecules, but is dominated by the long-range interaction between charged particles. If a number of neutrals is present, with density $n_0$, we can define the *degree of ionization* as the ratio $\alpha = n_e/(n_0 + Zn_i)$. When this ratio is close to one, the influence of inter-particle collisions involving neutrals is negligible.

Let us consider the equilibrium process and show how a neutral gas can be transformed into a plasma, when its temperature exceeds a certain value. This critical temperature will depend on the ionization energy of the atoms or molecules of the gas. In order to understand the transition from a neutral gas to a plasma, we assume a gas in equilibrium at a given temperature $T$, and consider two atomic quantum states with energies $E_0$ and $E_i$. Assuming a Boltzmann equilibrium, we can say that the ratio $r$ between the density of atoms in these two states is given by

$$r \equiv \frac{n_i}{n_0} = \frac{g_i}{g_0} \exp\left(-\frac{E}{T}\right) , \tag{1.18}$$

where $E = E_i - E_0$ is the energy difference between the two states and the temperature $T$ is written in energy units. We have also used the *degeneracy factors*, $g_i$ and $g_0$, which determine the number of quantum states with the same energy, $E_i$ and $E_0$. Let us now assume that $E_i$ is the lowest energy level of the ionized ion-electron pair, and $E_0$ is the ground state of the neutral atom. The difference $E$ will then give the ionization energy. The above relation will then allow us to say that the number of ions will equal the number of neutrals ($r = 1$) when the temperature attains the critical temperature

$$T_{cr} = \frac{E}{\ln(g_i/g_0)} . \tag{1.19}$$

This gives the temperature above which ionization becomes dominant ($T > T_{cr}$) and a plasma replaces the neutral gas. It is particularly useful to write a formula for the degree of ionization defined above, assuming single ionized ions, $Z = 1$, as

$$\alpha = \frac{r}{1+r} = \frac{(g_i/g_0)\exp(-E/T)}{1 + (g_i/g_0)\exp(-E/T)} . \tag{1.20}$$

It is interesting to note that the transition between the neutral state $\alpha = 0$ and the plasma state $\alpha = 1$ occurs in a temperature range $\Delta T$ centred around the critical temperature $T_{cr}$ given by

$$\Delta T = \frac{1}{(d\alpha/dT)_{cr}} = \frac{4T_{cr}}{\ln(g_i/g_0)} . \tag{1.21}$$

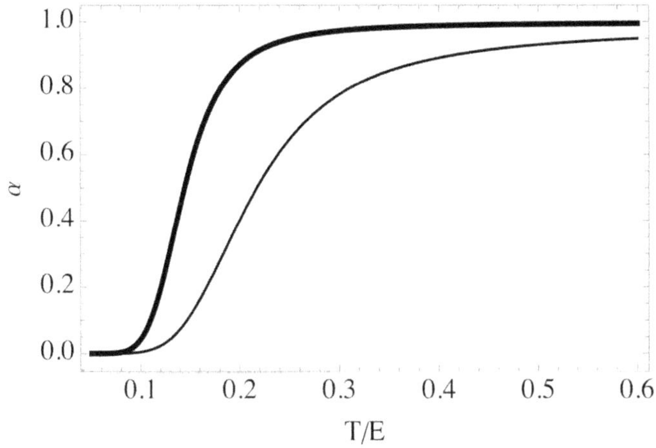


**FIGURE 1.2**
Degree of ionization $\alpha$ as a function of the temperature $T$, for two different values of the degeneracy ratio: $(g_i/g_0) = 10^3$, in bold, and $(g_i/g_0) = 10^2$.

When taking the derivative of $\alpha$, we have assumed that the degeneracy factors $g_i$ and $g_0$ evolve very slowly with the temperature, and therefore their contribution to the derivative can be neglected. In general, the ionization state is much more degenerate than the ground state, $g_i/g_0 \gg 1$, such that we have $\Delta T \ll T_{cr}$. This means that the transition between a neutral gas and a plasma can be quite sudden, when we change the temperature, as expected in a typical *phase transition*. This is illustrated in Figure 1.2. The degeneracy ratio can be determined from Quantum Mechanics and is given by

$$\frac{g_i}{g_0} = \left( \frac{2\pi m_e T}{h^2} \right)^{3/2} \frac{1}{n_i}, \tag{1.22}$$

where $m_e$ is the electron mass and $h$ is Planck's constant.

This expression can be easily understood by means of a simple argument. If the ground state is non-degenerate, we have $g_0 = 1$. On the other hand, the energy spectrum of the ionized states is a continuum. The degeneracy factor of the free-electron states can be estimated as the ratio between the phase-space volume occupied by a thermal electron and the elementary volume $h^3$. The phase-space volume occupied by a thermal electron of momentum $p = \sqrt{2m_e T}$ is given by $(2m_e T)^{3/2} V_0$, where $V_0 = 1/n_e$ is the volume occupied by a single electron. Assuming that $n_e n_i$, this allows us to write

$$g_i = (2m_e T)^{3/2} (1/n_i) h^{-3} = \left( \frac{2m_e T}{h^2} \right)^{3/2} \frac{1}{n_i}, \tag{1.23}$$

which only differs from (1.22) by a numerical factor of order one. Using this expression in eq. (1.21), we can then derive the *Saha equation* as

$$r \equiv \frac{n_i}{n_0} = \left( \frac{m_e T}{2\pi\hbar^2} \right)^{3/2} \frac{\exp(-E/T)}{n_i} , \tag{1.24}$$

If we now write the energy and the temperature, $E$ and $T$, in electron-Volt ($1eV = 1.6 \times 10^{-19} Joule$), and the ion density $n_i$ in number per cubic metre ( $m^{-3}$), we can rewrite this as

$$r = 3 \times 10^{27} \frac{T^{3/2}}{n_i} \exp\left( -\frac{E}{T} \right) . \tag{1.25}$$

For instance, in the case of Hydrogen, the ionization energy is $E = 13.6\, eV$. If a given volume of Hydrogen is at such a pressure that $n = n_0 + n_i = 10^{16}\, cm^{-3}$, then from eq. (1.19) we get $T_{cr} = 0.5\, eV$. We can obtain plasma at even lower temperatures, if we use Cesium or Potassium vapours, with ionization energies $E = 3.89$ and $4.34\, eV$. Plasmas with temperatures around $1\, eV$ can therefore be obtained. Modern methods, using plasma out of equilibrium, can significantly lower these values, as it occurs when we resonantly create highly excited Rydberg states or ionize a laser-cooled gas [7, 10].

## 1.4 Intense Field Ionization

The Saha equation was originally derived to determine the temperature of stellar atmospheres, starting from the spectroscopy observation of their ionization degrees. But, although it explains in a very clear and suggestive way the neutral gas-plasma transition, it cannot be applied to most plasmas observed in the laboratory or in nature, which are often out of equilibrium.

For instance, in the upper atmosphere of the Sun, called solar corona, plasmas are not in complete thermal equilibrium with radiation, because they are optically thin, and photons can travel through the medium without much interaction. Ionization is then provided by electron-neutral collisions, and a new equilibrium state, called *coronal equilibrium*, can occur when ionization and recombination rates exactly compensate.

In the laboratory, a common process of plasma production involves the creation of an initial bunch of electrons, accelerated by applied electric fields. These initial electrons can subsequently ionize the atoms by inelastic collisions. The resulting electron population is directly heated by the applied electric fields, and for this reason, their temperature is usually larger than that of the ions and neutral atoms with which it interacts.

Of particular interest for the main topic of this book is the process of ionization by intense fields, associated with intense laser pulses. The intensity threshold for field ionization can be estimated by assuming that ionization occurs when the incident laser field $E_0$ is equal to or larger than the electric field that is binding the electron inside the atom. Using this simple criterium is then easy to establish the *threshold intensity*, $I_a$, such that

$$I_a = \frac{\epsilon_0}{2} c |E_a|^2 , \tag{1.26}$$

where the atomic field $E_a$ is the electric field that binds the electrons inside the atom, typically at the *Bohr radius*, $r_a$, as defined by

$$E_a = \frac{e}{4\pi\epsilon_0} \frac{1}{r_a^2} , \quad r_a = 4\pi\epsilon_0 \frac{\hbar^2}{me^2} . \tag{1.27}$$

It is also sometimes useful to write this field in terms of the *reduced Compton wavelength*, $r_C = \lambda_C / 2\pi$, as

$$E_a = \frac{e}{4\pi\epsilon} \frac{\alpha^2}{r_C^2} , \quad r_C = \frac{\hbar}{mc} , \tag{1.28}$$

where we have introduced the fine structure constant $\alpha \simeq 1/137$, defined as

$$\alpha = \frac{1}{4\pi\epsilon_0} \frac{e^2}{\hbar c} , \quad r_a = \frac{r_C}{\alpha} . \tag{1.29}$$

Another useful elementary distance is the *classical electron radius*, $r_e$, defined as the distance at which its rest energy $mc^2$ equals its self-electrostatic energy. We have

$$r_c = \frac{1}{4\pi\epsilon_0} \frac{e^2}{mc^2} . \tag{1.30}$$

These different elementary scales are related, from the largest $r_e$ to the shortest $r_C$, by the fine structure constant as

$$r_e = \alpha \, r_C = \alpha^2 r_a . \tag{1.31}$$

The above atomic field $E_a$ can also be related to the so-called *Schwinger limit*, which is the critical field of quantum electrodynamics $E_S$ at which vacuum creation of electron-positron pairs spontaneously occurs. It is given by

$$E_S = \frac{m^2 c^3}{e\hbar} = 1.32 \times 10^{18} \, V/m. \tag{1.32}$$

Using the above definitions, we can easily get

$$E_S = \frac{e}{4\pi\epsilon_0} \frac{1}{\alpha r_c^2} , \quad E_S = \frac{E_a}{\alpha^3} . \tag{1.33}$$

The intense laser plasma interactions explore the region of electric fields above $E_a$ and well below $E_S$. We write these quantities in numbers as

$$r_a = 5.292 \times 10^{-11} \, m , \quad E_a = 5.1 \times 10^9 \, V/m , \quad I_a = 3.51 \times 10^{16} \, W/cm^2 . \tag{1.34}$$

From here, we conclude that atom ionization occurs for laser intensities above threshold, $I_0 > I_{th}$, or equivalently, $E_0 > E_a$. The electron velocity will oscillate with an amplitude given by $v_0 = eE_0/m\omega$, where $\omega$ is the laser frequency. This is sometimes written in normalized form as

$$v_0/c = 0.85\lambda_\mu^2\sqrt{I_{18}}\,, \tag{1.35}$$

where $I_{18}$ is the normalized intensity, written in $10^{18} W/cm^2$, and $\lambda_\mu$ is the laser wavelength $\lambda = 2\pi c/\omega$ written in micrometers. We can see that we are in the relativistic domain if $v_0/c \sim 1$, which occurs at intensities of the order of $I_0 \sim 10^{18}\, W/cm^2$, at a wavelength of nearly $\lambda \sim 1\,\mu m$. Notice that, between the threshold field $E_a$ and the Schwinger limit $E_S$, there are 12 orders of magnitude in terms of laser intensity.

It should however be noticed that ionization will eventually occur, with some probability, at intensities well below threshold. This can be due to absorption of a single photon, if its energy is equal to the ionization energy of the atom, or to multi-photon absorption if the photon energy is much smaller. Tunnel ionization is also possible. The various kinds of ionization regimes are often characterized by a dimensionless parameter, called the *Keldish parameter*, $\gamma_K$. It can be defined as the square root of the ratio between the binding energy of the atom $U_i$ and the ponderomotive energy of the incident wave, $U_p$ [11, 12], or

$$\gamma_K = \sqrt{\frac{U_i}{U_p}}\,, \quad U_p = \frac{e^2 E_0^2}{2m\omega^2}\,. \tag{1.36}$$

Alternatively, we can define it as the ratio between another critical field, $E_K$, sometimes called the Keldish field, and the laser electric field $E_0$. We have

$$\gamma_K = \frac{E_K}{E_0}\,, \quad E_K = \frac{\omega}{e}\sqrt{2mU_i}\,. \tag{1.37}$$

For weak fields, such that $\gamma_K \gg 1$, we are in the tunneling regime. For $\gamma_K \simeq 1$ we are in the intermediate regime. The intense field regime corresponds to $\gamma_K \ll 1$. But, a description of the different ionization regimes is outside the scope of the present work. Our aim here is not to determine the conditions for plasma formation, but to study the plasma specific properties as a physical medium.

# 2

# *Single Particle Motion*

A characteristic feature of plasma physics is that we need to describe it at three different levels that most of the time superpose: single particle motion, kinetic processes and fluid behavior. In this chapter, we focus on single particle trajectories and summarize the well-known results from classical Mechanics and Electrodynamics, in order to fix the notation and to establish links between the relevant equations. For more details, see the basic references [13, 14] and [15, 16]. In subsequent Chapters, these basic particle results will be used to built the kinetic description of a plasma as a gas of interacting particles, and the fluid description when particles lose their identity and are replaced by a continuum.

## 2.1 Equations of Motion

We start with a simple description of the classical equations of motion, in their different versions: Lagrangian, Hamiltonian and Newtonian. It is known that the equations of motion can be derived from an *action principle*, which states that the trajectory of a particle between two instants, $t_1$ and $t_2$, corresponds to a stationary value (minimum or maximum) of the *action integral* $S$, defined as

$$S = \int_{t_1}^{t_2} \mathcal{L}(\mathbf{v}, \dot{\mathbf{v}}, t)\, dt \quad \rightarrow \quad \delta S = 0\,, \tag{2.1}$$

where $\mathbf{v} = d\mathbf{r}/dt$ is the particle velocity. The function inside the integral is called the *Lagrangian*. For a particle with mass $m$ (we always define mass as the rest mass), moving freely in the absence of fields, the Lagrangian is

$$\mathcal{L}_0(\mathbf{v}, \dot{\mathbf{v}}, t) = -\frac{mc^2}{\gamma}\,, \quad \gamma = \frac{1}{\sqrt{1 - v^2/c^2}}\,, \tag{2.2}$$

where $c$ is the speed of light in vacuum, and $\gamma$ is the relativistic gamma-factor. Taking an arbitrary deviation $\delta\mathbf{r}$ with respect to the true trajectory, and assuming that it vanishes at the two extremes of the time interval, $\delta\mathbf{r}(t_1) =$

DOI: 10.1201/9781003514701-2

$\delta\mathbf{r}(t_2) = 0$, we can write the action principle as

$$\delta S = \int_{t_1}^{t_2} \left( \frac{\partial \mathcal{L}}{\partial \mathbf{r}} \delta \mathbf{r} + \frac{\partial \mathcal{L}}{\partial \mathbf{v}} \frac{\partial}{\partial t} \delta \mathbf{r} \right) dt = 0 \,, \tag{2.3}$$

After integrating by parts the second term, we arrive at

$$\frac{\partial \mathcal{L}}{\partial \mathbf{r}} - \frac{\partial}{\partial t} \frac{\partial \mathcal{L}}{\partial \mathbf{v}} = 0 \,. \tag{2.4}$$

This is the *Euler-Lagrange equation*. Let us now define the *canonical momentum*,

$$\mathbf{p} = \frac{\partial \mathcal{L}}{\partial \mathbf{v}} \,, \tag{2.5}$$

and use the *Legendre transformation*

$$\mathcal{L}(\mathbf{v}, \dot{\mathbf{v}}, t) = \mathbf{p} \cdot \mathbf{v} - H(\mathbf{r}, \mathbf{p}, t) \,, \tag{2.6}$$

where the *Hamiltonian* $H(\mathbf{r}, \mathbf{p}, t)$ is a new function to be specified. Introducing this in the action principle (2.3), we get

$$\delta S = \int_{t_1}^{t_2} \left[ \left( \frac{d\mathbf{r}}{dt} - \frac{\partial H}{\partial \mathbf{p}} \right) \cdot \delta \mathbf{p} + \left( \mathbf{p} \cdot \frac{d}{dt} \delta \mathbf{r} - \frac{\partial H}{\partial \mathbf{r}} \cdot \delta \mathbf{r} \right) \right] dt = 0 \,, \tag{2.7}$$

where no restrictions were imposed on the momentum variations $\delta\mathbf{p}$. Integrating the third term by parts, we then arrive at

$$\frac{d\mathbf{r}}{dt} = \frac{\partial H}{\partial \mathbf{p}} \,, \quad \frac{d\mathbf{p}}{dt} = - \frac{\partial H}{\partial \mathbf{r}} \,. \tag{2.8}$$

These are the *canonical equations* of motion, which describe the evolution of the particle position and momentum. We note that, for a free particle, the canonical momentum is

$$\mathbf{p} \equiv \frac{\partial \mathcal{L}_0}{\partial \mathbf{v}} = m\gamma \mathbf{v} \,, \tag{2.9}$$

Now, using the Legendre transformation (2.6), we get for the free Hamiltonian

$$H_0(\mathbf{r}, \mathbf{p}, t) = \mathbf{p} \cdot \mathbf{v} - \mathcal{L}_0 = m\gamma c^2 \,, \tag{2.10}$$

which identifies the Hamiltonian with the energy of the particle $\mathcal{E}$, written in terms of its momentum, as $H_0 = \mathcal{E}_0 = mc^2 \sqrt{1 + (p/mc)^2}$.

Let us now consider the motion of a particle in the presence of an electromagnetic field. For this purpose, we add an interaction term to the Lagrangian (2.2), such that the new Lagrangian becomes

$$\mathcal{L} = \mathcal{L}_0 + \mathcal{L}_{int} \,, \quad \mathcal{L}_{int} = q(\mathbf{A} \cdot \mathbf{v} - V) \,, \tag{2.11}$$

where $q$ is the particle charge ($q = -e$ for electrons, and $q = Ze$ for $Z$ times ionized positive ions ), $V$ and $\mathbf{A}$ are the scalar and vector potentials. We can immediately see that the new term modifies the canonical momentum, which becomes

$$\mathbf{P} = \frac{\partial \mathcal{L}}{\partial \mathbf{v}} = \mathbf{p} + q\mathbf{A} \,. \tag{2.12}$$

We keep the notation $\mathbf{p} = m\gamma\mathbf{v}$ for the *kinetic momentum*. Using eq. (2.6), we can now write the new Hamiltonian, as

$$H(\mathbf{r}, \mathbf{P}, t) = \mathbf{p} \cdot \mathbf{v} - \mathcal{L}_0 = m\gamma c^2 + qV \,, \tag{2.13}$$

Comparing this with eq. (2.10), we obtain the expression

$$H(\mathbf{r}, \mathbf{P}, t) = \sqrt{m^2c^4 + (\mathbf{P} - q\mathbf{A})^2c^2} + qV \,, \tag{2.14}$$

where the potentials $V$ and $\mathbf{A}$ can be explicitly dependent on $\mathbf{r}$ and $t$. It is particularly useful to take the weakly relativistic limit, valid for $v^2/c^2 \ll 1$. Developing the square root, and noticing that the rest energy $mc^2$ can be dropped because it is a constant, we obtain

$$H(\mathbf{r}, \mathbf{P}, t) = \frac{1}{2m}(\mathbf{P} - q\mathbf{A})^2 + qV \,. \tag{2.15}$$

A similar expression will reappear later for the covariant Hamiltonian in the fully relativistic regime.

Let us now consider the equations of motion in the Newtonian form. For this purpose, we use the Lagrangian equation (2.4). Noting that

$$\frac{\partial \mathcal{L}}{\partial \mathbf{r}} = \nabla \mathcal{L} - q\nabla(\mathbf{A} \cdot \mathbf{v}) - q\nabla V \,, \tag{2.16}$$

using the vectorial identity

$$\nabla(\mathbf{a} \cdot \mathbf{b}) = (\mathbf{a} \cdot \nabla)\mathbf{b} + (\mathbf{b} \cdot \nabla)\mathbf{a} + \mathbf{a} \times (\nabla \times \mathbf{b}) + \mathbf{b} \times (\nabla \times \mathbf{a}) \,, \tag{2.17}$$

where $\mathbf{a}$ and $\mathbf{b}$ are arbitrary vectors, and noting that $\mathbf{v}$ is independent of $\mathbf{r}$, we get

$$\frac{d}{dt}(\mathbf{p} + q\mathbf{A}) = \frac{\partial \mathcal{L}}{\partial \mathbf{r}} = q(\mathbf{v} \cdot \nabla)\mathbf{A} + q\mathbf{v} \times (\nabla \times \mathbf{A}) - qV \,. \tag{2.18}$$

This simplifies if we introduce the electric and magnetic fields, defined as

$$\mathbf{E} = -\nabla V - \frac{\partial \mathbf{A}}{\partial t} \,, \quad \mathbf{B} = \nabla \times \mathbf{A} \,, \tag{2.19}$$

and we arrive at the equation of motion, written in terms of the kinetic momentum $\mathbf{p} = m\gamma\mathbf{v}$, as

$$\frac{d\mathbf{p}}{dt} = q\left(\mathbf{E} + \mathbf{v} \times \mathbf{B}\right) \,. \tag{2.20}$$

The term on the r.h.s. is known as the *Lorentz force*. It is also useful to state the evolution of the total (kinetic + rest) energy $\mathcal{E} = m\gamma c^2$. Using this equation of motion, we can easily get

$$\frac{d\mathcal{E}}{dt} = q\mathbf{E} \cdot \mathbf{v} \,. \tag{2.21}$$

The magnetic field **B** is absent in this expression, meaning that it only changes the direction of motion, not the particle energy. This is of fundamental importance for particle acceleration.

## 2.2 Covariant Equations

### 2.2.1 Spacetime

It is known from Relativity that the laws of physics can be written in identical form in any inertial frame of reference. This means that they are *covariant*. It is therefore important to write the previous equations of motion in their covariant form. For that purpose, we introduce the spacetime 4-vector position $x$, defined by

$$x^\mu \equiv (ct, \mathbf{r}) \,, \quad \mu = (0, 1, 2, 3) \,, \tag{2.22}$$

where the component $x^0$ represents time and the other three the vector position $\mathbf{r} \equiv (x^1, x^2, x^3)$. The particle trajectory in 4-spacetime defines a *worldline*, and a 4-vector position defines an *event*. The distance between two infinitesimally close events is the interval $ds$, such that

$$ds^2 = c^2 dt^2 - d\mathbf{r}^2 \,. \tag{2.23}$$

When these events belong to the same particle trajectory, we can write the interval as

$$ds = c d\tau \,, \quad d\tau = \frac{dt}{\gamma} \,. \tag{2.24}$$

Here, $\tau$ is the particle *proper time*. Equation (2.23) can be defined by an internal product, as

$$ds^2 = dx_\mu dx^\mu = g_{\mu\nu} dx^\nu dx^\mu \,, \tag{2.25}$$

where the *metric tensor* $g_{\mu\nu}$ defines the geometric properties of the 4-spacetime. In Special Relativity, the spacetime is flat and is called the *Minkowski spacetime*, as defined by the metric

$$g_{\mu\nu} = g^{\mu\nu} = diag.(1, -1, -1, -1) \,. \tag{2.26}$$

We also have $g^{\nu}_{\mu} = \delta^{\nu}_{\mu}$. It is used to lower or raise the indices. The particle 4-*velocity* can be defined as

$$u^{\mu} = \frac{dx^{\mu}}{d\tau} \equiv \left( c\frac{dt}{d\tau}, \frac{d\mathbf{r}}{d\tau} \right) = \gamma(c, \mathbf{v}) \,. \tag{2.27}$$

We can also define the 4-*momentum*, as

$$p^{\mu} = mu^{\mu} = (\mathcal{E}/c, \mathbf{p}) = m\gamma(c, \mathbf{v}) \,. \tag{2.28}$$

The module of this quantity is a constant, as shown by

$$p^2 \equiv p_{\mu} p^{\mu} = \frac{\mathcal{E}^2}{c^2} - \mathbf{p}^2 = m^2 c^2 \,. \tag{2.29}$$

The derivative of the 4-momentum with respect to the proper time is the 4-*force*, $f^{\mu} = dp^{\mu}/d\tau$. The Lorentz force, defined in eq. (2.20) can be seen as the space component of a 4-force, and the covariant Lorentz equation of motion takes the form

$$\frac{dp^{\mu}}{d\tau} = qF^{\mu}_{\nu} u^{\nu} \,. \tag{2.30}$$

Here, $F^{\mu}_{\nu}$ is the *electromagnetic tensor*, and can be defined as the rotational of a 4-*potential*

$$F^{\mu\nu} = \partial^{\mu} A^{\nu} - \partial^{\nu} A^{\mu} \,, \quad A^{\mu} \equiv (V/c, \mathbf{A}) \,, \tag{2.31}$$

where $\partial^{\mu} = \partial/dx^{\mu}$ are the partial derivatives with respect to the coordinates $x^{\mu}$. In eq. (2.30), we obviously have $F^{\mu}_{\nu} = g_{\nu\rho} F^{\rho\mu}$. We can write the electromagnetic tensor explicitly, in terms of the electric and magnetic fields, as

$$F^{\mu\nu} = \frac{1}{c} \begin{bmatrix} 0 & -E^1 & -E^2 & -E^3 \\ E^1 & 0 & cB^3 & -cB^2 \\ E^2 & -cB^3 & 0 & cB^1 \\ E^3 & cB^2 & -cB^1 & 0 \end{bmatrix} . \tag{2.32}$$

We can see that $F^{\mu\nu} = -F^{\nu\mu}$. It is known that the fields created by electric currents $\mathbf{J}$ and charge distributions $\rho$, are determined by the *Maxwell's equations*, which take the covariant form

$$\partial_{\nu} F^{\mu\nu} = -\mu_0 J^{\mu} \,, \quad \partial \tilde{F}^{\mu\nu} = 0 \,, \tag{2.33}$$

where $J^{\mu} \equiv (c\rho, \mathbf{J})$ is the 4-current, and $\mu_0$ is the permeability of vacuum. Here, we have used the *dual* of the electromagnetic tensor $F^{\mu\nu}$, defined by

$$\tilde{F}^{\mu\nu} = \frac{1}{2} e^{\mu\nu\epsilon\sigma} F_{\epsilon\sigma} \,, \tag{2.34}$$

where, $e^{\mu\nu\epsilon\sigma}$ is the *Levi-Civita tensor*. Its components are equal to zero unless all its four indices are different from each other, $\mu \neq \nu \neq \epsilon \neq \sigma$, and are equal to 1 (or to −1) for pair (or even) permutations of (0, 1, 2, 3). We have $e^{\mu\nu\epsilon\sigma} = -e_{\mu\nu\epsilon\sigma}$. Eqs. (2.33), can be written in explicit three-dimensional form as

$$\nabla \cdot \mathbf{D} = \rho \quad , \quad \nabla \times \mathbf{H} = \frac{\partial \mathbf{D}}{\partial t} + \mathbf{J} \tag{2.35}$$

where we have used $\mathbf{B} = \mu_0 \mathbf{H}$, and defined the displacement vector $\mathbf{D} = \epsilon_0 \mathbf{E}$, with the vacuum permittivity $\epsilon_0 = 1/\mu_0 c^2$. This is completed by

$$\nabla \cdot \mathbf{B} = 0 \quad , \quad \nabla \times \mathbf{E} = -\frac{\partial \mathbf{B}}{\partial t}, \tag{2.36}$$

These four Maxwell's equations relate the electric and magnetic fields **E** and **H** (or the displacement and magnetic induction fields, **D** and **B**) to the charge and current distributions $\rho$ and **J**, assuming that these sources are located in vacuum. To these expressions we should add the continuity equation,

$$\frac{\partial \mathbf{J}}{\partial t} + \nabla \rho = 0 \,. \tag{2.37}$$

It can be rewritten in covariant form as

$$\partial_\mu J^\mu = 0 \,. \tag{2.38}$$

For a cloud of particles with density $\rho$, moving with the same velocity $u^\mu$, the 4-current takes the form

$$J^\mu = \rho \frac{dx^\mu}{d\tau} = \rho u^\mu \,. \tag{2.39}$$

A single particle, moving with a trajectory $x^\mu(\tau)$, can be described by the current

$$J^\mu = q \frac{dx^\mu}{d\tau} \delta[\mathbf{r} - \mathbf{r}(\tau)]_{t = x^0(\tau)} \,, \tag{2.40}$$

The particle charge $q = \int J^0(\mathbf{r}, t) d\mathbf{r}$ is an invariant.

### 2.2.2 Field Transformations

It is now useful to consider the *Lorentz transformations* relating two different inertial frames and to identify the field invariants. The 4-position $x \equiv (ct, \mathbf{r})$, defined in the lab frame $K$, transforms into a new 4-position $x' \equiv (ct', \mathbf{r}')$, defined in another frame $K'$, moving with velocity **V** with respect to the lab. These transformations are

$$x'^\mu = \Lambda^\mu_\nu x^\nu \,, \quad x^\nu = \Lambda'^\nu_\mu x'^\mu \,, \tag{2.41}$$

where $\Lambda^\mu_\nu \Lambda'^\nu_\mu = 1$, are such that

$$ct' = \gamma\,(ct - \beta \mathbf{n} \cdot \mathbf{r}) \,, \quad \mathbf{r}' = \mathbf{r} + (\gamma - 1)(\mathbf{n} \cdot \mathbf{r})\mathbf{n} - \gamma V t\, \mathbf{n} \,. \tag{2.42}$$

Here, we have used: the normalized velocity $\beta = V/c$, the relativistic gamma-factor $\gamma = 1/\sqrt{1 - \beta^2}$ and the unit vector along the direction of motion

$\mathbf{n} = \mathbf{V}/|V|$ . This is sometimes called the *Lorentz boost*. In standard form, when $K'$ moves along the $x$-axis and $\mathbf{n} = \mathbf{e}_x$, these formulae reduce to

$$ct' = \gamma(ct - \beta x)\,, \quad x' = \gamma(x - \beta ct)\,, \quad y' = y\,, \quad z' = z\,. \tag{2.43}$$

We can see that the coordinates are boosted in the negative $x$-direction. The Lorentz transformation can also sometimes be described as a *rotation* by an imaginary angle $(i\zeta)$, such that $\beta = \tanh\zeta$. As we can see, $\gamma = \cosh\zeta$, and these expressions become

$$ct' = ct\cosh\zeta - x\sinh\zeta\,, \quad x' = x\cosh\zeta - ct\sinh\zeta\,, \quad y' = y\,, \quad z' = z\,. \tag{2.44}$$

We can also consider the velocity transformation, given by

$$\mathbf{v}' = \frac{1}{(1 - \mathbf{V}\cdot\mathbf{v})}\left[\frac{\mathbf{v}}{\gamma} - \mathbf{V} + \frac{\gamma}{(1+\gamma)c^2}(\mathbf{V}\cdot\mathbf{v})\mathbf{V}\right]. \tag{2.45}$$

This expression transforms the velocity $\mathbf{v} = d\mathbf{r}/dt$, defined in the lab frame, into the velocity $\mathbf{v}' = d\mathbf{r}'/dt'$ in the moving frame. For $\mathbf{v}$ and $\mathbf{V}$ both parallel to the axis $\mathbf{e}_x$, this reduces to the well-known *velocity formula*

$$v' = \frac{v - V}{(1 - Vv/c^2)}\,. \tag{2.46}$$

Of particular relevance are the quantities that remain constant under a Lorentz transformation, the so-called *relativistic invariants*. As already mentioned, the interval is a scalar, and therefore an invariant, $ds'^2 = ds^2$. The same with the module of the 4-momentum, $p_\mu p^\mu = m^2c^2$. As for the electromagnetic tensor, it transforms as

$$F'^{\mu\nu} = \Lambda^\mu_\epsilon \Lambda^\nu_\sigma F^{\epsilon\sigma}\,. \tag{2.47}$$

Writing explicitly the electric and magnetic fields, it corresponds to

$$\mathbf{E}' = (\mathbf{E}\cdot\mathbf{n})\mathbf{n} + \gamma\left[\mathbf{n}\times(\mathbf{E}\times\mathbf{n}) + \mathbf{V}\times\mathbf{B}\right],$$

$$\mathbf{B}' = (\mathbf{B}\cdot\mathbf{n})\mathbf{n} + \gamma\left[\mathbf{n}\times(\mathbf{B}\times\mathbf{n}) - (\mathbf{V}\times\mathbf{E})/c^2\right]. \tag{2.48}$$

This immediately shows that the fields parallel to $\mathbf{n}$ are invariants, $\mathbf{E}'_\parallel = \mathbf{E}_\parallel$, and $\mathbf{B}'_\parallel = \mathbf{B}_\parallel$. We can also identify two important field invariants, defined as

$$F_{\mu\nu}F^{\mu\nu} = -2\left(E^2 - c^2B^2\right), \quad F_{\mu\nu}\tilde{F}^{\mu\nu} = -4\left(\mathbf{E}\cdot\mathbf{B}\right). \tag{2.49}$$

They imply that, if the field amplitudes are equal at a given inertial frame, which means $E^2 = c^2B^2$, they will stay equal in all inertial frames, $E'^2 = c^2B'^2$. Similarly, if they are orthogonal in one frame, $(\mathbf{E}\cdot\mathbf{B}) = 0$, the same will be true in every frame, $(\mathbf{E}'\cdot\mathbf{B}') = 0$.

### 2.2.3 Covariant Hamiltonian

Let us now consider the Hamiltonian equations for particles moving in an electromagnetic field, written in covariant form. They will use the canonical momentum $P^\mu = p^\mu + qA^\mu$. Starting with the Hamiltonian (2.13), we can write the canonical equations as

$$\frac{d\mathbf{r}}{dt} = \frac{\partial H}{\partial \mathbf{P}} = mc^2 \frac{\partial \gamma}{\partial \mathbf{P}}, \quad \frac{d\mathbf{P}}{dt} = -\frac{\partial H}{\partial \mathbf{r}}, \tag{2.50}$$

with $\gamma = H/mc^2$. This leads to

$$\frac{d\mathbf{r}}{dt} = \frac{\mathbf{p}}{m\gamma} = \mathbf{v}, \quad \frac{d\mathbf{P}}{dt} = -q\left(\mathbf{v} \cdot \frac{\partial \mathbf{A}}{\partial \mathbf{r}} - \frac{\partial V}{\partial \mathbf{r}}\right), \tag{2.51}$$

Noting that $dt/d\tau = \gamma$, we can easily transform the first of these equations into a covariant equation, using the proper time $\tau$, as

$$\frac{dx^\mu}{d\tau} = p^\mu \equiv (c\gamma, \mathbf{p}), \tag{2.52}$$

But the second of eqs. (2.51), which is equivalent to the Lorentz equation (2.20), is clearly not covariant. To circumvent the difficulty, we define a *covariant Hamiltonian*, $\mathcal{H}$, as

$$\mathcal{H} = \frac{1}{2m}(P_\mu - qA_\mu)(P^\mu - qA^\mu) = \frac{p_\mu p^\mu}{2m}, \tag{2.53}$$

This is formally identical to the classical non-relativistic Hamiltonian of a free particle. Using eq. (2.14) and introducing the 4-canonical momentum $P^\mu = (H/c, \mathbf{P})$, we conclude that this new Hamiltonian, sometimes called the *super-Hamiltonian*, is a constant of motion

$$\mathcal{H} \equiv \mathcal{H}(x^\mu, P^\mu) = \frac{1}{2}mc^2. \tag{2.54}$$

The covariant equations, equivalent to eqs. (2.50), can then be written as

$$\frac{dx^\mu}{d\tau} = \frac{\partial \mathcal{H}}{\partial P_\mu}, \quad \frac{dP_\mu}{d\tau} = -\frac{\partial \mathcal{H}}{\partial x^\mu}, \tag{2.55}$$

## 2.3 Static Fields

Specific examples of field configurations will be explored in this and in subsequent Sections. We first apply the particle equations of motion to the simplest case of a constant and uniform field. In this case, the field tensor $F^{\mu\nu}$ is independent of $x^\mu$. Starting from the covariant Lorentz equation (2.30), we can easily get, after integration

$$u^\mu(\tau) = \exp\left(\frac{q\tau}{m}F^\mu_\nu\right) u^\nu(0). \tag{2.56}$$

A second integration would then give the particle trajectory. But this formal integration is not very useful. In order to understand the particle behavior in constant and uniform fields, we need to consider a number of more specific examples, and derive more explicit solutions.

### i) Static Magnetic Field B

We first take $\mathbf{E} = 0$, and reduce eqs. (2.20)–(2.21) to

$$\frac{d\mathbf{u}}{dt} = \frac{q}{m}(\mathbf{v} \times \mathbf{B}) \,, \quad \frac{d\mathcal{E}}{d\tau} = 0 \,. \tag{2.57}$$

where $\mathbf{u} = \gamma\mathbf{v}$ and $\mathcal{E} = m\gamma c^2$, with $\gamma$ constant. This can be written as

$$\frac{d\mathbf{v}}{dt} = s\omega_c\,(\mathbf{v} \times \mathbf{b}) \,, \quad \omega_c = \frac{|q|B}{\gamma m} \,, \tag{2.58}$$

where $s$ is the sign of the charge $q$, $\mathbf{b} = \mathbf{B}/B$ defines the direction of the magnetic field and $\omega_c$ is the *cyclotron frequency*. For electrons, we have $s = -1$ and $|q| = e$. If the magnetic field is along $z$, as $\mathbf{b} = \mathbf{e}_z$, we immediately get $v_z = const.$ and

$$\frac{d}{dt}(v_x + iv_y) = -is\omega_c(v_x + iv_y) \,. \tag{2.59}$$

After integration, we get

$$(v_x + iv_y) = \alpha \exp(-is\omega_c t) \,, \tag{2.60}$$

which is formally identical to the solution (2.56). If we write the constant of integration as $\alpha = v_{\perp 0} \exp(i\varphi)$, we obtain

$$v_x = v_{\perp 0} \cos(s\omega_c t + \varphi) \,, \quad v_y = v_{\perp 0} \sin(s\omega_c t + \varphi) \,, \tag{2.61}$$

where $v_{\perp 0} = (v_x^2 + v_y^2)^{1/2}$ is a constant. A second integration then leads to the particle trajectory

$$\mathbf{r}(t) = \mathbf{r}_0 + (v_z t)\,\mathbf{e}_z + s\,r_L\,(i\mathbf{e}_x - \mathbf{e}_y) \exp(-is\omega_c t + \varphi) \,, \tag{2.62}$$

where $r_L$ is the *Larmor radius*, defined as

$$r_L = \frac{v_{\perp 0}}{\omega_c} = \frac{m\gamma}{|q|B} v_{\perp 0} \,, \tag{2.63}$$

This quantity is also sometimes called the *cyclotron radius*, or the *gyroradius*.

### ii) Static Electric Field E

We now consider $\mathbf{B} = 0$, and the equations of motion give

$$\frac{d\mathbf{u}}{dt} = \frac{q}{m}\mathbf{E} \,, \quad \mathbf{u} = \mathbf{v}\gamma \,. \tag{2.64}$$

This shows that the velocity component perpendicular to the electric field stays constant, $\mathbf{u}_\perp = const.$, but the relativistic factor varies in time, due to parallel acceleration. This gives

$$u_\parallel = \gamma v_\parallel = \frac{q}{m} E t + u_{\parallel 0}, \quad \gamma = \sqrt{\gamma_0^2 + (at/c)^2}, \tag{2.65}$$

where $\gamma_0 = \gamma(t=0)$, and $a = qE/m$ is a constant acceleration. If we take the field parallel to the $x$-axis, we get

$$\frac{dx}{dt} = \frac{qE}{m\gamma} t = \frac{at}{\sqrt{\gamma_0^2 + (at/c)^2}}. \tag{2.66}$$

After integration, we obtain

$$x(t) = \frac{c^2}{a} \sqrt{\gamma_0^2 + (at/c)^2}. \tag{2.67}$$

A constant should be added to this expression, in order to satisfy the initial condition $x(t=0) = x_0$. On the other hand, the velocity along the perpendicular direction changes in time because, although there is no force in that direction, the value of $\gamma$ also changes, while $\mathbf{u}_\perp$ remains constant. This allows us to state that

$$\frac{d\mathbf{r}_\perp}{dt} = \frac{\mathbf{u}_{\perp 0}}{\gamma_0} \frac{1}{\sqrt{1 + (at/\gamma_0 c)^2}}, \tag{2.68}$$

which gives, after integration

$$\mathbf{r}_\perp(t) = \frac{\mathbf{p}_{\perp 0} c}{qE} \sinh^{-1} \left( \frac{at}{\gamma_0 c} \right), \tag{2.69}$$

with $\mathbf{p}_{\perp 0} = m\mathbf{u}_{\perp 0}$, and $\gamma_0 = (1 + u_{\parallel 0}^2 + u_{\perp 0}^2)^{1/2}$. The solution of the equations of motion is completed. But we should also note that eqs. (2.67) and (2.69) can be written in parametric form. For this purpose, we take $\mathbf{p}_{\perp 0}$ in the $y$ direction, and get

$$\sinh \left( \frac{ay}{u_{\perp 0} c} \right) = \frac{at}{\gamma_0 c}. \tag{2.70}$$

Using eq. (2.67), this then leads to

$$x = \frac{\gamma_0 c^2}{a} \cosh \left( \frac{ay}{u_{\perp 0} c} \right), \tag{2.71}$$

which defines a *catenary* in the $xy$ plane.

### iii) Static Parallel Fields, E ∥ B

Let us consider static electric and magnetic fields **E** and **B**, pointing in the same $z$-direction. The electric field has no an influence on the cyclotron motion, and we are reduced to a superposition of the two previous cases: constant acceleration in the field direction, plus cyclotron motion in the perpendicular plane. Therefore, the Lorentz equation decomposes into

$$\frac{d\mathbf{u}_\parallel}{dt} = \frac{q}{m}\mathbf{E}\,, \quad \frac{d\mathbf{u}_\perp}{dt} = \frac{q}{m}\left(\mathbf{v}_\perp \times \mathbf{B}\right)\,, \tag{2.72}$$

with $\mathbf{v} = \mathbf{u}/\gamma$, where the relativistic factor varies in time due to the parallel acceleration. Using our previous results, we can write the solution for the parallel motion as

$$\mathbf{r}_\parallel = z(t)\,\mathbf{e}_z\,, \quad z(t) = \frac{c^2}{a}\sqrt{\gamma_0^2 + (at/c)^2}\,, \tag{2.73}$$

with $a = qE/m$. And, for the perpendicular motion, we have

$$\frac{d}{dt}\left(u_x + iu_y\right) = -is\omega_c(t)\left(u_x + iu_y\right)\,, \quad \omega_c(t) = \frac{|q|B}{m\gamma(t)}\,. \tag{2.74}$$

This leads to

$$\mathbf{r}_\perp(t) = r_L(t)\exp(-i\varphi)\,, \quad \frac{d\varphi}{dt} = s\omega_c(t)\,, \tag{2.75}$$

with $r_L = \mathbf{v}_{\perp 0}/\omega(t)$, Integration of the phase equation, then gives

$$t = \frac{\gamma_0 c}{a}\sinh\left(\frac{E}{cB}\varphi\right)\,. \tag{2.76}$$

Using eq. (2.73), we also get

$$z = \frac{\gamma_0 c}{a}\cosh\left(\frac{E}{cB}\varphi\right)\,. \tag{2.77}$$

which completes eq. (2.76). Such a trajectory corresponds to helical motion with a radius $r_L$ that increases with time, as illustrated in Figure 2.1.

### iv) Static Perpendicular Fields, $\mathbf{E} \perp \mathbf{B}$

This case can be transformed into one of the first two cases, using appropriate Lorentz transformations. We have noted that $(\mathbf{E}\cdot\mathbf{B}) = 0$ is an invariant, which means that the fields are orthogonal in any inertial frame. Let us assume that the first invariant of eq. (2.49) is nonzero, $E^2 \neq c^2B^2$. We can then make a Lorentz transformation to a reference frame moving with velocity $\mathbf{V}_D$ with respect to the lab, such that

$$\mathbf{V}_D = \frac{(\mathbf{E} \times \mathbf{B})}{B^2}\,. \tag{2.78}$$

This quantity is called the *drift velocity*. Using the field transformations defined by eqs. (2.48), we can easily get

$$\mathbf{E}'_\parallel = 0\,, \quad \mathbf{E}'_\perp = \gamma\left(\mathbf{E} + \mathbf{V}_D \times \mathbf{B}\right) = 0\,. \tag{2.79}$$

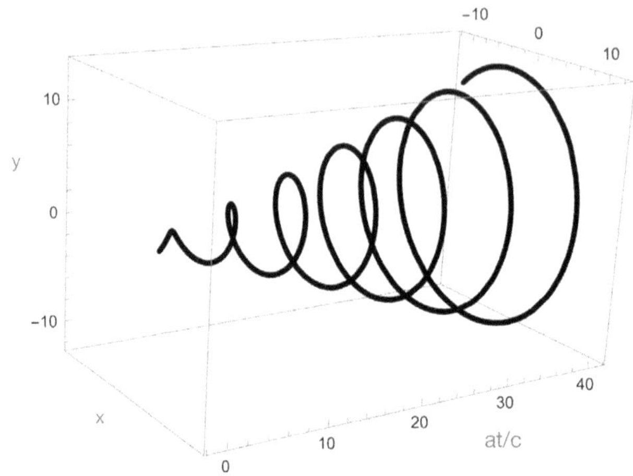


**FIGURE 2.1**
Particle trajectory in the field of parallel static fields, **E** and **B**, for $\gamma_0 = 1$.

We conclude that, in the new frame, the electric field is equal to zero, $\mathbf{E}' = 0$, as in case i), and that we are reduced to a purely magnetic field configuration, such that

$$\mathbf{B}' = \frac{1}{\gamma}\mathbf{B}, \quad \gamma = \frac{1}{\sqrt{1 - E^2/c^2B^2}}. \tag{2.80}$$

We notice that this new frame is only meaningful for real values of $\gamma$, which implies that $E^2 < c^2B^2$. This means that we need to define another Lorentz transformation, valid for the opposite case of $E^2 > c^2B^2$, to a frame moving with a new drift velocity, defined by

$$\mathbf{V}'_D = \frac{(\mathbf{E} \times \mathbf{B})}{E^2} c. \tag{2.81}$$

In this moving frame, we are now reduced to a purely electric field configuration as

$$\mathbf{B}" = 0, \quad \mathbf{E}" = \frac{1}{\gamma'}\mathbf{E}, \quad \gamma' = \frac{1}{\sqrt{1 - c^2B^2/E^2}}. \tag{2.82}$$

Let us finally consider the singular case, where neither of these Lorentz transformations is possible, because the fields have the same amplitude, $E^2 = c^2B^2$. Noting that the vector $(\mathbf{v} \times \mathbf{B})$ is in the $(-\mathbf{E})$ direction, the equations of motion for the different velocity components are

$$\frac{du_x}{dt} = \frac{q}{m} E \left(1 - \frac{v_z}{c}\right), \quad \frac{du_y}{dt} = 0, \quad \frac{du_z}{dt} = \frac{q}{m}\frac{v_x}{c} E, \tag{2.83}$$

with $\mathbf{E} = E\mathbf{e}_x$ and $\mathbf{B} = (E/c)\mathbf{e}_y$. Furthermore, the particle energy $\mathcal{E}$ satisfies the equation

$$\frac{d\mathcal{E}}{dt} = qEv_x, \tag{2.84}$$

These equations show the existence of two constants of motion: $p_y$, and $\alpha \equiv (\mathcal{E} - cp_z)$. Using $\mathcal{E} = mc^2\sqrt{1 + u^2/c^2}$, we then get

$$\mathcal{E} = \frac{\alpha}{2}\left(1 + \frac{\epsilon^2 + p_x^2 c^2}{\alpha^2}\right), \quad cp_z = -\frac{\alpha}{2}\left(1 - \frac{\epsilon^2 + p_x^2 c^2}{\alpha^2}\right), \tag{2.85}$$

where the constant quantity $\epsilon^2 = (m^2c^4 + p_y^2c^2)$ was also used. To complete our solution, we now have to calculate $p_x$. Using the first of eqs. (2.83), we obtain

$$\mathcal{E}\frac{dp_x}{dt} = qE(\mathcal{E} - cp_z) \equiv qE\alpha\,. \tag{2.86}$$

Replacing here the above expressions for $\mathcal{E}$ and $p_z$, we obtain a closed equation for $p_x$, which can be integrated in parametric form.

## 2.4 Motion in a Wave

As another important example, we consider a charged particle moving in the field of a plane electromagnetic wave in vacuum. For this purpose, we assume a linearly polarized wave, propagating along the $x$-direction. The corresponding electric and magnetic fields can be described as

$$\mathbf{E}(x,t) = E_0\mathbf{e}_y\cos(kx - \omega t)\,, \quad \mathbf{B}(x,t) = B_0\mathbf{e}_z\cos(kx - \omega t)\,, \tag{2.87}$$

where $E_0$ and $B_0 = kE_0/\omega$ are the amplitudes, $k$ is the wavenumber and $\omega = kc$ the wave frequency. The particle motion is determined by

$$\frac{dp_x}{dt} = \frac{qEk}{\omega}v_y\,, \quad \frac{dp_y}{dt} = qE\left(1 - \frac{k}{\omega}v_x\right), \quad \frac{dp_z}{dt} = 0\,, \tag{2.88}$$

where $\mathbf{p} = m\gamma\mathbf{v}$. We assume a reference frame where the particle is at rest in the absence of the wave. The energy of the particle, $\mathcal{E} = m\gamma c^2$, evolves according to

$$\frac{d\mathcal{E}}{dt} = qEv_y = \frac{\omega}{k}\frac{dp_x}{dt}\,, \quad \frac{d\gamma}{dt} = \frac{\omega}{kmc^2}\frac{dp_x}{dt}\,. \tag{2.89}$$

From these equations, we can conclude that there are two constants of motion

$$\mathcal{I}_1 = p_z\,, \quad \mathcal{I}_2 = \mathcal{E} - \frac{\omega}{k}p_x\,. \tag{2.90}$$

Using the auxiliary variable $\eta = x - ct$, we can write the second equation of motion (2.88) as

$$\frac{dp_y}{dt} = \frac{qE_0k}{\omega}\cos(k\eta)\left(\frac{\omega}{k} - v_x\right) = -\frac{qE_0k}{\omega}\cos(k\eta)\frac{d\eta}{dt}\,, \tag{2.91}$$

with the obvious solution

$$p_y = \frac{qE_0}{\omega} \sin(k\eta) \,. \tag{2.92}$$

Replacing this in the first equation of motion, we get

$$\gamma \frac{dp_x}{dt} = \frac{q^2 E_0^2}{m\omega^2} k \cos(k\eta) \sin(k\eta) \,. \tag{2.93}$$

This can also be written as

$$\left( \mathcal{I}_2 + \frac{\omega p_x}{kmc^2} \right) \frac{dp_x}{dt} = \frac{q^2 E_0^2}{m\omega^2} \frac{k}{2} \sin(2k\eta) \,. \tag{2.94}$$

On the other hand, the evolution of the energy is dictated by

$$\frac{d\mathcal{E}}{dt} = \frac{qE_0}{m} \cos(k\eta) \frac{p_y}{\gamma} \,, \tag{2.95}$$

or

$$\gamma \frac{d\gamma}{dt} = \frac{qE_0}{m^2 c^2} \cos(k\eta) p_y \,, \tag{2.96}$$

This then leads to the well-known solution [16]

$$x(\eta) = \frac{a_0^2}{8k\gamma_0^2} \sin(2k\eta) \quad , \quad y(\eta) = -\frac{a_0}{k\gamma_0} \cos(k\eta) \tag{2.97}$$

where we have used

$$a_0 = \frac{eE_0}{\omega mc} \quad , \quad \gamma_0 = \sqrt{1 + a_0^2/2} \tag{2.98}$$

We notice that, for a weak field such that $a_0 \ll 1$, the electron response is essentially linear, with an oscillating transverse motion nearly parallel to the wave electric field, and a second-order longitudinal correction oscillating with double frequency, in the direction of wave propagation. In contrast, in the limit of intense wave fields, $a_0 \gg 1$, the wave motion becomes nearly independent of the electric field amplitude, as given by

$$x(\eta) \simeq \frac{1}{4k} \sin(2k\eta) \quad , \quad y(\eta) = -\frac{\sqrt{2}}{k} \cos(k\eta) \tag{2.99}$$

The nonlinear term is now of the same order as the linear one, and the particle trajectory describes the famous *figure 8*, as illustrated in Figure 2.2. This nonlinear response to the plane wave field can be seen as a symbol of relativistic plasma physics, which is dominated by single particle, as well as, its collective nonlinearities.

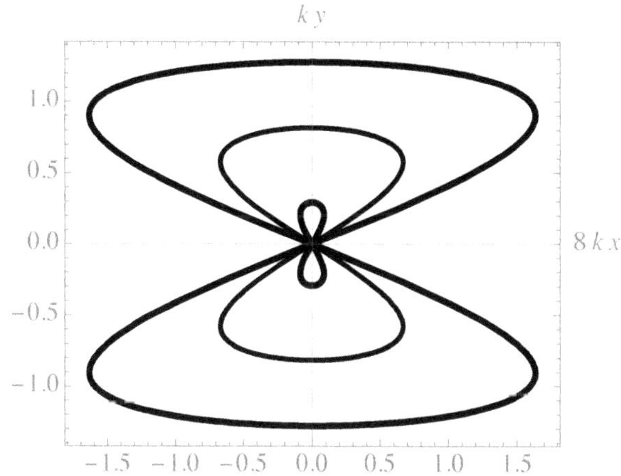


**FIGURE 2.2**
Particle trajectory in the field of an electromagnetic plane wave in vacuum, for increasing values of normalized field amplitude $a_0 = 0,3$ (inner curve), $a_0 = 1$, and $a_0 = 3$ (outer curve).

## 2.5 Slowly Varying Fields

It is now important to discuss some of the main features of particle motion in non-uniform and slowly varying static fields. For this purpose, we start with a short review of the concept of *action-angle variables*. These are pairs of canonical variables, valid for Hamiltonians that are periodic in some variable $r$, and can be written (ignoring the other degrees of freedom) as $H(r,p) = \mathcal{E}$, where $\mathcal{E}$ is a constant. If $H$ is periodic in $r$, then its conjugate momentum $p = p(r, \mathcal{E})$ is also periodic. We can then integrate $p$ along a periodic orbit, as

$$J = \frac{1}{2\pi} \oint p \, dr \, . \tag{2.100}$$

In such conditions, it is possible to replace the periodic Hamiltonian $H(r,p)$ by another Hamiltonian that only depends on $J$ and is independent of the periodic variable, $H(J) = \mathcal{E}$. Such a replacement, from the initial pair of variables $(r,p)$ to a new pair $(\theta, J)$, can be performed using a canonical transformation. The generating function of this transformation is $F \equiv F(r,J)$, and the angular variable is determined by $\theta = \partial F / \partial J$. The new equations of motion are

$$\frac{d\theta}{dt} = \frac{\partial H}{\partial J} = \omega(J) \, , \quad \frac{dJ}{dt} = -\frac{\partial H}{\partial \theta} = 0 \, , \tag{2.101}$$

where the angular frequency $\omega$ only depends on the action integral $J$, as defined by (2.100). The equations of motion are then trivially given by

$$\theta(t) = \omega(J) + \theta_0\,, \quad J = const.\,, \tag{2.102}$$

If, instead of one degree of freedom, we have a system with several independent degrees of freedom, the corresponding Hamiltonian $H(r_i, p_i)$, with $i = 1, 2, ...N$. It will be called *integrable*, if it is possible to define $N$ different pairs of similar action-angle variables, such that the equations of motion, written in terms of these variables, reduce to

$$\theta_i(t) = \omega_i(J_i) + \theta_0\,, \quad J_i = const. \tag{2.103}$$

The motion becomes *multiperiodic*.

An important property of the action integral is that it remains constant if the frequency $\omega$ changes slowly in time, or in other words, adiabatically. So, even for a time-varying frequency, the action $J$ is still a constant, and is called an *adiabatic invariant*. To illustrate this property, we consider the cyclotron motion of a particle that is moving over a region of slowly varying magnetic field, $\mathbf{B} = B\,\mathbf{e}_z$, such that the rate of change of the cyclotron frequency is negligible over one cycle, $\omega_c \gg (dB/dz)(1/Bv_z)$. Given the periodicity of the cyclotron motion in the transverse plane, we define the action integral as

$$J_c = \oint \mathbf{P}_\perp \cdot d\mathbf{s}\,, \tag{2.104}$$

where $\mathbf{P}_\perp = \mathbf{p}_\perp + q\mathbf{A}$ is the canonical momentum, and $d\mathbf{s}$ is the line element of the orbit in the perpendicular plane. Using the *Stokes theorem*, this can be rewritten as

$$J_c = \oint \mathbf{p}_\perp \cdot d\mathbf{s} + q \int_S (\mathbf{B}\cdot\mathbf{n})\, dS\,. \tag{2.105}$$

In the last term, the line integral was transformed into a surface integral, and $\mathbf{n}$ is the unit vector perpendicular to the surface. It can easily be shown that this surface integral is (apart from a minus sign) equal to the magnetic flux $\Phi_B$. We then get

$$J_c = \pi m\gamma\omega_c r_L^2 = q\Phi_B\,, \quad \Phi_B = \pi B r_L^2\,. \tag{2.106}$$

We conclude that both $J_c$ and the magnetic flux $\Phi_B$ are adiabatic invariants. This leads to the so-called *magnetic mirror* configuration, where an increasing magnetic field reflects the charged particles (e.g. electrons), confining them in the lower magnetic field region. This can occur in the Earth magnetic field, as well as in magnetic machines built in the laboratory. Writing the magnetic flux in terms of the particle energy, $\mathcal{E} = m\gamma c^2$, and noting that this quantity remains constant in a static magnetic field, we can write for a particle moving in the $z$-direction

$$\frac{d}{dz}\Phi_B = 0\,, \quad \frac{d}{dz}\left(\frac{v_\perp^2}{B}\right) = 0\,. \tag{2.107}$$

Defining a *pitch angle* $\alpha$ in velocity space, such at $v_\perp = v \sin \alpha$, we conclude from here that

$$\frac{\sin^2 \alpha(z)}{B} = const. \tag{2.108}$$

This shows that, for increasing values of $B(z)$, the particle will be reflected at some point $z = z_{max}$, defined by the condition $\sin \alpha(z_{max}) = 1$, or $\alpha = \pi/2$. At this point, the particle motion in the field direction stops, $v_z = 0$, and no further motion is allowed for $z > z_{max}$. The total particle energy is converted into circular energy at $z_{max}$, and the Larmor radius becomes

$$r_L(z_{max}) = \frac{mc}{|q|B} \left(\gamma^2 - 1\right)^{1/2} . \tag{2.109}$$

If, in a magnetic mirror configuration, we have a particle trajectory with some angle $\alpha_0$ in the regions of lower magnetic field, $B_{min}$, only when the inequality $\sin^2 \alpha_0 \geq (B_{min}/B_{max})$ is satisfied, will this particle remain confined. All particles with a lower pitch angle (or, equivalently, a larger velocity $v_z$ at the magnetic minimum) will be lost. This defines a *loss cone* in particle velocity space.

Trapped particles can be found between two magnetic field maxima, such as between the North and South poles of the Earth magnetic field, or in magnetic confinement machines, such as *magnetic bottles* and *tokamaks* [17]. In this case, the particle will oscillate back and forth in the $z$-direction. This allows us to introduce a new adiabatic invariant, called the *longitudinal invariant*, defined as

$$J_z = \oint \mathbf{v} \cdot d\mathbf{s} \equiv \oint v_z ds \,, \tag{2.110}$$

where integration is taken over the oscillation between two magnetic mirrors. This new invariant can then be written in terms of the parallel kinetic energy $m\gamma v_z^2$.

Similarly, a third invariant can eventually be assigned to the slow drift motion around the axis of symmetry of a magnetic configuration, thus allowing for a complete particle confinement in three dimensions. But confinement can be lost due to field perturbations, like those associated with turbulence, eventually involving a stochastic motion. Simple but relevant examples of stochastic motion in a broad wave spectrum will be discussed next.

## 2.6 Mappings

A discrete-time description of particle motion can sometimes be used, that corresponds to a stroboscopic sampling of their trajectories. In particular, this is useful to understand particle transport and particle acceleration processes. Two important examples are considered here, corresponding to the standard map and the Fermi-Ulam map.

### 2.6.1 Standard Map

We use a covariant formulation of the relativistic motion of a particle in the field of a large spectrum of electrostatic waves with amplitude $V_0$, equally spaced in frequency. We start from eqs. (2.51)–(2.52), and assume a 4-potential of the form

$$A^\mu \equiv (V/c, \mathbf{A} = 0) \ , \quad V = V_0 \sum_{n=-\infty}^{\infty} \sin(kx - n\Delta\omega t) \, , \tag{2.111}$$

where the 4-wavevectors are

$$k_n^\mu \equiv \left(\frac{\omega_n}{c}, \mathbf{k}\right) = \left(n\frac{\Delta\omega}{c}, k, 0, 0\right) . \tag{2.112}$$

Replacing this in the canonical Hamiltonian (2.53), we can see that $\mathcal{H}(x^\mu, P^\mu)$ is not explicitly dependent on the variables $x^2 = y$ and $x^3 = z$. This means that we have two constants of motion

$$P^2 = P_y = const. \, , \quad P^3 = P_z = const. \tag{2.113}$$

For simplicity, we assume that these constants are equal to zero, $P^2 = P^3 = 0$. We then use a reduced form of the Hamiltonian function, such that

$$\mathcal{H}(x^0, x^1; P^0, P^1) = \frac{1}{2m}\left\{\left[P^0 - \frac{qV_0}{c}\sum_n \sin(k_\nu x^\nu)\right]^2 - (P^1)^2\right\} , \tag{2.114}$$

Now, due to the pseudo-Euclidian nature of the Minkowski metric, we have $P^0 = P_0$ and $P^1 = -P_1$, and can write the canonical equations of motion (2.55) as

$$\frac{dx^0}{d\tau} = \frac{\partial\mathcal{H}}{\partial P_0} = \frac{1}{m}\left[P^0 - \frac{qV_0}{c}\sum_n \sin(kx^1 - n\frac{\Delta\omega}{c}x^0)\right] ,$$

$$\frac{dx^1}{d\tau} = \frac{\partial\mathcal{H}}{\partial P_1} = \frac{P^1}{m} , \tag{2.115}$$

and

$$\frac{dP_0}{d\tau} = -\frac{\partial\mathcal{H}}{\partial x^0} = \gamma\frac{qV_0}{c}\frac{\Delta\omega}{c}\sum_n n\cos(kx^1 - n\frac{\Delta\omega}{c}x^0) ,$$

$$\frac{dP_1}{d\tau} = -\frac{\partial\mathcal{H}}{\partial x^1} = \gamma q V_0 k \sum_n \cos(kx^1 - n\frac{\Delta\omega}{c}\omega x^0) , \tag{2.116}$$

where $\tau$ is the proper time, and $\gamma$ is the relativistic factor given by

$$\gamma = \frac{P_0}{mc} + \frac{qV_0}{mc^2}\sum_n \sin(kx^1 - n\frac{\Delta\omega}{c}x^0) , \tag{2.117}$$

Now, we use the identity

$$\sum_n \cos(kx^1 - n\frac{\Delta\omega}{c}x^0) = T\cos(kx^1)\sum_n \delta(t - nT)\,, \tag{2.118}$$

where $t = x^0/c$ and $T = 2\pi/\Delta\omega$. We can then write

$$\frac{dP_1}{d\tau} = \gamma q E_0 \cos(kx^1)\sum_n \delta(t - nT)\,, \tag{2.119}$$

with $E_0 = V_0 k$. This shows that the particle momentum changes by instantaneous kicks. Integrating over a time interval between two consecutive kicks, and using $d\tau = dt/\gamma$, we get

$$P_{1,n+1} = P_{1,n} + qE_0 T\cos(kx_n^1)dt\,, \tag{2.120}$$

where $x_n^1$ is the value of $x$ for $t = t_n$. This value can be obtained through the integration

$$x_{n+1}^1 = x_n^1 + \int_{t_n}^{t_n+T} \frac{P_1}{m\gamma}dt\,, \tag{2.121}$$

from where we get

$$x_{n+1}^1 = x_n^1 + \frac{T}{m}\frac{P_{1,n+1}}{\sqrt{1 + (P_{1,n+1}/c)^2}}\,. \tag{2.122}$$

Simplifying the notation to

$$x = kx^1\,, \quad p = \frac{P_1}{mc}\,, \quad K = \frac{qE_0 T}{mc}\,, \tag{2.123}$$

and using $kcT = 2\pi kc/\Delta\omega = 1$, we retrieve from eqs. (2.120) and (2.122) the *relativistic standard map*

$$p_{n+1} = p_n + K\sin x_n$$

$$x_{n+1} = x_n + \frac{p_{n+1}}{\sqrt{1 + p_{n+1}^2}}\,, \tag{2.124}$$

It describes the relativistic particle motion as a sequence of discrete points plotted in the $(x, p)$ phase space. It was first proposed by Chernikov [18] (see also [19]). If we take the case of non-relativistic motion, corresponding to $\gamma = 1$, we obtain the standard map in its original version, as

$$p_{n+1} = p_n + K\sin x_n$$

$$x_{n+1} = x_n + p_{n+1}\,. \tag{2.125}$$

This is sometimes called the *Chirikov-Taylor map* [20], and has been used to study several different aspects of chaotic motion (see [21, 22]). It can also

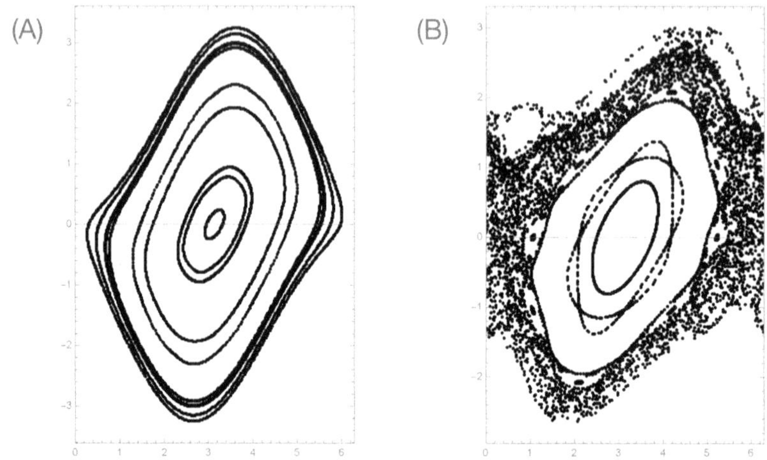


**FIGURE 2.3**
Standard Map $(p, x)$, for $K = 0.85$: (A) Relativistic map as defined by eq. (2.124) versus (B) non-relativistic map, eq. (2.125).

be applied to the magnetic field lines in toroidal machines used in confined magnetic fusion, such as *tokamaks* and *stellarators* [23, 24].

Although very similar, these two versions of the standard map (relativistic and non-relativistic) have quite different properties. This is illustrated in Figure 2.3, where we compare the two maps for the same value of the parameter $K$, showing that the relativistic motion is much less chaotic.

An interesting property of a map is related to the existence of *fixed points*, satisfying $(x_{n+1} = x_n \, , p_{n+1} = p_n)$. For the standard map (2.124), there are two sequences of fixed points, one stable $(s)$ and the other unstable $(u)$, defined by

$$(x_\ell, p_\ell)_s = (0, 2\pi\ell) \ , \quad (x_\ell, p_\ell)_u = (\pi/2, 2\pi\ell) \ , \tag{2.126}$$

with $\ell = (0, \pm 1, \pm 2, ...)$. We can see that, in the non-relativistic standard map, there is an an infinity of equidistant fixed points in the $(x, p)$ phase-plane. In contrast, for the relativistic map of eqs. (2.124), only two fixed points survive, one stable $(0, 0)$ and the other unstable $(\pi/2, 0)$. Another interesting property is related to the existence of *accelerator modes*, which correspond to an unlimited increase of the momentum $p$. This is achieved when we choose initial values such that

$$p_i = 0 \, , \quad K \sin x_i = 2\pi\ell \tag{2.127}$$

with $\ell$ integer. Eqs. (2.125) show that, for each iteration, the particle momentum $p$ increases by a quantity $2\pi\ell$, while the position $x$ (modulo $\pi$) stays the same. In contrast, for the relativistic map (2.124) no accelerator modes can be found.

### 2.6.2 Fermi-Ulam Map

Fermi acceleration was first proposed in 1948 [25] and is still considered the dominant mechanism for the production of *cosmic rays* [26, 27]. This mechanism assumes that charged particles (mainly protons) are confined in a magnetic bottle between moving magnetic clouds. As described above in Section 2.5, the particles will be reflected at some point $z_{max}$, which now moves and increases the kinetic energy of the particle. The process is then repeated indefinitely, thus explaining the extremely high energies observed in cosmic rays.

A simple mechanical model, first proposed by Ulam [22, 28], can be used to describe this repetitive process of acceleration: A particle moving between two parallel walls, one fixed, at $z = z_1$, and the other oscillating periodically around the origin, as $z = z_2(t) = \delta z \sin(\omega t)$. Assuming elastic collisions with the wall, this leads to the so-called *Fermi-Ulam map*.

In alternative, we use an electrostatic model [29]. We assume an electron moving in the field of a standing electrostatic field, $\mathbf{E} = E_0 \mathbf{e}_z \cos(kz) \sin(\omega t)$. The electron motion along $z$ is described by the Hamiltonian

$$H(p,z) = m\gamma c^2 + V(z)\sin(\omega t)\,, \quad \gamma = \sqrt{1 + (p/mc)^2}\,, \tag{2.128}$$

with the potential function

$$V(z) = V_0 \sin(kz)\,, \quad V_0 = -\frac{eE}{k}. \tag{2.129}$$

We can therefore write the canonical equations as

$$\frac{dz}{dt} = \frac{\partial H}{\partial p} = \frac{p}{m\gamma}\,, \quad \frac{dp}{dt} = -\frac{\partial H}{\partial z} = -\sin(\omega t)\frac{\partial V}{\partial z}. \tag{2.130}$$

Integrating these equations between $z_0$ and $z_1 = z_0 + L$, we obtain

$$z_1 - z_0 = \frac{p_1}{m\gamma_1}(t_1 - t_0) = L\,, \tag{2.131}$$

and, assuming that the field is only applied at the period positions $a = z_1 + nL$, we get

$$p_1 - p_0 = \frac{eE_0}{\omega}\int \sin(\theta)\cos[kz(\theta)]\delta(\theta-\theta_0)d\theta = \frac{eE_0}{\omega}\sin(\theta_0)\cos(kz_0)\,, \tag{2.132}$$

where we have defined $\theta = \omega t$. Now, defining a new variable $u$, and a new parameter $M$, such that

$$u = \frac{p}{eE_0/\omega} \equiv \frac{p}{mca_0}\,, \quad 2\pi M = \frac{L}{a_0 c}\,, \tag{2.133}$$

and generalizing to a succession of steps with length $L$, we obtain

$$u_{n+1} = u_n + \sin\theta_n\,,$$

$$\theta_{n+1} = \theta_n + \frac{2\pi M}{u_{n+1}} \sqrt{1 + a_0^2 u_{n+1}^2}\,. \tag{2.134}$$

This is the relativistic Fermi-Ulam map. Comparing with the relativistic standard map (2.125) we note an interesting inversion of the relativistic factor. It is also interesting to consider the limit of very weak electric fields, such that $a_0^2 \ll 1$. In this limit, the gamma factor tends to one, and we are reduced to

$$I_{n+1} = I_n + K \sin\theta_n$$

$$\theta_{n+1} = \theta_n + I_{n+1}, \tag{2.135}$$

where we have used a new variable $I = Ku$, and a new constant $K = 2\pi M$. We can then see that, in this limit, the Fermi-Ulam map coincides with the non-relativistic standard map of eq. (2.125).

# 3

# *Plasma Equations*

We now consider the basic plasma equations, which allow to describe the medium at various levels of approximation: kinetic and fluid. We first assume that the medium is made of charged particle populations, described by a variety of *kinetic equations*, established at different levels of approximation. On a second step, we take averages over the kinetic description, and derive *fluid equations* that describe the medium as a superposition of continuous fluids, one for each species. Under certain conditions, this fluid description can be reduced to a single fluid, which leads to the simplest plasma description, as a single magnetodynamical fluid. We are then in the land of *magnetohydrodynamics* or MHD. The present chapter is therefore devoted to the study of kinetic and fluid plasma evolution equations. We also consider the influence of collisions between particles, and the way they can influence the plasma evolution towards thermal equilibrium.

## 3.1 BBGKY Hierarchy

Let us describe a plasma as a system with a large number, $N \gg 1$, of point particles. Following the principles of classical Statistical Physics, we represent the state of this system in *phase-space*, an ideal space with a large number of dimensions, representing the coordinates and momenta of each particle. For simplicity, we only discuss a generic single particle species, the extension to multiple species being straightforward. For each particle $i$, we define 6 dimensions, corresponding to 3 dimensions for position $\mathbf{r}_i$, and 3 other dimensions for momentum $\mathbf{p}_i$. The probability density defined in this $6N$-dimensional phase-space is represented by the function

$$\rho(\mathbf{r}_1, \mathbf{p}_1, ...\mathbf{r}_j, \mathbf{p}_j, ...\mathbf{r}_N, \mathbf{p}_N, t) \equiv \rho(\Gamma, t) \,, \tag{3.1}$$

and the volume element $d\Gamma$, in simplified notation, by

$$d\Gamma = d1\, d2 \ldots dN \,, \quad di \equiv d\mathbf{r}_i d\mathbf{p}_i \,. \tag{3.2}$$

By definition, this probability density is normalized according to

$$\int \rho(\Gamma, t)\, d\Gamma = 1 \,, \tag{3.3}$$

DOI: 10.1201/9781003514701-3

where integration extends to the entire phase-space. It is also useful to define a reduced *one-particle* distribution function, $f_1(1)$, as

$$f_1(1) = N \int \rho(\Gamma, t)\, d\Gamma_{N-1}\,, \quad d\Gamma_{N-1} = d2\,...dN\,. \tag{3.4}$$

More generally, we can define a reduced *s-particle* distribution function as

$$f_s(1, ...s) = V_s \int \rho(\Gamma, t)\, d\Gamma_{N-s}\,, \quad d\Gamma_{N-s} = d(s+1)\,...dN\,, \tag{3.5}$$

with $V_s = N!/(N-s)!$, valid for $s \geq 1$. In order to obtain evolution equations for these quantities, we write their time derivative as

$$\frac{\partial f_s}{\partial t} = V_s \int \frac{\partial \rho}{\partial t}\, d\Gamma_{N-s}\,, \tag{3.6}$$

It is known that the distribution function $\rho(\Gamma, t)$ satisfies a conservation law in phase-space, determined by the *Liouville equation*

$$\frac{d\rho}{dt} \equiv \frac{\partial \rho}{\partial t} + \sum_{i=1}^{N} \{\rho,\, H\}_i = 0\,, \tag{3.7}$$

where $H$ is the Hamiltonian of the system. Here, use the *Poisson brackets*

$$\{\rho,\, H\}_i = \frac{\partial \rho}{\partial \mathbf{r}_i} \cdot \frac{\partial H}{\partial \mathbf{p}_i} - \frac{\partial \rho}{\partial \mathbf{p}_i} \cdot \frac{\partial H}{\partial \mathbf{r}_i}\,. \tag{3.8}$$

The Liouville equation can be written in a more compact form as

$$\frac{\partial \rho}{\partial t} = L\rho\,, \quad L \equiv \sum_{j=1}^{N} \{H,\, \}_j = 0\,, \tag{3.9}$$

where $L$ is called the *Liouville operator* or *Liouvillian*. We should note that, in general, this operator is a sum of terms $L_j$ that only depend on the coordinates and momenta of one particle, and terms $L'_{ij}$ that depend on the coordinates and momenta of two particles, and represent two particle interactions, such that

$$L = \sum_{i=1}^{N} L_i + \sum_{i}^{N} \sum_{j \geq i}^{N} L'_{ij}\,, \tag{3.10}$$

When we replace these results in eq. (3.6), we obtain

$$\frac{\partial f_s}{\partial t} = V_s \int \left[ \sum_{i=1}^{N} L_i + \sum_{i}^{N} \sum_{j \geq i}^{N} L'_{ij} \right] \rho(\Gamma, t)\, d\Gamma_{N-s}\,, \tag{3.11}$$

The first term can be separated into $s$ plus $(N-s)$ terms as

$$\sum_{i=1}^{N} L_i = \sum_{i=1}^{s} L_i + \sum_{i=s+1}^{N} L_i \,. \tag{3.12}$$

It is obvious that the first $s$ terms in the sum over $L_j$ are not affected by the integration over $d\Gamma_{N-s}$, and we simply get

$$V_s \sum_{i=1}^{s} L_i \int \rho(\Gamma, t) d\Gamma_{N-s} = \sum_{i=1}^{s} L_i f_s \,, \tag{3.13}$$

On the other hand, the contribution of the second term in the sum in eq. (3.12) is exactly equal to zero. This can be shown taking the time derivative of the normalization condition (3.3), which gives

$$\frac{\partial}{\partial t} \int \rho(\Gamma, t)\, d\Gamma = \int \left( \sum_{i=1}^{N} L_i + \sum_{i}^{N} \sum_{j \geq i}^{N} L'_{ij} \right) \rho(\Gamma, t)\, d\Gamma = 0 \,. \tag{3.14}$$

This result is perfectly general, and is independent of the details of the external potentials, or the interaction forces. This leads us to conclude that, in order to satisfy this equality, all the terms have to be separately equal to zero, which means that

$$\int L_i \rho(\Gamma, t) di = 0 \,, \quad \int L'_{ij} \rho(\Gamma, t) didj = 0 \,, \tag{3.15}$$

The first of these expressions can then be used to demonstrate the validity of neglecting the sum between $i = s+1$ and $N$, in eq. (3.12).

Let us now turn to the second expression to draw conclusions on the contributions of the second term in (3.6). We can identify three distinct situations:

i) First, if the indices $i$ and $j$ are smaller than $s$, $(i,j) \leq s$, the operators $L'_{ij}$ can be taken out of the integral, leading to

$$\sum_{i=1}^{s} \sum_{j>i} L'_{ij} f_s \,,$$

ii) Second, if the indices $i$ and $j$ are larger than $s$, $(i,j) > s$, the integral is zero, according to eq. (3.15).

iii) Finally, if $i$ is smaller than $s$, but $j$ is larger, $(i \leq s, j > s)$, we can write for questions of symmetry that

$$V_s \int \sum_{i=1}^{s} \sum_{j=s+1}^{N} L'_{ij}(\Gamma, t)\, d\Gamma_{N-s} = \sum_{i=1}^{s} \int L'_{i,s+1} f_{s+1} d(s+1) \,,$$

Using all of these considerations, we can then write the evolution equation

(3.6) in its final form, as

$$\frac{\partial f_s}{\partial t} = \left[ \sum_{i=1}^{s} \left( L_i + \sum_{j>i}^{s} L'_{ij} \right) \right] f_s + + \sum_{i=1}^{s} \int L'_{i,s+1} f_{s+1} d(s+1) \,. \qquad (3.16)$$

This is the famous *BBGKY hierarchy*, collectively honoring the pioneering work of Bogoliubov [30], Born and Green [31], Kirkwood [32] and Yvon [33]. For more details, see for instance [2, 6, 34]. This represents a chain of equations, connecting the evolution of the reduced distribution function $f_s$ in terms of $f_{s+1}$ by an integral over $d(s+1) \equiv d\mathbf{r}_{s+1} d\mathbf{p}_{s+1}$.

This infinite chain of equations is exactly equivalent to the Liouville equation (3.9) from where we started, but gives a detailed description of the different particle distributions, $f_s$, which is physically very useful. We can write the first two equations of this chain, in a more explicit form, as

$$\left( \frac{\partial}{\partial t} - L_1 \right) f_1 = \int L'_{12} f_2 \, d\mathbf{r}_2 d\mathbf{p}_2 \,, \qquad (3.17)$$

and

$$\left( \frac{\partial}{\partial t} - L_1 - L_2 \right) f_2 = L'_{12} f_2 + \int \left( L'_{13} + L'_{23} \right) f_3 \, d\mathbf{r}_3 d\mathbf{p}_3 \,, \qquad (3.18)$$

where $f_1 \equiv f_1(\mathbf{r}_1, \mathbf{p}_1)$ and $f_2 \equiv f_2(\mathbf{r}_1, \mathbf{p}_1, \mathbf{r}_2, \mathbf{p}_2)$. The first of these equations can eventually be transformed into a closed equation for the one-particle distribution function $f_1$, if we find a way to represent the two-particles distribution function $f_2$, in terms of $f_1$. In particular, this leads to the *Boltzmann equation*, where the term in $f_2$ is written as a *collision integral* and is expressed in terms of products of $f_1$. Other methods can also be used to close the chain of equations, to be considered later, the most drastic one (but nevertheless important) leading to the *Vlasov equation*. This equation plays a crucial role in Plasma Physics.

## 3.2 Klimontovich Equation

Let us now consider a different, but equivalent representation of a system of $N$ particles, where the extended $6N$-dimensional phase-space is avoided. We have seen that the dynamical state of a particle (let us call it particle 1) is determined by the evolution of its position and momentum, the so-called *particle trajectory*, $\mathbf{r}_1(t)$ and $\mathbf{p}_1(t)$. This single particle can be represented in 6-dimensional phase-space $(\mathbf{r}, \mathbf{p})$, using *Dirac delta functions*, as

$$N_1(\mathbf{r}, \mathbf{p}, t) = \delta[\mathbf{r} - \mathbf{r}_1(t)] \, \delta[\mathbf{p} - \mathbf{p}_1(t)] \,. \qquad (3.19)$$

This defines a kind of *microscopic density* or *microdensity*. If we have a number $n_\alpha \gg 1$ of identical particles of species $\alpha$ per unit volume, then the microscopic density of this species is given by

$$N_\alpha(\mathbf{r}, \mathbf{p}, t) = \sum_{j=1}^{n_\alpha} \delta[\mathbf{r} - \mathbf{r}_j(t)]\, \delta[\mathbf{p} - \mathbf{p}_j(t)]\,. \tag{3.20}$$

It is obvious that the integration of this quantity over phase-space $(\mathbf{r}, \mathbf{p})$ gives the number of particles $n_\alpha$. In the usual situation of a plasma with two particle species, we have $\alpha = e$ for electrons, and $\alpha = i$ for ions. If we know exactly what are the trajectories of all the particles in the medium, then the quantity $N_\alpha(\mathbf{r}, \mathbf{p}, t)$ will completely define the plasma evolution. Of course, in order to determined these trajectories we need to solve the equations of motion

$$\frac{d\mathbf{r}_j}{dt} = \mathbf{v}_j \equiv \frac{\mathbf{p}_j}{\gamma_j m_\alpha}\,, \quad \frac{d\mathbf{p}_j}{dt} = q_\alpha \left[\mathbf{E}_m(\mathbf{r}_j, t) + \mathbf{v}_j \times \mathbf{B}_m(\mathbf{r}_j, t)\right] \equiv \mathbf{F}_m(\mathbf{r}_j, t)\,, \tag{3.21}$$

where the electric and magnetic fields $\mathbf{E}_m$ and $\mathbf{B}_m$ included in the Lorentz force are the microscopic values, defined at the exact particle positions $\mathbf{r}_j$, and therefore depend on the position and velocities of all the other particles. They are indeed determined by Maxwell's equations where the charge and current densities are determined microscopically as

$$\rho_m(\mathbf{r}, t) = \sum_\alpha q_\alpha \int N_\alpha(\mathbf{r}, \mathbf{p}, t) d\mathbf{p}\,, \quad \mathbf{J}_m(\mathbf{r}, t) = \sum_\alpha q_\alpha \int \mathbf{v} N_\alpha(\mathbf{r}, \mathbf{p}, t) d\mathbf{p}\,. \tag{3.22}$$

In order to obtain the plasma equations, we need to calculate the time derivative of the microdensities (3.20). For that purpose, we use the derivative of an implicit function as

$$\frac{d}{dt} f\left(g(t)\right) = \frac{dg}{dt}\frac{df}{dg}\,,$$

which allows us to write

$$\frac{\partial N_\alpha}{\partial t} = \sum_{j=1}^{n_\alpha} \left( \frac{d\mathbf{r}_j}{dt} \cdot \frac{\partial}{\partial \mathbf{r}_j} + \frac{d\mathbf{p}_j}{dt} \cdot \frac{\partial}{\partial \mathbf{p}_j} \right) \delta[\mathbf{r} - \mathbf{r}_j(t)]\, \delta[\mathbf{p} - \mathbf{p}_j(t)]\,. \tag{3.23}$$

Using the obvious relation

$$\frac{\partial}{\partial x} f(x - y) = -\frac{\partial}{\partial y} f(x - y)\,,$$

we can rewrite this expression as

$$\frac{\partial N_\alpha}{\partial t} = -\sum_{j=1}^{n_\alpha} \left( \mathbf{v}_j \cdot \frac{\partial}{\partial \mathbf{r}} + \mathbf{F}_m(\mathbf{r}_j, t) \cdot \frac{\partial}{\partial \mathbf{p}} \right) \delta[\mathbf{r} - \mathbf{r}_j(t)]\, \delta[\mathbf{p} - \mathbf{p}_j(t)]\,. \tag{3.24}$$

We can now use the well-known property of the Dirac $\delta$-function

$$x\,\delta(x-y) = y\,\delta(x-y)\,,$$

which allows us to replace $\mathbf{r}_j$ and $\mathbf{v}_j$ by $\mathbf{r}$ and $\mathbf{v}$, inside the bracket. This transforms the above equation into

$$\frac{\partial N_\alpha}{\partial t} + \mathbf{v} \cdot \frac{\partial}{\partial \mathbf{r}} N_\alpha + \mathbf{F}_m \cdot \frac{\partial}{\partial \mathbf{p}} N_\alpha = 0\,. \tag{3.25}$$

This equation is known as the *Klimontovich equation* [35]. It describes the evolution of the microdensity (3.20) in a reduced 6-dimensional phase-space, and contains all the possible information on the kinetic state of the plasma species $\alpha$. Notice that this state also depends on the state of the other species, which appear implicitly in the microscopic force. It should be stressed that this equation determines the conservation of the microscopic density in the reduced phase-space $(\mathbf{r}, \mathbf{p})$. This is obvious if we recognize that the differential operator acting on the microdensity $N_\alpha$ is nothing but the total time derivative

$$\frac{D}{Dt} \equiv \frac{\partial}{\partial t} + \frac{d\mathbf{r}}{dt} \cdot \frac{\partial}{\partial \mathbf{r}} + \frac{d\mathbf{p}}{dt} \cdot \frac{\partial}{\partial \mathbf{p}}\,. \tag{3.26}$$

This means that the Klimontovich equation can be written as a conservation equation, of the form

$$\frac{D}{Dt} N_\alpha(\mathbf{r}, \mathbf{p}, t) = 0\,. \tag{3.27}$$

It states the conservation of the number of particles along time. And, because it is an exact equation, it has to be equivalent to the Liouville equation for the system of charged particles, although written in a one-particle phase-space. This is an exact equation, but is not written in covariant form, as discussed later. In general, it is not very practical, because it envolves a quantity that is not a regular function, and contains an excess of information about the medium. Simplified and more useful versions of the plasma kinetic equations will be described next.

We now replace the singular quantity $N_\alpha$ by a regular function, $f_\alpha$, defined as the statistical average of the microdensity

$$f_\alpha(\mathbf{r}, \mathbf{v}, t) = \langle N_\alpha(\mathbf{r}, \mathbf{p}, t) \rangle\,, \tag{3.28}$$

This is a one-particle distribution function for the particle species $\alpha$. The averaging is taken over a *statistical ensemble*, made of an infinite number of different realizations of the same dynamical system. We also define macroscopic fields, which are mean-values of the electric and magnetic fields, as $\mathbf{E} = \langle \mathbf{E}_m \rangle$ and $\mathbf{B} = \langle \mathbf{B}_m \rangle$. This allow us to write

$$N_\alpha = f_\alpha + \delta N_\alpha\,, \quad \mathbf{E}_m = \mathbf{E} + \delta\mathbf{E}\,, \quad \mathbf{B}_m = \mathbf{B} + \delta\mathbf{B}\,, \tag{3.29}$$

where the quantities $\delta N_\alpha$, $\delta\mathbf{E}$ and $\delta\mathbf{B}$, are deviations with respect to the mean-values. Taking an average over the Klimontovich equation (3.25), we get an equation for the distribution function $f_\alpha$ as

$$\left(\frac{d}{dt} + \mathbf{v}\cdot\frac{\partial}{\partial\mathbf{r}} + \mathbf{F}\cdot\frac{\partial}{\partial\mathbf{p}}\right) f_\alpha = -\left\langle \delta\mathbf{F}\cdot\frac{\partial}{\partial\mathbf{p}}\delta N_\alpha\right\rangle \equiv \left(\frac{\partial f_\alpha}{\partial t}\right)_{coll.} . \tag{3.30}$$

where $\mathbf{F} = q_\alpha(\mathbf{E}+\mathbf{v}\times\mathbf{B})$ is the macroscopic Lorentz force, and $\delta\mathbf{F} = (\mathbf{F}_m - \mathbf{F})$ represents the microscopic deviations with respect to this mean-value. In order to calculate the collision term, we need to determine the fluctuations $\delta N_\alpha$. An equation for this quantity can be established if we subtract eq. (3.30) form the exact Klimontovich equation (3.25). The result is

$$\left(\frac{d}{dt} + \mathbf{v}\cdot\frac{\partial}{\partial\mathbf{r}} + \mathbf{F}\cdot\frac{\partial}{\partial\mathbf{p}}\right) \delta N_\alpha = \left\langle \delta\mathbf{F}\cdot\frac{\partial}{\partial\mathbf{p}}\delta N_\alpha\right\rangle - \mathbf{F}\cdot\frac{\partial}{\partial\mathbf{p}} f_\alpha . \tag{3.31}$$

In general, the calculation of the fluctuations $\delta N_\alpha$ is difficult, and several theoretical approaches have been proposed to solve it. Some of them will be discussed later.

## 3.3 Vlasov Equation

Fortunately, in many different situations, we can neglect the influence of collisions, and drop the last term in eq. (3.30). We are then reduced to the well-known Vlasov equation

$$\left(\frac{d}{dt} + \mathbf{v}\cdot\frac{\partial}{\partial\mathbf{r}} + \mathbf{F}\cdot\frac{\partial}{\partial\mathbf{p}}\right) f_\alpha = 0 . \tag{3.32}$$

Notice that this is also an equation of conservation, which can be rewritten as

$$\frac{D}{Dt} f_\alpha(\mathbf{r},\mathbf{p},t) = 0 . \tag{3.33}$$

For this reason, it has been also called a *one-particle Liouville equation.* Although this equation is not written in a covariant form it can easily be transformed into a covariant equation, if we use the covariant Hamiltonian, or super-Hamiltonian, $\mathcal{H}$. For that purpose we consider the motion of a particle of species $\alpha$, moving in an electromagnetic field. As shown in the previous chapter, this super-Hamiltonian is given by

$$\mathcal{H}(x^\mu, P_\mu) = \frac{1}{2m}\left[\left(P^0 - q_\alpha\frac{V}{c}\right)^2 - (\mathbf{P} - q_\alpha\mathbf{A})^2\right] = \frac{1}{2}mc^2 , \tag{3.34}$$

where we have used the 4-position, $x^\mu \equiv (ct, \mathbf{r})$ and 4-momentum, $P^\mu \equiv (H/c, \mathbf{P})$. Let us then consider a generic dynamical variable $f_\alpha(\tau, \mathbf{r}, \mathbf{P})$, where

$\tau$ is the particle proper time such that $d\tau = dt/\gamma$, where $\gamma = (1 - v^2/c^2)^{-1/2}$. If we identify this with the one-particle distribution function, and if there are no collisions, we can state the conservation law

$$\frac{df_\alpha}{d\tau} \equiv \frac{\partial f_\alpha}{\partial \tau} + \frac{\partial f_\alpha}{\partial \mathbf{r}} \cdot \frac{d\mathbf{r}}{d\tau} + \frac{\partial f_\alpha}{\partial \mathbf{P}} \cdot \frac{d\mathbf{P}}{d\tau} = 0 \,. \tag{3.35}$$

Using the canonical equations of motion

$$\frac{d\mathbf{r}}{\partial \tau} = \frac{\partial \mathcal{H}}{\partial \mathbf{P}} \,, \quad \frac{d\mathbf{P}}{\partial \tau} = -\frac{\partial \mathcal{H}}{\partial \mathbf{r}} \,, \tag{3.36}$$

and replacing this in eq. (3.35), we obtain

$$\frac{\partial f_\alpha}{\partial \tau} + \{f_\alpha, \mathcal{H}\} = 0 \,, \tag{3.37}$$

where we have used the *Poisson bracket*

$$\{f_\alpha, \mathcal{H}\} = \frac{\partial f_\alpha}{\partial \mathbf{r}} \cdot \frac{\partial \mathcal{H}}{\partial \mathbf{P}} - \frac{\partial f_\alpha}{\partial \mathbf{P}} \cdot \frac{\partial \mathcal{H}}{\partial \mathbf{r}} \,. \tag{3.38}$$

We can easily recognize that, if the fields in the Hamiltonian are determined by the charges and currents of all the particles existing in the system, then eq. (3.37) coincides with the Vlasov equation (3.32). This can be shown by writing the explicit expression for the Poisson bracket. Noting that

$$\frac{\partial \mathcal{H}}{\partial \mathbf{r}} = -q_\alpha(\gamma \mathbf{E} + \mathbf{u} \times \mathbf{B}) \,, \quad \frac{\partial \mathcal{H}}{\partial \mathbf{P}} = \frac{\partial \mathbf{r}}{\partial \tau} = \mathbf{u} \,, \tag{3.39}$$

where $\mathbf{u} = \gamma \mathbf{v} = \mathbf{p}/m_\alpha$, we can easily transform eq. (3.37) into

$$\frac{\partial f_\alpha}{\partial \tau} + \mathbf{u} \cdot \frac{\partial f_\alpha}{\partial \mathbf{r}} + \frac{q_\alpha}{m_\alpha} (\gamma \mathbf{E} + \mathbf{u} \times \mathbf{B}) \cdot \frac{\partial f_\alpha}{\partial \mathbf{u}} = 0 \,. \tag{3.40}$$

where $\gamma = (1 + u^2/c^2)^{1/2}$. If instead, we want to write the distribution function in terms of the laboratory time $t$, as $f_\alpha = f_\alpha(t, \mathbf{r}, \mathbf{p})$, we get

$$\frac{\partial f_\alpha}{\partial t} + \mathbf{v} \cdot \frac{\partial f_\alpha}{\partial \mathbf{r}} + q_\alpha (\mathbf{E} + \mathbf{v} \times \mathbf{B}) \cdot \frac{\partial f_\alpha}{\partial \mathbf{p}} = 0 \,, \tag{3.41}$$

which is the explicit form of eq. (3.32). As mentioned, this is a simple but pivotal equation, that will be used to analyse several processes in most of this book.

It seems, at first sight that, by ignoring the collision integral we have eliminated all the correlations between particles. But this is not the case if we calculate the fields self-consistently, using the charge and current distributions associated with the distribution functions $f_\alpha$, as

$$\rho(\mathbf{r}, t) = \sum_\alpha q_\alpha \int f_\alpha(\mathbf{r}, \mathbf{p}, t) d\mathbf{p} \,, \quad \mathbf{J}(\mathbf{r}, t) = \sum_\alpha q_\alpha \int \mathbf{v} f_\alpha(\mathbf{r}, \mathbf{p}, t) d\mathbf{p} \,. \tag{3.42}$$

The use of self-consistent fields will allow us to retain particle correlations that are associated with the collective fields. The only correlations still absent in the Vlasov equations are those due to short-range particle interactions that occur inside the Debye sphere, as discussed later.

## 3.4 Thermal Equilibrium

Let us now consider thermal equilibrium. For an unspecified particle species, this should correspond to a distribution function such that $f_0(\mathbf{p})d\mathbf{p}$ is the number of particles in the momentum interval between $\mathbf{p}$ and $\mathbf{p} + d\mathbf{p}$, where we define the element of volume in momentum space as $d\mathbf{p} = dp_x dp_y dp_z$. In the absence of fields, the relativistic version of the *Maxwell-Boltzmann equilibrium* should take the form

$$f_0(\mathbf{p}) = C \exp\left[-\frac{H(\mathbf{p})}{T}\right], \quad H(\mathbf{p}) = mc^2\sqrt{1 + p^2/m^2c^2}\,, \tag{3.43}$$

where $C$ is a normalization constant. It is also useful to add the equivalent expression

$$f_0(\mathbf{p}) = C \exp\left[-\mu\gamma(p)\right], \quad \gamma(p) = \sqrt{1 + p^2/m^2c^2}\,, \quad \mu = \frac{mc^2}{T}\,. \tag{3.44}$$

The normalization constant should be chosen such that

$$\int f_0(\mathbf{p})d\mathbf{p} = 1\,. \tag{3.45}$$

This integration leads to

$$\frac{4\pi}{(mc)^3}C\int_0^\infty p^2 dp \exp\left[-\mu\gamma(p)\right] = 1\,. \tag{3.46}$$

Using $\gamma$ as an auxiliary variable, such that $dp = m^2c^2(\gamma/p)d\gamma$, we obtain

$$4\pi C\int_0^\infty e^{-\mu\gamma}(\gamma^2 - 1)^{1/2}\gamma d\gamma = (mc)^{-3}\,. \tag{3.47}$$

Integrating by parts and taking into account that

$$\int (\gamma^2 - 1)^{1/2}\gamma d\gamma = \frac{1}{3}(\gamma^2 - 1)^{3/2}\,,$$

we can easily transform the above expression into

$$\frac{4\pi}{3\mu}C\int_1^\infty e^{-\mu\gamma}(\gamma^2 - 1)^{3/2}d\gamma = (mc)^{-3}\,. \tag{3.48}$$

We now use the definition of the *modified Bessel function* of the second kind, as (see [36], chapter 11)

$$K_\nu(z) = \frac{\sqrt{\pi}(z/2)^\nu}{(\nu - 1/2)!}\int_1^\infty e^{-zt}(t^2 - 1)^{\nu - 1/2}dt\,,$$

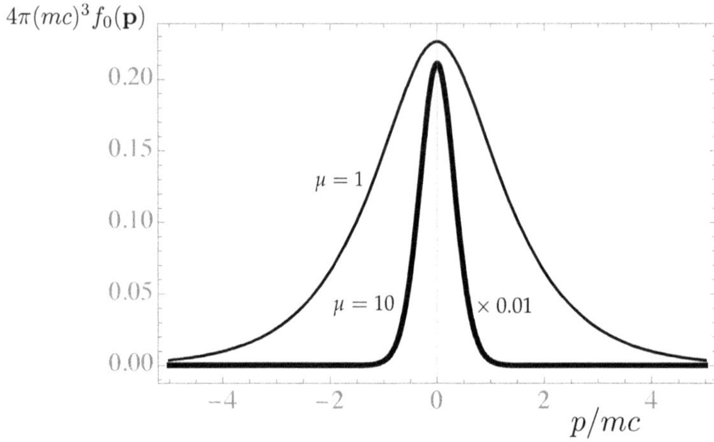


**FIGURE 3.1**
Maxwell-Jüttner distribution, in normalized form, $4\pi(mc)^3 f_0(\mathbf{p})$, as a function of the normalized momentum $p/mc$, for different values of the thermal parameter $\mu = mc^2/T$.

This allows us to write the normalization constant in terms of the function $K_2(z)$, with $z = \mu$, as

$$C = \frac{\mu}{4\pi(mc)^3 K_2(\mu)} . \tag{3.49}$$

The equilibrium distribution (3.44) becomes

$$f_0(\mathbf{p}) = \frac{1}{4\pi(mc)^3} \frac{\mu}{K_2(\mu)} \exp\left[-\mu\gamma(p)\right] . \tag{3.50}$$

This is known as the *Maxwell-Jüttner distribution*, or simply the *relativistic Maxwell*, first derived by Jüttner in 1911 [37]. Writing it more explicitly in terms of the temperature, we have

$$f_0(\mathbf{p}) = \frac{1}{4\pi T m^2} \frac{1}{K_2(mc^2/T)} \exp\left[-\frac{H(\mathbf{p})}{T}\right] , \tag{3.51}$$

This is illustrated in Figure 3.1. It is also useful to derive an approximate expression of this equilibrium distribution at low temperatures, such that $\mu \gg 1$. For that purpose, we use the assymptotic expansion of the Bessel function $K_2(\mu)$, valid in this limit, as

$$K_2(\mu) = e^{-\mu}\sqrt{\frac{\pi}{2\mu}}\left[1 + \frac{1}{2\mu}\right]$$

Taking the dominant term, and using $\gamma \simeq (1 + p^2/2m^2c^2)$, we can then reduce the equilibrium distribution to

$$f_0(\mathbf{p}) = \frac{m^{3/2}}{(2\pi T)^{3/2}} \exp\left[-\frac{p^2}{2m}\right] . \tag{3.52}$$

This means that we are reduced to the *Maxwell-Boltzmann distribution* as expected. Noting that $\mathbf{p} = m\mathbf{v}$, we need to multiply this expression by $m^3$, in order to get the equilibrium distribution written in terms of velocity, $f_0(\mathbf{v})$, normalized as $\int f_0(\mathbf{v})d\mathbf{v} = 1$.

## 3.5 Fluid Equations

Instead of the above plasma kinetic description, providing the detailed momentum distribution of the different plasma particles, $f_\alpha(\mathbf{p})$, we can use a fluid description based on macroscopic quantities. For that purpose, we start from the Vlasov equation (3.32), multiply it by successive powers of $\mathbf{p}$, and integrate over the momentum space. This leads to a hierarchy of macroscopic plasma equations, which can be interrupted by some simplifying assumption. The first of these equations results from a direct integration of eq. (3.32) over the momentum space. For that purpose, we define the mean density $n_\alpha$, and the mean velocity $\mathbf{v}_\alpha$ of the particle species $\alpha$, as

$$n_\alpha = \int f_\alpha(\mathbf{r}, \mathbf{p}, t)d\mathbf{p} , \quad \mathbf{v}_\alpha = \frac{1}{n_\alpha} \int \mathbf{v} f_\alpha(\mathbf{r}, \mathbf{p}, t)d\mathbf{p} . \tag{3.53}$$

This leads to the *continuity equation*

$$\frac{\partial n_\alpha}{\partial t} + \nabla(n_\alpha \mathbf{v}_\alpha) = 0 . \tag{3.54}$$

It states that the particle number is conserved. We should notice that the force term in the Vlasov equation (3.32) gives no contribution to this equation. This can be easily understood if we use an integration by parts, leading to

$$\int \left(\mathbf{F} \cdot \frac{\partial f_\alpha}{\partial \mathbf{p}}\right) d\mathbf{p} = - \int f_\alpha \left(\frac{\partial}{\partial \mathbf{p}} \cdot \mathbf{F}\right) d\mathbf{p} . \tag{3.55}$$

The second integral gives the mean value of the divergence of the Lorentz force $\mathbf{F}$ over the momentum space. Using the divergence theorem, we can easily verify that this averaged divergence is zero.

Next, we multiply eq. (3.32) by $\mathbf{p}$, and integrate over the momentum space. We then get the *momentum conservation equation*, of the form

$$\frac{d\mathbf{p}_\alpha}{dt} + \frac{1}{n_\alpha} \nabla \cdot \bar{\bar{P}}_\alpha = q_\alpha \left(\mathbf{E} + \mathbf{v}_\alpha \times \mathbf{B}\right) . \tag{3.56}$$

Here, we have used the mean particle momentum $\mathbf{p}_\alpha$

$$\mathbf{p}_\alpha = \frac{1}{n_\alpha} \int \mathbf{p}\, f_\alpha(\mathbf{r}, \mathbf{p}, t) d\mathbf{p} \,. \tag{3.57}$$

and introduced the pressure tensor $\bar{\bar{P}}_\alpha$, defined as

$$\bar{\bar{P}}_\alpha = \int (\mathbf{p} - \mathbf{p}_\alpha)\,(\mathbf{v} - \mathbf{v}_\alpha)\, f_\alpha(\mathbf{r}, \mathbf{p}, t) d\mathbf{p} \,. \tag{3.58}$$

The pressure term in eq. (3.56) has to be determined by a third equation of the fluid hierarchy, which is obtained from the Vlasov equation multiplying it by the tensor $\mathbf{pp}$, and integrating in momentum space. This new equation will involve the *heat flow tensor* $\bar{\bar{Q}}_\alpha$, defined as

$$\bar{\bar{Q}}_\alpha = \int (\mathbf{p} - \mathbf{p}_\alpha)\,(\mathbf{v} - \mathbf{v}_\alpha)\,(\mathbf{v} - \mathbf{v}_\alpha)\, f_\alpha(\mathbf{r}, \mathbf{p}, t) d\mathbf{p} \,. \tag{3.59}$$

In alternative, we can use an *ansatz* on the pressure tensor in eq. (3.56), thus breaking the sequence of coupled fluid equations of increasing order. For instance, in isotropic plasmas we can use a simple *equation of state* of the form

$$\bar{\bar{P}}_\alpha = \bar{\bar{1}}\, C(T_\alpha) n_\alpha^{g_\alpha} \,, \tag{3.60}$$

where, for each species, the constant $C(T_\alpha)$ and a polytropic exponent $g_\alpha$ are to be defined, as shown later.

## 3.6 MHD

The simplest possible description of a plasma corresponds to a single conducting fluid, moving in the presence of a magnetic field. This description was used for the first time by Alfvén in 1942 [38]. This is called *Magnetohydrodynamics*, or MHD in short. In such a description, the charged particle species (electrons and ions) lose their identity, and are assumed to move together. Of course, this is only valid for slow processes, much slower than $1/\omega_{pe}$.

Here, we present two different versions of MHD, one containing the basic physics concepts and the other, a more formal covariant description.

### 3.6.1 Basic MHD Concepts

We start with the definition of the single fluid plasma density $\rho$, and the corresponding current, $\mathbf{J}$, as

$$\rho \equiv \sum_\alpha \rho_\alpha = \sum_\alpha m_\alpha n_\alpha \,, \quad \mathbf{J} = \sum_\alpha q_\alpha n_\alpha \mathbf{v}_\alpha \,, \tag{3.61}$$

where $\alpha$ represents the different plasma particle species and $\mathbf{v}_\alpha$ is the mean velocity of each species. Obviously, this single fluid satisfies a continuity equation, of the form

$$\frac{\partial \rho}{\partial t} + \nabla \cdot \rho \mathbf{v} = 0 . \tag{3.62}$$

where $\mathbf{v}$ is the velocity of the center of mass of this single fluid. In terms of the plasma particle species, we could define it as

$$\mathbf{v} = \frac{1}{\rho} \sum m_\alpha n_\alpha \mathbf{v}_\alpha . \tag{3.63}$$

This continuity equation is completed by the equation of motion, which determines the center of mass velocity $\mathbf{v}$ in the presence of a magnetic field $\mathbf{B}$, as

$$\rho \frac{d\mathbf{v}}{dt} = -\nabla \cdot \bar{\bar{P}} + \mathbf{J} \times \mathbf{B} , \tag{3.64}$$

where $\bar{\bar{P}}$ is the pressure tensor. In its simplest version, we can use $\bar{\bar{P}} = P \bar{\bar{1}}$, where the pressure $P$ is a scalar. It can be related to the fluid density $\rho$ by an equation of state. The standard form of the adiabatic equation of state is given by $(d/dt)(P\rho^{-\gamma}) = 0$, where $\gamma$ is the adiabatic index.

In order to determine the magnetic field $\mathbf{B}$ in the presence of the current $\mathbf{J}$, we need to use Maxwell's equations. For sufficiently slow motions of the fluid, we can neglect the displacement current and write

$$\nabla \times \mathbf{B} = \mu_0 \mathbf{J} , \quad \nabla \cdot \mathbf{B} = 0 . \tag{3.65}$$

The current $\mathbf{J}$ in a conducting medium with conductivity $\sigma$ can be written in the fluid rest frame as $\mathbf{J} = \sigma \mathbf{E}'$, where $\mathbf{E}'$ is the electric field in that frame. For $v \ll c$ this is related to the electric field in the lab frame as $\mathbf{E}' = \mathbf{E} + \mathbf{v} \times \mathbf{B}$. Replacing this in the Faraday equation, we get

$$\nabla \times \mathbf{E} = \frac{1}{\sigma} \nabla \times \mathbf{J} - \nabla \times (\mathbf{v} \times \mathbf{B}) = -\frac{\partial \mathbf{B}}{\partial t} . \tag{3.66}$$

Notice that eq. (3.65) allows us to write

$$\nabla \times (\nabla \times \mathbf{B}) \equiv -\nabla^2 \mathbf{B} = \mu_0 \nabla \times \mathbf{J} . \tag{3.67}$$

From eq. (3.66) we then get

$$\frac{\partial \mathbf{B}}{\partial t} = \nabla \times (\mathbf{v} \times \mathbf{B}) + \frac{1}{\mu_0 \sigma} \nabla^2 \mathbf{B} . \tag{3.68}$$

This completes the set of equations for a basic MHD description of the medium, with eqs. (3.64)–(3.65), where the electric field is absent.

Let us discuss the physical meaning of eq. (3.68). It basically shows that the magnetic field changes in the medium by two distinct processes. One is the convective motion of the fluid, as determined by the center of mass velocity

**v**. The other is called the magnetic field diffusion, and is dictated by a finite conductivity $\sigma$. In order to understand *magnetic diffusion*, let us assume that convection is absent, $\mathbf{v} = 0$. In this case, eq. (3.68) reduces to a diffusion equation of the form

$$\frac{\partial \mathbf{B}}{\partial t} = D_m \nabla^2 \mathbf{B}\,, \quad D_m = \frac{1}{\mu_0 \sigma}\,, \tag{3.69}$$

where $D_m$ is the magnetic diffusion coefficient. In a medium with size $L$, the characteristic diffusion time $\tau_D$ is of order

$$\tau_D = \frac{L^2}{D_m} = \mu_0 \sigma L^2\,. \tag{3.70}$$

This shows that, when the conductivity is very high, and consequently the diffusion time-scale is very large, we can neglect the last term in eq. (3.68) and simply retain the convective process

$$\frac{\partial \mathbf{B}}{\partial t} = \nabla \times (\mathbf{v} \times \mathbf{B})\,. \tag{3.71}$$

This approximation is called the *ideal MHD*, where the effect of dissipation associated with a finite conductivity is not present. It can adequately describe processes much faster than $\tau_D$. In such an ideal case, the magnetic field **B** is simply determined by the fluid motion and moves along with it, because it has no time to diffuse. We can then say that we have a *frozen-in* magnetic field.

In order to understand the meaning of this frozen-in effect, we consider a surface S inside the fluid, limited by a closed contour $C$. The magnetic flux crossing this surface is

$$\phi_m = \int_S \mathbf{B} \cdot d\mathbf{S}\,, \tag{3.72}$$

where $d\mathbf{S}$ is a vector perpendicular to the element of surface $dS$. Let us consider a temporal variation of the flux, as determined by

$$\frac{d\phi_m}{dt} = \int_S \frac{d\mathbf{B}}{dt} \cdot d\mathbf{S} + \int_S \mathbf{B} \cdot \frac{d\mathbf{S}}{dt}\,, \tag{3.73}$$

The first term in this expression gives the variation directly due to a variation of the field. The second term represents a contribution to the variation of the boundary $C$, because we know that the flux does not depend on the exact shape of the surface $S$, only on its boundary. But, at the boundary, the temporal variation of the surface can be written in terms of the fluid velocity, as $d\mathbf{S}/dt = \mathbf{v} \times d\mathbf{s}$, where $d\mathbf{s}$ is a line element of the contour $C$. Which means that

$$\frac{d\phi_m}{dt} = \int_S \frac{d\mathbf{B}}{dt} \cdot d\mathbf{S} + \oint_C \mathbf{v} \cdot \mathbf{B} \times d\mathbf{s}\,. \tag{3.74}$$

Now, we can use eq. (3.71) and the *Stokes theorem* to calculate the first term, as

$$\int_S \frac{d\mathbf{B}}{dt}\cdot d\mathbf{S} = \int_S \nabla\times(\mathbf{v}\times\mathbf{B})\cdot\frac{d\mathbf{S}}{dt} = \oint_C (\mathbf{v}\times\mathbf{B})\cdot d\mathbf{s} = -\oint_C \mathbf{v}\cdot\mathbf{B}\times d\mathbf{s}\,. \quad (3.75)$$

This exactly compensates the second term of eq. (3.118), and the net result is

$$\frac{d\phi_m}{dt} = 0\,, \quad (3.76)$$

We can then conclude that the magnetic flux through any surface $S$ moving with the fluid is always constant, $\phi_m$ = const. This also means that the number of magnetic field lines that cross the surface is always the same. Or, in other words, the magnetic field lines are transported by the fluid.

The MHD equations are particularly suited to determine the plasma equilibrium in a given magnetic field configuration, such as the magnetic field around the Earth and planets, or a magnetic fusion machine. To illustrate the basic equilibrium conditions, we go back to eqs. (3.64) and consider the static situation where $d\mathbf{v}/dt = 0$. Using a scalar pressure $P$, we obtain

$$\nabla P = \mathbf{J}\times\mathbf{B} = \frac{1}{\mu_0}(\nabla\times\mathbf{B})\times\mathbf{B} = \frac{1}{\mu_0}\left[\mathbf{B}\cdot\nabla\mathbf{B} - \frac{1}{2}\nabla B^2\right]\,, \quad (3.77)$$

We then get

$$\nabla\left(P + \frac{B^2}{2\mu_0}\right) = \frac{1}{\mu_0}\mathbf{B}\cdot\nabla\mathbf{B}\,. \quad (3.78)$$

This is the plasma equilibrium equation for a given magnetic configuration. Let us consider the case where the derivatives along the magnetic lines are zero, or $\mathbf{B}\cdot\nabla\mathbf{B} = 0$. The equilibrium equation then shows that the kinetic pressure is exactly compensated by the magnetic pressure, as

$$\left(P + \frac{B^2}{2\mu_0}\right) = const\,. \quad (3.79)$$

An important quantity in the study of plasma equilibrium is the so-called $\beta$ *parameter*, which determines the ratio between the kinetic pressure and the magnetic pressure,

$$\beta = \frac{2\mu_0 P}{B^2}\,. \quad (3.80)$$

In a confined plasma, this parameter is always less than 1 and can be used as a *figure of merit* of a given magnetic confinement. It can easily be seen that, outside the plasma ($P = 0$), the magnetic field $B_{ext}$ is always larger than the field inside the plasma, $B_{int}$, because eq. (3.79) necessarily implies that

$$\frac{B_{ext}^2}{2\mu_0} = P + \frac{B_{int}^2}{2\mu_0}\,. \quad (3.81)$$

This shows that the plasma expels the magnetic field, which is a characteristic feature of *diamagnetic media*.

Before proceeding further, it is important to know that the MHD equilibrium equation, as stated in (3.79), can be formulated in terms of an integral equation. This integral formulation takes the form of a *virial theorem*. For that purpose, we define two pressure terms, $P_\perp$ and $P_\parallel$, such that

$$P_\perp = P + \frac{B^2}{2\mu_0}, \quad P_\parallel = P - \frac{B^2}{2\mu_0}, \tag{3.82}$$

which represent the pressure across and along the direction of the magnetic field. With these two quantities we can build the following tensor

$$T_{jk} = P_\perp \left( \delta_{jk} - \frac{B_j B_k}{B^2} \right) + P_\parallel \frac{B_j B_k}{B^2}. \tag{3.83}$$

This can also be written in matrix form, as

$$\bar{\bar{T}} = \begin{bmatrix} P_\perp & 0 & 0 \\ 0 & P_\perp & 0 \\ 0 & 0 & P_\parallel \end{bmatrix} \tag{3.84}$$

Using this tensor, it is then possible to rewrite the equilibrium equation (3.79) as the divergence of a tensor

$$\nabla \cdot \bar{\bar{T}} = 0, \quad or \quad \frac{\partial T_{jk}}{\partial x_j} = 0. \tag{3.85}$$

Using the following identity

$$\frac{\partial}{\partial x_j} x_k T_{jk} = T_{jk}\delta_{jk} + x_k \frac{\partial T_{jk}}{\partial x_j}, \tag{3.86}$$

and taking eq. (3.85) into account, we obtain

$$\frac{\partial}{\partial x_j} x_k T_{jk} = T_{jk}\delta_{jk}. \tag{3.87}$$

Let us now integrate this expression over the volume $V$ limited by a closed surface $S$. We get

$$\int_V \frac{\partial}{\partial x_j}(x_k T_{jk}) dV = \int_V T_{jk}\delta_{jk} dV. \tag{3.88}$$

Using the *divergence theorem* on the left-hand side of this equation, we also have

$$\oint_S (x_k T_{jk}) dS_j = \int_V T_{jk}\delta_{jk} dV. \tag{3.89}$$

We can now use the explicit form of the tensor $T_{jk}$ as defined above. We obtain

$$\oint_S \left[ P_\perp (\mathbf{r} \cdot d\mathbf{S}) + (P_\parallel - P_\perp) \frac{(\mathbf{r} \cdot \mathbf{B})(\mathbf{B} \cdot d\mathbf{S})}{B^2} \right] = \int_V (2P_\perp + P_\parallel)\, dV \,. \tag{3.90}$$

Now, using the definitions of $P_\perp$ and $P_\parallel$, as given in (3.82), we obtain an expression relating the magnetic field $\mathbf{B}$ and the plasma pressure, as

$$\oint_S \left[ \left( P + \frac{B^2}{2\mu_0} \right) (\mathbf{r} \cdot d\mathbf{S}) - \frac{1}{\mu_0} \frac{(\mathbf{r} \cdot \mathbf{B})(\mathbf{B} \cdot d\mathbf{S})}{B^2} \right] = \int_V \left( 3P + \frac{B^2}{2\mu_0} \right) dV \,. \tag{3.91}$$

This states the MHD equilibrium in the form of a virial theorem, relating the pressure and the magnetic forces.

### 3.6.2 Covariant MHD

Let us now consider the covariant form of MHD equations. In this case, we replace the fluid velocity $\mathbf{v}$ defined in the three-dimensional space by the 4-velocity $u^\alpha = (\gamma, \gamma \mathbf{v})$ defined in spacetime, where now $\gamma = 1/\sqrt{1 - v^2/c^2}$ is the relativistic factor of the moving fluid. If we are in the fluid rest frame, we simply have $(u^0 = 1, u^j = 0)$. We can define the plasma energy-momentum tensor $T^{\mu\nu}$ as

$$T^{\mu\nu} = P g^{\mu\nu} + h u^\mu u^\nu \,, \quad h = P + \epsilon \,, \tag{3.92}$$

where $g^{\mu\nu}$ represents the metric tensor, the scalar $P$ is the pressure, as above. We have also introduced the enthalpy $h$, defined as the sum of the pressure and the energy density. Knowledge of this tensor allows us to state the fluid equation of motion as [39]

$$\frac{\partial T^{\mu\nu}}{\partial x^\nu} \equiv \partial_\nu T^{\mu\nu} = 0 \,. \tag{3.93}$$

Notice the formal analogy with eq. (3.85), where the divergence was taken in the three-dimensional space for a static equilibrium configuration. In contrast, here we have defined the divergence in spacetime, and included the temporal evolution of the fluid. In a more general formulation, we should also take into account the interaction of the fluid with an applied electromagnetic field, and include a source term, such that

$$\partial_\nu T^{\mu\nu} = F^{\mu\nu} J_\nu \,. \tag{3.94}$$

where $J_\nu$ is the 4-vector current and $F^{\mu\nu}$ is the electromagnetic field tensor. A detailed discussion of relativistic MHD is outside the scope of the present work. Complementary information can be found in the papers [40, 41, 42].

## 3.7 Collisions

We now consider the influence of particle collisions in the kinetic plasma equations. As we have seen, they can adequately be written as

$$\left(\frac{d}{dt} + \mathbf{v} \cdot \frac{\partial}{\partial \mathbf{r}} + \mathbf{F} \cdot \frac{\partial}{\partial \mathbf{p}}\right) f_\alpha = \left(\frac{\partial f_\alpha}{\partial t}\right)_{coll} . \tag{3.95}$$

where we now focus in the collision term appearing on the right side. A simple, and sometimes useful, approach proposed by Bhatnagar, Gross and Krook [43] consists in the use of a phenomenological collision frequency $\nu_c$, such that

$$\left(\frac{\partial f_\alpha}{\partial t}\right)_{coll} = -\nu_c \left(f_\alpha - f_{\alpha 0}\right) , \tag{3.96}$$

where $f_{\alpha o}$ is the equilibrium distribution for the particle species $\alpha$. It describes the decay toward equilibrium on a time scale of $1/\nu_c$, which is usually associated with the particle collisions. The dominant collisional effects are associated with particles moving at distances $r$ smaller that the Debye length, or $r < \lambda_D$. Each particle, seen as a test particle, moves under the influence of multiple collisions along a random trajectory in momentum space. This can be described by a Fokker-Planck equation, of the generic form

$$\left(\frac{\partial f_\alpha}{\partial t}\right)_{coll} = \frac{\partial}{\partial p_i}\left[-A_i f_\alpha + \frac{1}{2}\frac{\partial}{\partial p_j}\left(B_{ij} f_\alpha\right)\right] , \tag{3.97}$$

where $A_i$ are the components of a *friction vector* and $B_{ij}$ the components of a *diffusion tensor*, to be defined. We should comply with the conservation of the particle number, momentum and energy, such that

$$\int (1; \mathbf{p}; p^2) \left(\frac{\partial f_\alpha}{\partial t}\right)_{coll} d\mathbf{p} = 0 \tag{3.98}$$

The first condition is immediately satisfied by eq. (3.97), and the other two lead to

$$A_i = -\nu_c \left(p_i - p_{i0}\right) , \quad B_{ij} = 2\nu_c \frac{T}{m\gamma c^2} \delta_{ij} , \tag{3.99}$$

where $\mathbf{p}_0$ is a drift momentum, $T$ the equilibrium temperature and $\gamma$ the thermal relativistic factor. In this case, we have a collision term of the form

$$\left(\frac{\partial f_\alpha}{\partial t}\right)_{coll} = \nu_c \frac{\partial}{\partial \mathbf{p}} \cdot \left[\left(\mathbf{p} - \mathbf{p}_0\right) f_\alpha + \frac{T}{m\gamma c^2}\frac{\partial}{\partial \mathbf{p}} f_\alpha\right] , \tag{3.100}$$

The validity of these simple model will be discussed in the following.

### 3.7.1 Coulomb Collisions

Let us examine the contribution of individual particle collisions to the plasma collective behavior. Charged particles interact via Coulomb collisions, and our starting point is the cross-section of binary collisions. We then discuss the global effect of binary collisions and the difficulties associated with the description of a collisional plasma.

Let us then consider a charged particle of charge $q_1$ and mass $m_1$, colliding with another particle, with charge $q_2$ and mass $m_2$. Starting with a non-relativistic formulation, the equations of motion of each particle under the Coulomb field created by the other, can be written as

$$m_1 \frac{d^2 \mathbf{r}_1}{dt^2} = \frac{q_1 q_2}{4\pi\epsilon_0} \frac{\mathbf{r}}{|r|^3}, \quad m_2 \frac{d^2 \mathbf{r}_2}{dt^2} = -\frac{q_1 q_2}{4\pi\epsilon_0} \frac{\mathbf{r}}{|r|^3}, \tag{3.101}$$

where $\mathbf{r} = \mathbf{r}_1 - \mathbf{r}_2$ is the vector distance between the two particles. Defining the center-of-mass vector position $\mathbf{R}$, and the reduced mass of this system as

$$\mathbf{R} = \frac{m_1 \mathbf{r}_1 + m_2 \mathbf{r}_2}{m_1 + m_2}, \quad m = \frac{m_1 m_2}{m_1 + m_2}, \tag{3.102}$$

we can transform the equations of motion into

$$\frac{d^2 \mathbf{R}}{dt^2} = 0, \quad m \frac{d^2 \mathbf{r}}{dt^2} = \frac{q_1 q_2}{4\pi\epsilon_0} \frac{\mathbf{r}}{|r|^3}. \tag{3.103}$$

The first equation indicates that the center-of-mass moves at constant speed, while the second one shows that the force applied to the particle with reduced mass $m$ is equal to the force applied to the first particle. This equation of motion can be derived by a Lagrangian of the form $\mathcal{L} = T(r) - V(r)$. Noting that the Coulomb potential $V(r) = q_1 q_2 / r$ is a central field, which only depends on the absolute value of the distance $r$, we can use polar coordinates in the plane defined by the center of mass and the velocity of the reduced particle, and write the Lagrangian as

$$\mathcal{L}(r, \dot{r}, \dot{\theta}) = \frac{1}{2} m \left( \dot{r}^2 + r^2 \dot{\theta}^2 \right) - V(r). \tag{3.104}$$

Noting that this Lagrangian is independent of the angular variable ($\partial \mathcal{L} / \partial \theta = 0$), we conclude that the moment conjugate $p_\theta$ is constant

$$p_\theta = \frac{\partial \mathcal{L}}{\partial \dot{\theta}} = m r^2 \dot{\theta} = const. \tag{3.105}$$

This constant of motion is precisely the angular momentum of the incident reduced particle, $L \equiv p_\theta$. The temporal evolution of the angle $\theta$ is then determined by the equation

$$\dot{\theta} \equiv \frac{d\theta}{dt} = \frac{L}{m r^2}. \tag{3.106}$$

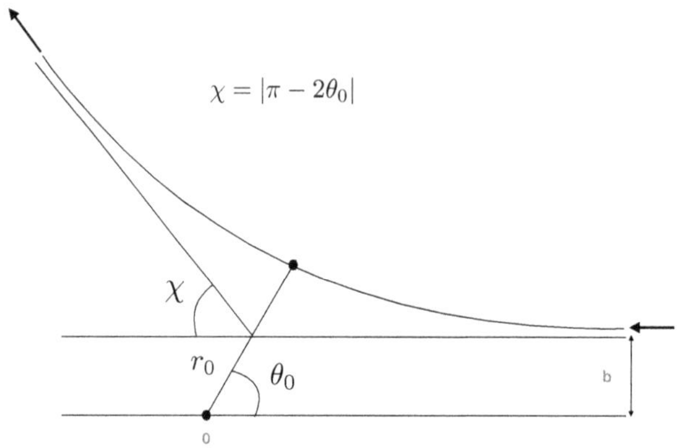


**FIGURE 3.2**
Geometry of Coulomb collisions: Deflection angle $\chi$ and nearest position $(r_0, \theta_0)$.

We can write the total energy of the particle as

$$\mathcal{E} = T(r) + V(r) = \frac{1}{2}\left(m\dot{r}^2 + \frac{L^2}{mr^2}\right) + \frac{q_1 q_2}{4\pi\epsilon_0 r} . \tag{3.107}$$

From this expression we derive the temporal evolution of the distance $r$, as given by

$$\frac{dr}{dt} = \sqrt{\frac{2}{m}\left[\mathcal{E} - U(r)\right]}, \quad U(r) = V(r) + \frac{L^2}{2mr^2} . \tag{3.108}$$

Eliminating time from eqs. (3.105) and (3.107), we obtain the particle trajectory in the plane $(r, \theta)$, as

$$d\theta = \frac{L}{r^2}\frac{dr}{\sqrt{(2/m)\left[\mathcal{E} - U(r)\right]}} . \tag{3.109}$$

This result shows that the incident (reduced) particle approaches the center of mass up to a minimum distance $r = r_0$, where its energy equals the effective potential, $\mathcal{E} = U(r)$, and $\theta = \theta_0$. This angle is determined by the integral

$$\theta_0 = \int_{r_0}^{\infty} \frac{L}{r^2}\frac{dr}{\sqrt{(2/m)[\mathcal{E} - U(r)]}} . \tag{3.110}$$

The trajectory is symmetric with respect to this point $(r_0, \theta_0)$ (see Figure 3.2). Therefore, the two asymptotes to the trajectory make an angle $\chi$ with respect to the initial direction of motion, called the *deflection angle*, defined as

$$\chi = |\pi - 2\theta_0| . \tag{3.111}$$

It is useful to write the constants of motion, $\mathcal{E}$ and $L$, in terms of the initial velocity $v_0 \equiv v(r \to \infty)$, and of the impact parameter $b$, defined in the figure. For that purpose, it should be noticed that, when $r \to \infty$, the angle $\theta$ tends to zero, according to $\theta \simeq b/r$. From that we get in this limit, $\dot{\theta} = b\dot{r}/r^2$. This allows us to state that

$$\mathcal{E} = \frac{1}{2} m v_0^2 \,, \quad L \equiv m r^2 \dot{\theta} = m b v_0 \,. \tag{3.112}$$

Replacing this in eq. (3.110), we transform the integral into

$$\theta_0 = \int_0^{x_0} \frac{dx}{\sqrt{1 - x^2 - 2x(b_0/b)}} \,. \tag{3.113}$$

where we have used a new variable $x = b/r$, and defined $x_0 = b/r_0$. We have also introduced a new parameter $b_0$ defined by the relation

$$b_0 = \frac{rV(r)}{mv_0^2} = \frac{q_1 q_2}{4\pi\epsilon_0 m v_0^2} \,. \tag{3.114}$$

The physical meaning of this parameter will become apparent soon. Integration of eq. (3.113) then leads to

$$\theta_0 = \left[ tan^{-1} \sqrt{\frac{x_0(x_0 - x)}{(1 + x x_0)}} \right]_0^{x_0} = 2\, tan^{-1}(x_0) \,, \tag{3.115}$$

Or, simply, $tan(\theta_0/2) = x_0$. This is equivalent to write $\tan\theta_0 = 2x_0/(1 - x_0^2)$. But, on the other hand, eq. (3.113) shows that $(1 - x_0^2) = 2x_0(b_0/b)$. Using the above definition of the deflection angle, we arrive at the conclusion that

$$\tan\left(\frac{\chi}{2}\right) = \frac{b_0}{b} \,. \tag{3.116}$$

This is a very useful result. It shows that an impact parameter $b = b_0$ corresponds to a deflection angle of $\chi = \pi/2$. This means that $b_0$ is the limit between *soft collisions* leading to small deviations ($\chi < \pi/2$), and *hard collisions* leading to strong deviations ($\chi > \pi/2$), of the particle trajectory.

Let us consider now scattering of a flux of a large number of incident particles by a single test particle, having all the same velocity but different impart parameters $b$. The incident particles will be deflected along different angles $\chi$. Let us define $\Gamma_0 = nv$ as the flux of incident particles, where $n$ is the particle density. Then, the number of particles scattered per unit time in the interval $(\chi, \chi + d\chi)$ will be $dN = 2\pi\Gamma_0\, bdb$, and using eq. (3.116) we can then write the differential cross-section $d\sigma = dN/\Gamma_0$ in the form

$$d\sigma = 2\pi b db = 2\pi b_0 cotan(\chi/2) \frac{db}{d\chi} d\chi \,, \tag{3.117}$$

Or, equivalently

$$d\sigma = \pi b_0^2 \frac{\cos(\chi/2)}{\sin^3(\chi/2)} d\chi \,, \tag{3.118}$$

Finally, defining the element of solid angle $d\Omega = 2\pi \sin\theta d\theta$, and using (3.111), we obtain

$$d\sigma = \sigma(\chi) d\chi \,, \quad \sigma(\chi) = \frac{b_0^2}{\sin^4(\chi/2)} = \left( \frac{q_1 q_2}{4\pi\epsilon_0 \mathcal{E} \sin^2(\chi/2)} \right)^2 . \tag{3.119}$$

This is the famous *Rutherford formula*. For relativistic particles, we should replace the energy $\mathcal{E}$ by $p^2/2m$, and use the relativistic momentum.

### 3.7.2 Coulomb Logarithm

Let us now calculate the force **F** acting on the test particle, due to the incident particle flux. Given the angular symmetry of the Coulomb collision process, this will be equal to

$$\mathbf{F} = -\frac{d}{dt} \sum_{i=1}^{N_0} \mathbf{p}_i = -\frac{d}{dt} \sum_{i=1} p_{zi} \mathbf{e}_z \,, \tag{3.120}$$

for an incident beam along the $z$-axis. Noting that we are dealing with elastic scattering, the absolute value of a particle momentum stays constant. So, if $\mathbf{p}'$ is the momentum after collision, we will have $|\mathbf{p}'| = |\mathbf{p}|$, and $\delta p \equiv |\mathbf{p}' - \mathbf{p}| = 2p \sin(\chi/2)$. This leads to a variation in the $z$-direction given by $\delta p_z = -\delta p \sin(\chi/2)$. More explicitly, we have

$$\delta p_z = -2p \sin^2(\chi/2) = -2p \frac{b_0^2}{b^2 + b_0^2} \,. \tag{3.121}$$

The total force defined in eq. (3.120) can then be written as

$$\mathbf{F} = -\mathbf{e}_z \int_0^\infty dN \delta p_z db = -\mathbf{e}_z \left( 4\pi \mathcal{E} b_0^2 \right) \int_0^\infty \frac{bdb}{b^2 + b_0^2} \,, \tag{3.122}$$

This integral has a logarithmic divergence, which can be avoided if we introduce an upper cutoff, and integrate between $b = 0$ and $b = b_{max}$, where the upper limit of the impact parameter $b_{max}$ is to be defined. To adequately choose this limit, we note that the Coulomb potential created by a point particle is shielded at a distance of the order of the Debye length. We can then assume that there are no collisions for $b > \lambda_D$. This means that we can replace the above integral by

$$\mathcal{I} = \int_0^{\lambda_D} \frac{bdb}{b^2 + b_0^2} = \ln \left( 1 + \frac{\lambda_D^2}{b_0^2} \right)^{1/2} \,, \tag{3.123}$$

When we have $\lambda_D \gg b_0$, this is approximately given by

$$\mathcal{I} = \ln \Lambda \,, \quad \Lambda \equiv \frac{\lambda_D}{b_0} = \frac{4\pi\epsilon_0 \lambda_D}{q_1 q_2/\mathcal{E}} \,, \tag{3.124}$$

where, for electron-ion collisions, we have $q_1 = e$ and $q_2 = Ze$, and the particle kinetic energy is $\mathcal{E}$.

This integral is known as the *Coulomb logarithm*, and plays a crucial role in plasma kinetic theory. As a typical example of a magnetic fusion plasma, we can take $T = 10\,KeV$ and $n = 10^{14}\,cm^{-3}$, and use $Z = 1$. This leads to $\Lambda = 10^7$ and $\ln \Lambda = 16$. This, and other typical examples show that, in quite general plasma conditions, we have $\lambda_D \gg b_0$. In this case, most of particle collisions occur in the region of very large impact parameters and small deflection angles, $\chi \simeq 2b_0/b \ll 1$. In order to understand this statement, we divide the integral of eq. (3.122) in two parts, $\mathcal{I} = \mathcal{I}_{prox.} + \mathcal{I}_{dist.}$, such that

$$\mathcal{I}_{prox.} = \int_0^{b_1} \frac{bdb}{b^2 + b_0^2} = \ln \left( 1 + \frac{b_1^2}{b_0^2} \right)^{1/2} = \ln \sqrt{5} \tag{3.125}$$

and

$$\mathcal{I}_{dist.} = \int_{b_1}^{\lambda_D} \frac{bdb}{b^2 + b_0^2} = \ln \left( 1 + \frac{\lambda_D^2}{b_1^2} \right)^{1/2} = \ln \frac{\lambda_D}{2b_0} \,, \tag{3.126}$$

where we have used $b_1 = 2b_0$. This shows that, for $\lambda_D \gg b_0$, we get for the total force

$$\frac{F_{prox.}}{F_{dist.}} = \frac{\mathcal{I}_{prox.}}{\mathcal{I}_{dist.}} \simeq \frac{1}{\ln \Lambda} \ll 1 \,. \tag{3.127}$$

This means that the total contribution of distant collisions to the interaction of one particle with all the other particles in a plasma is nearly $\ln \Lambda$ times larger than the contribution of close collisions. Therefore, the Coulomb interactions of particles in a plasma are usually much more a collective effect than a result of single particle binary collisions. This is clearly different from the case of a neutral gas of atoms or molecules, where kinetic effects are simply due to close collisions, therefore justifying the name *plasma*, chosen by Langmuir to define the *ionized gas*.

We are now in position to calculate the average momentum and energy exchanged between a particle and the medium. We have considered above a flux of incident particles with the same momentum $\mathbf{p}$. We should now take the momentum distribution into account. For incident particles of a given species $\beta$, their density is

$$n_\beta = \int f_\beta(\mathbf{p}) d\mathbf{p} \,. \tag{3.128}$$

We can then establish the force on a single particle of species $\alpha$ by a number $dn_\beta = f_\beta d\mathbf{p}$, of particles of species $\beta$, with momenta between $\mathbf{p}'$ and $(\mathbf{p} + d\mathbf{p}')$, as

$$d\mathbf{F}_\alpha(\mathbf{p}) = -\mathbf{e}_z \left( \frac{m q_\alpha^2 q_\beta^2}{4\pi\epsilon_0 |\mathbf{p} - \mathbf{p}'|^2} \right) \ln \Lambda \, dn_\beta(\mathbf{p}') \,, \tag{3.129}$$

Replacing $\mathbf{e}_z$ by the unit vector along the relative momentum, and integrating in momentum space, we obtain

$$\mathbf{F} = -\frac{m_\alpha q_\alpha^2 q_\beta^2}{4\pi\epsilon_0} \int \frac{\mathbf{p} - \mathbf{p}'}{|\mathbf{p} - \mathbf{p}'|^3} \ln\Lambda\, f_\beta(\mathbf{p}') d\mathbf{p}' \,, \tag{3.130}$$

For a plasma near equilibrium at a temperature $T$, the Coulomb logarithm is approximately given by

$$\ln\Lambda \simeq \ln \frac{\lambda_D}{< b_0 >} = \ln\left(\frac{4\pi\epsilon_0\lambda_D}{q_\alpha q_\beta/3T}\right) \,, \tag{3.131}$$

and can be taken out of the integral. We can make a more suggestive interpretation of the above force, by considering an effective charge density $\rho_\beta$, defined as

$$\rho_\beta(\mathbf{p}) = \frac{q_\alpha^2 q_\beta^2}{\epsilon_0} f_\beta(\mathbf{p}) \,, \tag{3.132}$$

and the analogue of an effective electrostatic potential

$$\varphi_\beta(\mathbf{p}) = -\frac{1}{4\pi\epsilon_0} \int \frac{\rho_\beta(\mathbf{p})}{|\mathbf{p} - \mathbf{p}'|} d\mathbf{p}' \,. \tag{3.133}$$

Replacing this in eq. (3.130), and summing over all plasma species $\beta$, we obtain

$$\mathbf{F}_\alpha(\mathbf{p}) = -m_\alpha \sum_\beta \ln\Lambda\, \mathbf{E}_\beta(\mathbf{p}) \,, \tag{3.134}$$

where $\mathbf{E}_\beta(\mathbf{p})$ is defined as the gradient of the above effective potential, taken in momentum space, as

$$\mathbf{E}_\beta(\mathbf{p}) = -\nabla\varphi_\beta(\mathbf{p}) = \frac{1}{4\pi\epsilon_0} \int \frac{(\mathbf{p} - \mathbf{p}')}{|\mathbf{p} - \mathbf{p}'|^3} \rho_\beta(\mathbf{p}) d\mathbf{p}' \,. \tag{3.135}$$

A similar calculation can be made for the rate of energy exchanged by particle collisions $\mathcal{E}_\alpha$. The result is

$$\frac{d}{dt}\mathcal{E}_\alpha(\mathbf{p}) = -\frac{m_\alpha q_\alpha^2}{4\pi\epsilon_0} \sum_\beta q_\beta^2 \int \frac{\mathbf{p} \cdot (\mathbf{p} - \mathbf{p}')}{|\mathbf{p} - \mathbf{p}'|^3} f_\beta(\mathbf{p}') d\mathbf{p}' \,, \tag{3.136}$$

As before, this can be rewritten in terms of the effective potential (3.133), as

$$\frac{d}{dt}\mathcal{E}_\alpha(\mathbf{p}) = -m_\alpha \sum_\beta \ln\Lambda \left[\mathbf{p} \cdot \mathbf{E}_\beta(\mathbf{p}) - \varphi_\beta(\mathbf{p})\right] \,. \tag{3.137}$$

### 3.7.3 Relaxation Times

When the distribution functions $f_\beta(\mathbf{p})$ are isotropic, the effective collision field $\mathbf{E}_\beta(\mathbf{p})$ is parallel to the direction of $\mathbf{p}$. Then, using eq. (3.134) we can calculate the momentum exchange in the same direction. However, it is also useful to know the momentum exchange in the perpendicular direction. Taking the time derivative of $p^2 = p_\parallel^2 + p_\perp^2$, we have

$$\frac{dp^2}{dt} = 2m_\alpha \frac{d\mathcal{E}_\alpha}{dt}, \quad \frac{dp_\parallel^2}{dt} = 2\mathbf{p} \cdot \frac{d\mathbf{p}}{dt}. \tag{3.138}$$

From this, we get

$$\frac{dp_\perp^2}{dt} = 2m_\alpha n \frac{d\mathcal{E}_\alpha}{dt} - 2\mathbf{p} \cdot \frac{d\mathbf{p}}{dt}. \tag{3.139}$$

Using eqs. (3.134) and (3.137), we can then derive

$$\frac{dp_\perp^2}{dt} = 2\sum_\beta \ln\Lambda\, \varphi_\beta(\mathbf{p}). \tag{3.140}$$

Let us now define the parallel and perpendicular relaxation times $\tau_\parallel$ and $\tau_\perp$, such that

$$\frac{dp_\parallel}{dt} = -\frac{p}{\tau_\parallel}, \quad \frac{dp_\perp^2}{dt} = -\frac{p^2}{\tau_\parallel}, \tag{3.141}$$

Comparing this with our previous results, we can easily establish the expression of the relaxation times, as

$$\tau_\parallel = \frac{p}{\sum_\beta \ln\Lambda\, E_\beta(\mathbf{p})}, \quad \tau_\perp = \frac{m_\alpha \mathcal{E}_\alpha}{2\sum_\beta \ln\Lambda\, \varphi_\beta(\mathbf{p})}, \tag{3.142}$$

Let us consider the explicit case of a non-relativistic Maxwellian distribution

$$f_\beta(\mathbf{p}) = n_\beta \left(\frac{m_\beta}{2\pi T_\beta}\right)^{1/2} \exp\left(-\frac{p^2}{2m_\beta T_\beta}\right). \tag{3.143}$$

We can calculate the explicit values of $\tau_\parallel$ and $\tau_\perp$. Assuming that the plasma is made of electrons ($q = -e$) and ions ($q = Ze$), we obtain the collision times associated with an electron (as a test particle) colliding with electrons and ions. In the first case, we only use the term $\beta \equiv$ electrons, and in the second $\beta \equiv$ ions. We obtain

$$\tau_{ee} = \frac{6\pi\sqrt{3}\epsilon_0^2 m_e^2 m_e^{1/2} T_e^{3/2}}{n_e e^4 \ln\Lambda}, \tag{3.144}$$

with $\tau_{ee} = \tau_{\parallel ee} = \tau_{\perp ee}$. These results are valid when the test particle moves at thermal velocity, $v = \sqrt{3T_e/m_e}$. We also get: $\tau_{ee} = (Z/2)\tau_{\parallel ei} = Z\tau_{\perp ei}$. A similar calculation for ion-ion collisions, with $n_i = Zn_e$, leads to

$$\tau_{ii} = \tau_{ee}\frac{1}{Z^3}\left(\frac{m_i}{m_e}\right)^{1/2}\left(\frac{T_i}{T_e}\right)^{3/2}, \tag{3.145}$$

This means that, in the usual case of $Z = 1$ and $T_i = T_e$, we have $\tau_{ii} = \tau_{ee}\sqrt{m_i/m_e}$. For the inverse situation of a test ion colliding with the electrons, we get

$$\tau_{ie} = \tau_{ee}\frac{1}{Z^2}\frac{m_i}{m_e}, \tag{3.146}$$

These results show that it is possible to establish an ordering between the different relaxation times, such that

$$\tau_{ie} \gg \tau_{ii} \gg \tau_{ee}, \tag{3.147}$$

The relaxation times for the energy are of the same order of magnitude. This allows us to say that, if the initial conditions correspond to a plasma slightly out of equilibrium, the electron gas is the first to attain thermal equilibrium, at a temperature $T_e$, followed by the ions at a temperature $T_i$, and only much later the two populations attain global equilibrium at $T_e = T_i = T$.

Another useful way to characterize the plasma medium is through the concept of *mean free-path*, which corresponds to the distance traveled by a particle between two consecutive collisions. It can be defined, for electrons and ions, by

$$\lambda_e = v_{the}\tau_{ee}, \quad \lambda_i = v_{thi}\tau_{ii}, \quad \frac{\lambda_e}{\lambda_i} \simeq Z^3\left(\frac{T_e}{T_i}\right)^2, \tag{3.148}$$

where the thermal velocities were defined by $v_{th\alpha} = \sqrt{T_\alpha/m_\alpha}$. This expressions show that, for singly ionized ions ($Z = 1$), and for a plasma in equilibrium ($T_e = T_i$), the electron and ion mean free-paths are the same, even if their relaxation times are very different.

### 3.7.4 Resistivity

If we apply an external electric field $\mathbf{E}$ to a neutral plasma, electrons and ions are accelerated by the field and tend to move in opposite directions, thus creating a current $\mathbf{J}$. However, particle collisions will create a resistance. The relation between resistivity $\eta$ and current $\mathbf{J}$ is stated by the *Ohm's law*

$$\mathbf{E} = \eta\mathbf{J}, \quad \mathbf{J} = \sum_\alpha q_\alpha n_\alpha < \mathbf{v}_\alpha > . \tag{3.149}$$

A steady-state value of the current corresponds to an equilibrium between the particle acceleration produced by the field and the deceleration produced by collisions. This means that, in steady-state, the electron momentum will evolve according to:

$$\frac{d\mathbf{p}}{dt} = -e\mathbf{E} \simeq -m_e(< \mathbf{v}_i > - < \mathbf{v}_e >)\nu_{ei}, \tag{3.150}$$

where $n_e = n_i = n$, and $q_i = -q_e = e$. This is equivalent to write

$$\mathbf{E} = -\frac{m_e\nu_{ei}}{e^2 n}\mathbf{J}. \tag{3.151}$$

This immediately gives the plasma resistivity. Such a simple formulation is valid for isotropic plasmas, and also for parallel resistivity in a magnetized plasma. Using the above results for the collision frequencies, we can then write, for a plasma in equilibrium, and for an arbitrary ion charge $Z$, the plasma resistivity as

$$\eta = \sqrt{\frac{m_e}{3}} \frac{Ze^2 \ln \Lambda}{12\pi\epsilon_0^2 T_e^{3/2}} \,. \tag{3.152}$$

A more rigorous derivation introduces a numerical correction of order one [44]. A practical expression that includes this small correction is

$$\eta = 1.65 \times 10^{-9} Z \frac{\ln \Lambda}{T_e^{3/2}} \,. \quad (Ohm.m) \tag{3.153}$$

This is result known as the *Spitzer resistivity*. As an example, we consider a plasma with an electron temperature of $T_e = 1KeV$. Its resistivity is equal to $\eta = 1.8 \times 10^{-8}\, Ohm.m$, a value slightly higher than the resistivity of copper at room temperature. For $T_e = 50\, KeV$, the plasma resistivity would be 350 times smaller.

The decrease of plasma resistivity with temperature results from a decrease of the collision cross-sections with the particle velocity. This leads to another characteristic plasma effect, the *runaway electrons*. In order to understand it, we go back to the momentum equation (3.150), and write it for a single electron, in the absence of an exact balance between electric field acceleration and the collisions. We have, for a general expression

$$\frac{d\mathbf{p}}{dt} = -e\mathbf{E} - \sum_\beta \frac{\ln \Lambda}{m} \mathbf{E}_\beta(\mathbf{p}) \,, \tag{3.154}$$

where $\mathbf{E}_\beta(\mathbf{p})$ is the effective field defined in eq. (3.135). In a more explicit form, we can write this formula for the standard case of $Z = 1$, as

$$\frac{d\mathbf{p}}{dt} = -e\mathbf{E} + \frac{e^4 \ln \Lambda}{4\pi\epsilon_0^2} \frac{n_e \mathbf{v}}{m_e v^3} \,. \tag{3.155}$$

As we have seen, the second term represents the deceleration due to collisions, while the first gives the electric field acceleration. Noting that the collision term varies with the particle velocity as $1/v^3$, a critical velocity $v = v_{cr} < c$ can be defined, above which the first term becomes dominant and the electrons will be continuously accelerated by the field $\mathbf{E}$. This critical velocity is determined by

$$v_{cr}^2 = \frac{e^3 n_e \ln \Lambda}{4\pi\epsilon_0^2 m_e E} \,. \tag{3.156}$$

As an example, for a plasma density of $n_e = 10^{14} cm^{-3}$, and an electric field of $E = 10\, V/m$, typical values for a magnetic fusion plasma, all electrons with energy above $5\, KeV$ will runaway. In other words, they will be continuously accelerated, creating a fast beam that decouples from the thermal plasma. This *runaway effect* has been abundantly documented by experiments. We should however note that radiation processes such as Bremsstrahlung, and beam-plasma instabilities, to be discussed in later chapters, will lead to a decrease and eventual saturation of the runaway process.

# 4

# *Linear Waves*

We first consider the case of waves in plasmas in the linear approximation, assuming that their dispersive properties are independent of the wave amplitude, and only depend on the properties of the medium. Formally, this approximation is only valid for waves with infinitesimal amplitudes, but can be used as a basis for the understanding of waves in the general case, and stay approximately valid even for nonlinear wave phenomena, as discussed in subsequent chapters. Being an essential part of plasma theory, linear waves have abundantly been described in several books on plasma physics, specially focused in the non-relativistic version, such as [45, 46, 47, 48]. It should be recognized that the general framework of linear wave theory remains valid in relativistic plasmas. We start with a general derivation of the wave dispersion relation, and then discuss the cases of electrostatic and electromagnetic waves in both isotropic and magnetized media.

## 4.1 Dispersion Relation

We assume an infinite, homogeneous and stationary plasma. This allows us to ignore the influence of boundaries and to identify the characteristic oscillations that can take place inside the medium. We start with the Maxwell's equations, written in vectorial form, as

$$\nabla \times \mathbf{E} = -\frac{\partial \mathbf{B}}{\partial t}, \quad \nabla \cdot \mathbf{E} = \frac{\rho}{\epsilon_0},$$

$$\nabla \times \mathbf{H} = \epsilon_0 \frac{\partial \mathbf{E}}{\partial t} + \mathbf{J}, \quad \nabla \cdot \mathbf{B} = 0, \tag{4.1}$$

where $\mathbf{B} = \mu_0 \mathbf{H}$, and $\epsilon_0 \mu_0 = 1/c^2$. From here we can easily derive the general form of the propagation equation

$$\nabla \times (\nabla \times \mathbf{E}) + \frac{1}{c^2} \frac{\partial^2 \mathbf{E}}{\partial t} = -\mu_0 \frac{\partial \mathbf{J}}{\partial t}. \tag{4.2}$$

It states that the conductive current $\mathbf{J}$ is the source of the electric field $\mathbf{E}$. Notice that, by using $\mathbf{B} = \mu_0 \mathbf{H}$, we have neglected magnetization effects.

DOI: 10.1201/9781003514701-4

In plasmas this is globally valid, and we can explicitly treat magnetization in special cases when necessary. Using the vector identity $\nabla \times (\nabla \times \mathbf{a}) = \nabla(\nabla \cdot \mathbf{a}) - \nabla^2 \mathbf{a}$, valid for an arbitrary vector field $\mathbf{a}$, this equation can also be written as

$$\left(\nabla^2 - \frac{1}{c^2}\frac{\partial^2}{\partial t}\right)\mathbf{E} - \nabla\left(\nabla \cdot \mathbf{E}\right) = \mu_0 \frac{\partial \mathbf{J}}{\partial t}, \tag{4.3}$$

On the other hand, the current is related with the electric field by an expression of the form

$$\mathbf{J} = \sum_\alpha \int dt' \int d\mathbf{r}' \, \bar{\bar{\sigma}}_\alpha(\mathbf{r}, \mathbf{r}', t, t') \cdot \mathbf{E}(\mathbf{r}', t), \tag{4.4}$$

where we have assumed that $\mathbf{J} \equiv \mathbf{J}(\mathbf{r}, t)$. The conductivity tensor for particle species $\alpha$ (electrons or ions), $\bar{\bar{\sigma}}_\alpha$, can be derived from the plasma equations, as shown below. In the case of homogeneous and stationary plasmas, this can be rewritten as

$$\mathbf{J} = \sum_\alpha \int d\tau \int d\mathbf{s} \, \bar{\bar{\sigma}}_\alpha(\mathbf{s}, \tau) \cdot \mathbf{E}(\mathbf{r} - \mathbf{s}, t - \tau). \tag{4.5}$$

It is now useful to consider a double space and time Fourier transformation of the field, such that

$$\mathbf{E}(\mathbf{r}, t) = \int \frac{d\omega}{2\pi} \int \frac{d\mathbf{k}}{(2\pi)^3} \mathbf{E}(\omega, \mathbf{k}) \exp(i\mathbf{k} \cdot \mathbf{r} - i\omega t). \tag{4.6}$$

Noting that the current $\mathbf{J}$ is now determined by convolution integrals, we can replace eq. (4.5) by

$$\mathbf{J}(\mathbf{r}, t) = \sum_\alpha \int \frac{d\omega}{2\pi} \int \frac{d\mathbf{k}}{(2\pi)^3} \bar{\bar{\sigma}}_\alpha(\omega, \mathbf{k}) \cdot \mathbf{E}(\omega, \mathbf{k}) \exp(i\mathbf{k} \cdot \mathbf{r} - i\omega t). \tag{4.7}$$

This shows that, for each spectral field component $\mathbf{E}(\omega, \mathbf{k})$, we can derive from eq. (4.3) a relation of the form

$$\bar{\bar{\mathcal{D}}}(\omega, \mathbf{k}) \cdot \mathbf{E}(\omega, \mathbf{k}) = 0, \tag{4.8}$$

where we have introduced the *dispersion tensor*

$$\bar{\bar{\mathcal{D}}}(\omega, \mathbf{k}) = \left[\mathbf{k}\mathbf{k} - k^2 \bar{\bar{1}} + \frac{\omega^2}{c^2} \bar{\bar{\kappa}}(\omega, \mathbf{k})\right]. \tag{4.9}$$

Here, $\bar{\bar{1}}$ is the unit tensor, and $\bar{\bar{\kappa}} \equiv \bar{\bar{\kappa}}(\omega, \mathbf{k})$ is usually called the *dielectric tensor*, and is defined as

$$\bar{\bar{\kappa}} = \bar{\bar{1}} + \frac{i}{\omega \epsilon_0} \sum_\alpha \bar{\bar{\sigma}}_\alpha. \tag{4.10}$$

In order to satisfy eq. (4.8) for a non-vanishing electric field, the following condition should be satisfied

$$\det[\bar{\bar{\mathcal{D}}}(\omega, \mathbf{k})] = 0 . \tag{4.11}$$

This gives the *dispersion relation*, valid for a given wave component $(\omega, \mathbf{k})$, and establishes a relation between the wave frequency $\omega$ and the wavevector $\mathbf{k}$. If this relation is satisfied, the double Fourier transform of the field is reduced to a single transformation, given that

$$\mathbf{E}(\omega, \mathbf{k}) = 2\pi \, \mathbf{E}_{\mathbf{k}} \, \delta \left(\omega - \omega(\mathbf{k})\right) , \tag{4.12}$$

where $\omega = \omega(\mathbf{k})$ is a solution of the dispersion relation (4.11). Replacing this in eq. (4.6), we reduce it to

$$\mathbf{E}(\mathbf{r}, t) = \int \mathbf{E}_{\mathbf{k}} \exp\left[i\mathbf{k} \cdot \mathbf{r} - i\omega(\mathbf{k})t\right] \frac{d\mathbf{k}}{(2\pi)^3} . \tag{4.13}$$

The dielectric tensor can be obtained directly from the equation $\nabla \cdot \mathbf{E} = \rho/\epsilon_0$, making use of the continuity equation

$$\frac{\partial \rho}{\partial t} + \nabla \cdot \mathbf{J} = 0 , \tag{4.14}$$

This allows us to write, for the Fourier component of the charge density $\rho(\omega, \mathbf{k})$, an expression of the form

$$\rho = i\mathbf{k} \cdot \mathbf{E} = \frac{\mathbf{k}}{\omega} \sum_{\alpha} \bar{\bar{\sigma}}_{\alpha} \cdot \mathbf{E} , \tag{4.15}$$

which immediately leads to

$$i\mathbf{k} \cdot \bar{\bar{\epsilon}} \cdot \mathbf{E} = 0 , \quad \bar{\bar{\epsilon}} = \epsilon_0 \bar{\bar{\kappa}} . \tag{4.16}$$

We can also see that the displacement field, $\mathbf{D} = \bar{\bar{\epsilon}} \cdot \mathbf{E}$, satisfies the equation $\nabla \cdot \mathbf{D} = 0$. At this point, we should notice that, for each mode component $(\omega, \mathbf{k})$, we can define a unit polarization vector $\mathbf{e}_{\mathbf{k}} = \mathbf{E}_{\mathbf{k}}/|E_{\mathbf{k}}|$, satisfying the normalization $(\mathbf{e}_{\mathbf{k}}^* \cdot \mathbf{e}_{\mathbf{k}}) = 1$. Going back to eq. (4.8), and multiplying it by $\mathbf{e}_{\mathbf{k}}^*$, we obtain another possible representation of the dispersion relation in terms of the polarization vector, as

$$\mathcal{D}(\omega, \mathbf{k}) \equiv \mathbf{e}_{\mathbf{k}}^* \cdot \bar{\bar{\mathcal{D}}}(\omega, \mathbf{k}) \cdot \mathbf{e}_{\mathbf{k}} = 0 \tag{4.17}$$

Explicit expressions for the dispersion relation of plasma waves will be discussed next.

We conclude this introductory section by noting that the dispersion relation can also be derived in a more formal way, using the covariant formulation of Maxwell's equations. For that purpose, we replace eqs. (4.1) by

$$\partial_\mu F^{\mu\nu} = \mu_0 J^\mu , \quad \partial_\mu \left(e^{\mu\nu\sigma\delta} F_{\sigma\delta}\right) = 0 , \tag{4.18}$$

where $J^\mu$ is the 4-current, and

$$F^{\mu\nu} = \partial^\mu A^\nu - \partial^\nu A^\mu \,, \quad A^\mu = (V/c, \mathbf{A})\,. \tag{4.19}$$

We now consider wave perturbations, such that the 4-current and the 4-potential can be Fourier analysed as

$$J^\mu(x) = \int J^\mu e^{-i(k\cdot x)} \frac{d^4k}{(2\pi)^4}\,, \quad A^\mu(x) = \int A^\mu e^{-i(k\cdot x)} \frac{d^4k}{(2\pi)^4}\,. \tag{4.20}$$

Here, we use the internal product $(k \cdot x) \equiv k_\nu x^\nu = \mathbf{k}\cdot\mathbf{r} - \omega t$, with the 4-wavevector $k^\nu = (\omega/c, \mathbf{k})$. The quantity $d^4k$ represents integration over space and time components of $k^\nu$. We can also write, for the electromagnetic tensor

$$F^{\mu\nu}(k) = i\left(k^\nu A^\mu - k^\mu A^\nu\right)\,, \tag{4.21}$$

From Maxwell's equations (4.18), we then get

$$(k\cdot A)k^\mu - (k\cdot k)A^\mu = \mu_0 \sum_\alpha J^\mu_\alpha\,, \tag{4.22}$$

where $\alpha = (e, i)$ represent the electron and ion currents. Following the above steps, we then arrive at a relation of the form

$$\mathcal{D}^{\mu\nu}(k)A_\nu(k) = 0\,, \tag{4.23}$$

where

$$\mathcal{D}^{\mu\nu}(k) = (k\cdot k)\delta^{\mu\nu} - k^\mu k^\nu + \epsilon^{\mu\nu}(k)\,. \tag{4.24}$$

Here, $\delta^{\mu\nu}$ represents the unit tensor and $\epsilon^{\mu\nu}$ the dielectric tensor in the 4-dimensional spacetime. This dispersion relation can then be written as

$$\det[\mathcal{D}^{\mu\nu}(k)] = 0\,. \tag{4.25}$$

Comparison with the three-dimensional expressions above allows to retrieve the explicit expressions for these new tensors.

## 4.2 Fluid Description

Let us consider the basic waves that can be excited in isotropic plasmas. The simplest possible model that allows to identify these modes is based on the fluid plasma equations. A more accurate model, based on the kinetic theory will be discussed latter.

We first consider the case of high-frequency electrostatic waves, where the ion motion can be ignored, and focus on the electron plasma response. The relevant equations are those stating the conservation of mass and momentum of the electron fluid, and take the simple form

$$\frac{\partial n_e}{\partial t} + \nabla \cdot (n_e \mathbf{v}_e) = 0 \,, \quad \frac{d\mathbf{p}_e}{dt} + \frac{1}{n_e} \nabla P_e = -e\mathbf{E} \,, \tag{4.26}$$

where $n$ and $\mathbf{v}$ represent the mean electron density and velocity, respectively, and the magnetic force is ignored. If the electron fluid is initially at rest, the electron velocity is non-relativistic for infinitesimal wave amplitudes. The only relativistic effects are then related to the electron temperature $T_e$, which can take arbitrary values. For waves with frequency $\omega$ and wavevector $\mathbf{k}$ the perturbed quantities are, the electric field $\mathbf{E} = \mathbf{E}_\mathbf{k} \exp(i\mathbf{k}\cdot\mathbf{r} - i\omega t)$, the electron density

$$n_e = n_0 + n_\mathbf{k} \exp(i\mathbf{k} \cdot \mathbf{r} - i\omega t) \,, \tag{4.27}$$

where $n_0$ is the equilibrium density, and the electron momentum

$$\mathbf{p}_e = m_e \gamma_e \mathbf{v}_\mathbf{k} \exp(i\mathbf{k} \cdot \mathbf{r} - i\omega t) \,, \tag{4.28}$$

where the relativistic gamma factor $\gamma_e = \sqrt{1 + u_{th}^2/c^2}$ is a constant. We also assume an equation of state such that the electron pressure $P$ satisfies the simple relation

$$\nabla P_e = \frac{g_e}{m} \nabla n_e \,, \tag{4.29}$$

where the factor $g_e \equiv g_e(T_e)$ only depends on the temperature. After linearization, the fluid equations (4.26) then lead to

$$n_\mathbf{k} = n_0 \frac{\mathbf{k}}{\omega} \cdot \mathbf{v}_\mathbf{k} \,, \tag{4.30}$$

and

$$\omega^2 \mathbf{v}_\mathbf{k} = -i \frac{e\omega}{\gamma_e m_e} \mathbf{E}_\mathbf{k} + \frac{S_e^2}{\gamma_e} \mathbf{k}(\mathbf{k} \cdot \mathbf{v}_\mathbf{k}) \,, \tag{4.31}$$

where $S_e = \sqrt{g_e(T_e)/m_e}$ is the electron thermal velocity. For electrostatic waves, the electric field is determined by Poisson's equation,

$$\nabla \cdot \mathbf{E} = -\frac{e}{\epsilon_0}(n_e - n_0) \,, \tag{4.32}$$

which gives

$$i\mathbf{k} \cdot \mathbf{E}_\mathbf{k} = -\frac{e}{\epsilon_0} n_\mathbf{k} \,. \tag{4.33}$$

It can easily be found that the field $\mathbf{E}_\mathbf{k}$ is parallel to the wavevector $\mathbf{k}$, and to the electron velocity $\mathbf{v}_\mathbf{k}$, and we can write

$$E_\mathbf{k} = \frac{i}{k} \frac{e}{\epsilon_0} n_\mathbf{k} = \frac{i}{\omega} \frac{e n_0}{\epsilon_0} v_\mathbf{k} \,. \tag{4.34}$$

Using eq. (4.31), it is then possible to establish a dispersion relation of the form

$$\omega^2 = \frac{\omega_{pe}^2}{\gamma_e} + \frac{S_e^2}{\gamma_e} k^2 . \tag{4.35}$$

This reduces to the well-known results for non-relativistic plasmas, for $\gamma_e = 1$ and $g_e(T_e) = 3T_e$. In the general case, the correct values can only be found using a more accurate kinetic description, as discussed later (see [49, 50]). For our present purposes, we can use

$$\gamma_e \simeq \sqrt{1 + \frac{5}{2} \frac{T_e}{m_e c^2}} , \tag{4.36}$$

Assuming the weakly relativistic regime, such that $T_e \ll m_e c^2$, and get

$$\omega^2 \simeq \left(\omega_{pe}^2 + S_e^2 k^2\right) \left(1 - \frac{5}{2} \frac{T_e}{m_e c^2}\right) , \tag{4.37}$$

with $S_e^2 = 3T_e/m_e$. Or, in more explicit terms

$$\omega^2 \simeq \omega_{pe}^2 + \left(3 \frac{T_e}{m_e}\right) k^2 - \frac{5}{2} \frac{T_e}{m_e c^2} \omega_{pe}^2 - \frac{15}{2} \left(\frac{T_e}{m_e c^2}\right)^2 k^2 c^2 . \tag{4.38}$$

This is to be compared with the kinetic result [51]

$$\omega^2 \simeq \omega_{pe}^2 + \left(3 \frac{T_e}{m_e}\right) k^2 - \frac{5}{2} \frac{T_e}{m_e c^2} \omega_{pe}^2 - \frac{33}{2} \left(\frac{T_e}{m_e c^2}\right)^2 k^2 c^2 . \tag{4.39}$$

We can see that there is a nearly complete agreement, except for a factor of order one in the last term, which is a second-order correction on the temperature. Therefore, this simple model is accurate to order $T_e/m_e c^2$.

We now turn to the case of electromagnetic waves, where the electric field is transverse, $\mathbf{E_k} \cdot \mathbf{k} = 0$. In this case, we have

$$n_{\mathbf{k}} = 0 , \quad \mathbf{v_k} = -i \frac{e}{m_e \gamma_e \omega} \mathbf{E_k} . \tag{4.40}$$

Going back to the wave equation (4.3), and using the current $\mathbf{J} = -en_0 \mathbf{v}$, we can easily derive the dispersion relation for transverse waves, as

$$\omega^2 = \frac{\omega_{pe}^2}{\gamma_e} + k^2 c^2 . \tag{4.41}$$

In the weakly relativistic limit, we get the approximate expression

$$\omega^2 = \omega_{pe}^2 + k^2 c^2 - \frac{5}{2} \left(\frac{T_e}{m_e c^2}\right) \omega_{pe}^2 . \tag{4.42}$$

This shows a relativistic frequency shift of the plasma cut-off (defined by the condition: $\mathbf{k} = 0$), which increases with the temperature, in agreement with more accurate kinetic formulations [51]

$$\omega^2(k \to 0) \simeq \omega_{pe}^2 \left[ 1 - \frac{5}{2} \frac{T_e}{m_e c^2} \right] . \tag{4.43}$$

Finally, we consider low-frequency oscillations of the medium, where we need to include the ion contributions. In this case, eq. (4.31) is replaced by

$$\omega^2 \mathbf{v}_\alpha = i \frac{q_\alpha \omega}{\gamma_\alpha m_\alpha} \mathbf{E}_\mathbf{k} + \frac{S_\alpha^2}{\gamma_\alpha} \mathbf{k} (\mathbf{k} \cdot \mathbf{v}_\alpha) , \tag{4.44}$$

where the index $\alpha = (e, i)$ pertains to electrons and ions, and where we have used the thermal velocities $S_\alpha = \sqrt{g_\alpha(T_\alpha)/m_\alpha}$. For electrostatic modes, the electric field is now determined by

$$\mathbf{E}_\mathbf{k} = \frac{i}{k} \frac{ek}{\epsilon_0 \omega} (v_e - v_i) . \tag{4.45}$$

which allows to derive, from eq. (4.44), two coupled equations for the electron and ion velocities, as

$$v_e \left( \omega^2 - \frac{\omega_{pe}^2}{\gamma_e} - \frac{S_e^2}{\gamma_e} k^2 \right) + v_i \frac{\omega_{pe}^2}{\gamma_e} = 0 , \tag{4.46}$$

and

$$v_i \left( \omega^2 - \omega_{pi}^2 - S_i^2 k^2 \right) + v_e \omega_{pi}^2 = 0 , \tag{4.47}$$

where the ion plasma frequency is $\omega_{pi} = \omega_{pe}(m_i/m_e)^{1/2} \ll \omega_{pe}$. Notice that we have neglected the relativistic corrections to the ion motion, and assumed that $\gamma_i = 1$, which is justified by the ion much larger mass. This then leads to a dispersion relation of the form

$$\left( \omega^2 - \frac{\omega_{pe}^2}{\gamma_e} - \frac{S_e^2}{\gamma_e} k^2 \right) \left( \omega^2 - \omega_{pi}^2 - S_i^2 k^2 \right) - \frac{\omega_{pe}^2}{\gamma_e} \omega_{pi}^2 = 0 . \tag{4.48}$$

It is useful to consider this result in two opposite limits. First, in the high-frequency limit, such that $\omega^2 \gg \omega_{pi}^2$, this reduces to the dispersion relation of electron plasma waves, with a modified plasma frequency where the ion contributions are included, as determined by

$$\omega^2 = \omega_p^2 + \frac{S_e^2}{\gamma_e} k^2 , \quad \omega_p^2 = \frac{\omega_{pe}^2}{\gamma_e} + \omega_{pi}^2 . \tag{4.49}$$

Notice that the ratio between the electron and ion plasma frequencies is usually very large, $(\omega_{pe}/\omega_{pi}) \simeq \sqrt{m_i/m_e} \gg 1$. But the ion contribution to the total plasma frequency $\omega_p$ can eventually become relevant for very high temperatures, such that $\gamma_e \gg 1$. Going back to eq. (4.48) and taking the low-frequency limit we obtain the ion acoustic waves, which satisfy the following relation

$$\omega^2 = c_s^2 k^2 , \quad c_s^2 = \frac{1}{m_i} \left[ g_e(T_e) + g_i(T_i) \right] , \tag{4.50}$$

where $c_s$ is the *ion-acoustic velocity*, with $g_e(T_e) \simeq T_e$ and $g_i(T_i) \simeq T_i$. In the following, we give a more accurate description of waves in plasmas, based on the kinetic theory.

## 4.3 Transverse Waves

We now consider the case of electromagnetic waves in isotropic plasmas, based on a fully relativistic kinetic description of the electron gas. The ions are assumed immobile. Going back to the basic wave equation (4.3), using an electric field perturbation of the form $\mathbf{E_k} \exp(i\mathbf{k} \cdot \mathbf{r} - i\omega t)$, and noting that for transverse waves the condition $(\mathbf{k}.\mathbf{E_k}) = 0$ is verified, we obtain

$$\mathbf{E_k} = \frac{i\omega\mu_0}{(k^2 - \omega^2/c^2)} \mathbf{J_k} , \tag{4.51}$$

where the current perturbation $\mathbf{J_k}$ is determined by the kinetic expression for the electron current

$$\mathbf{J} = -en \int \mathbf{v} f(\mathbf{p}) \, d\mathbf{p} . \tag{4.52}$$

Here, $f(\mathbf{p})$ is the electron distribution function and $\mathbf{p} = m\gamma\mathbf{v}$. We skip the index $e$ for simplicity. This leads to

$$\mathbf{E_k} = -i \frac{e\omega\mu_0 n_0}{(k^2 - \omega^2/c^2)} \int \mathbf{v} f_\mathbf{k}(\mathbf{p}) \, d\mathbf{p} , \tag{4.53}$$

where the perturbed distribution function $f_\mathbf{k}(\mathbf{p})$ is determined by the Vlasov equation

$$\left( \frac{\partial}{\partial t} + \mathbf{v} \cdot \frac{\partial}{\partial \mathbf{r}} + \mathbf{F}_e \cdot \frac{\partial}{\partial \mathbf{p}} \right) f(\mathbf{r}, \mathbf{p}, t) = 0 , \tag{4.54}$$

and $F_e = -e(\mathbf{E} + \mathbf{v} \times \mathbf{B})$ is the Lorentz force. Assuming that the electron distribution function is defined as

$$f(\mathbf{r}, \mathbf{p}, t) = f_0(\mathbf{p}) + f_\mathbf{k}(\mathbf{p}) \exp(i\mathbf{k} \cdot \mathbf{r} - i\omega t) , \tag{4.55}$$

where $f_0(\mathbf{p})$ is the equilibrium distribution, and linearizing the Vlasov equation with respect to the perturbations, we obtain

$$f_\mathbf{k}(\mathbf{p}) = i \frac{e\mathbf{E_k}}{(\omega - \mathbf{k} \cdot \mathbf{v})} \cdot \frac{\partial f_0}{\partial \mathbf{p}} . \tag{4.56}$$

Replacing in eq. (4.53), this leads to a dispersion relation of the form

$$(k^2c^2 - \omega^2) = \omega_p^2 m \int \left( \mathbf{v}_\perp \cdot \frac{\partial f_0}{\partial \mathbf{p}_\perp} \right) \frac{d\mathbf{p}}{(1 - \mathbf{v} \cdot \mathbf{k}/\omega)}, \tag{4.57}$$

where $\mathbf{p}_\perp = m\gamma\mathbf{v}_\perp$ is the particle momentum in the direction of the electric field perturbation $\mathbf{E_k}$. In other words, $\mathbf{p}_\perp = \mathbf{p} \cdot \mathbf{e_k}$, where $\mathbf{e_k}$ is the unit polarization vector. This can also be written in the form

$$\frac{k^2c^2}{\omega^2} = \epsilon_t(\omega, \mathbf{k}), \tag{4.58}$$

where we have defined the function

$$\epsilon_t(\omega, \mathbf{k}) = 1 + \frac{\omega_p^2}{\omega^2} m \int \left( \mathbf{v}_\perp \cdot \frac{\partial f_0}{\partial \mathbf{p}_\perp} \right) \frac{d\mathbf{p}}{(1 - \mathbf{v} \cdot \mathbf{k}/\omega)}. \tag{4.59}$$

It is now useful to introduce polar coordinates, $\mathbf{r} \equiv (r, \theta, \varphi)$, such that the momentum is represented as

$$\mathbf{p} = p\,(\sin\theta\cos\varphi\,\mathbf{e}_x + \sin\theta\sin\varphi\,\mathbf{e}_y + \cos\theta\,\mathbf{e}_z), \tag{4.60}$$

and assume propagation along the $z$-axis, or $(\mathbf{k} \cdot \mathbf{p}) = kp\cos\theta$. The integral in the dispersion relation (4.57) can then be developed as

$$\int \frac{(\mathbf{v}_\perp \cdot \partial f_0/\partial \mathbf{p}_\perp)\, d\mathbf{p}}{(1 - \mathbf{v} \cdot \mathbf{k}/\omega)} = \omega \int_0^\infty p^2 dp \int_0^{2\pi} d\varphi \int_{-1}^{1} \frac{dx}{(m\gamma\omega - kpx)} \left( \mathbf{p}_\perp \cdot \frac{\partial f_0}{\partial \mathbf{p}_\perp} \right). \tag{4.61}$$

Integrating over the variable $x \equiv \cos\theta$, we obtain

$$\epsilon_t(\omega, \mathbf{k}) = 1 - \frac{\omega_p^2}{\omega} \int_0^\infty \frac{\partial f_0}{\partial p} \mathcal{I}_t(p) p^2 dp. \tag{4.62}$$

The auxiliary quantity $\mathcal{I}_t(p)$ is given by

$$\mathcal{I}_t(p) = 2\frac{m\gamma\omega}{kp} + \left( \frac{m^2\gamma^2\omega^2}{k^2p^2} - 1 \right) \ln \left[ \frac{m\gamma\omega - kp}{m\gamma\omega + kp} \right]. \tag{4.63}$$

We now use the Maxwell-Jüttner equilibrium distribution, given by the expression

$$f_0(p) = \frac{\mu \exp[-\mu\gamma(p)]}{4\pi (mc)^3 K_2(\mu)}, \tag{4.64}$$

with $\mu = (mc^2/T)$, and $\gamma \equiv \gamma(p) = \sqrt{1 + p^2/(m^2c^2)}$. Noting that

$$d\gamma = \frac{1}{2\gamma} \frac{p}{m^2c^2} dp, \tag{4.65}$$

we can write eq. (4.63) as [49]

$$\epsilon_t(\omega, \mathbf{k}) = 1 - \mu \frac{\omega_p^2}{2k^2c^2} \left\{ 1 - \frac{k^2c^2}{\omega^2} \frac{P(\mu)}{K_2(\mu)} - \left(1 - \frac{k^2c^2}{\omega^2}\right) \frac{\mu}{K_2(\mu)} \frac{\partial^2}{\partial \mu^2} \frac{P(\mu)}{\mu} \right\} . \tag{4.66}$$

where $P(\mu)$ is the *relativistic plasma dielectric function*, defined by

$$P(\mu) = \int_1^\infty \frac{d\gamma}{\sqrt{\gamma^2 - 1}} \frac{\exp(-\mu\gamma)}{\gamma^2 - (\gamma^2 - 1)k^2c^2/\omega^2}$$
$$-i\sigma \frac{\pi}{2} \frac{\omega}{kc} \exp\left[ -\mu \left(\frac{kc}{\omega}\right) \Big/ \sqrt{\frac{k^2c^2}{\omega^2} - 1} \right] . \tag{4.67}$$

Here, we have used the notation $\sigma = 0$ for $\Re(\omega^2) \geq k^2c^2$, and $\sigma = 1$ for $\Re(\omega^2) < k^2c^2$ and $\Im(\omega) = 0$. Notice that, in physical terms, the most relevant case is that of transverse waves with superluminal phase velocities, such that $\sigma = 0$. In general, this expression has to be solved numerically. Two important limiting cases can, however, be treated analytically. They are, the weakly relativistic and the ultra-relativistic plasma limits, as follows:

i) **Weakly-relativistic plasma limit**

In this case, we have

$$\gamma \simeq 1 + \frac{p^2}{2m^2c^2} , \quad \mathbf{p} \simeq m\mathbf{v} . \tag{4.68}$$

The dispersion relation (4.58)–(4.66) reduces to

$$\omega^2 = k^2c^2 - \omega_p^2 z F(z) , \quad z = \frac{\omega}{kc} \sqrt{\frac{\mu}{2}} , \tag{4.69}$$

where $F(z)$ is the non-relativistic plasma dispersion function [52], defined as

$$F(z) = \frac{1}{\sqrt{\pi}} \int_{-\infty}^\infty \frac{\exp(-u^2)}{u - z} du + i\sigma\sqrt{\pi} e^{-z^2} . \tag{4.70}$$

Noting that $\mu \gg 1$ for the weakly–relativistic plasma, we can use the asymptotic expansion of this function for large arguments

$$F(z) = -\frac{1}{z} \left(1 + \frac{3}{2} \frac{1}{z^2}\right) + i\sigma\sqrt{\pi} e^{-z^2} . \tag{4.71}$$

Note also that, for superluminal phase velocities the imaginary part of this function vanishes, and the dispersion relation reduce to

$$\omega^2 = k^2c^2 + \omega_p^2 \left(1 + \frac{3}{\mu} \frac{k^2c^2}{\omega^2}\right) . \tag{4.72}$$

We can see here that the superluminal properties are directly associated with a finite temperature. This superluminal effect was not included in the fluid expressions of Section 4.2.

ii) **Ultra-relativistic plasma limit**

In this case, we can use

$$\gamma \simeq \frac{p}{mc}, \quad \mathbf{p} \simeq m\gamma\mathbf{v} \simeq \frac{p}{c}\mathbf{v}. \tag{4.73}$$

The Maxwell-Jüttner equilibrium distribution (4.64) converts into

$$f_0(p) = \frac{c^3}{8\pi T^3} \exp[-\mu\gamma], \quad \mu\gamma \simeq \frac{pc}{T}, \tag{4.74}$$

The dispersion relation (4.58)–(4.66) then becomes

$$\omega^2 \left(1 - \frac{\mu\omega_p^2}{2k^2c^2}\right) = k^2c^2 \left\{1 + \frac{\mu\omega_p^2}{2k^2c^2} \frac{(\omega^2 - k^2c^2)}{2\omega kc} \ln\left[\frac{\omega - kc}{\omega + kc}\right]\right\}, \tag{4.75}$$

where $\mu \simeq 1$.

## 4.4 Landau Damping

Let us now consider electrostatic waves, and examine the problem of electron Landau damping in the relativistic regime. In this case, we replace Maxwell's equations by Poisson's equation. For perturbations evolving in space and time as $\exp(i\mathbf{k}\cdot\mathbf{r} - i\omega t)$, the electric field amplitude will satisfy the relation

$$E_\mathbf{k} = \frac{i}{k\epsilon}\rho_\mathbf{k} \equiv -\frac{ien_0}{k\epsilon_0} \int f_\mathbf{k}(\mathbf{p})dp_\parallel, \tag{4.76}$$

where $p_\parallel = \mathbf{p}\cdot\mathbf{k}/k$ is the electron momentum in the propagation direction. On the other hand, from the linearized Vlasov equation we get

$$f_\mathbf{k}(\mathbf{p}) = i\frac{eE_\mathbf{k}}{(\omega - \mathbf{k}\cdot\mathbf{v})} \frac{\partial f_0}{\partial p_\parallel}. \tag{4.77}$$

These two expressions lead to the dispersion relation

$$\epsilon_l(\omega, \mathbf{k}) = 1 + \frac{\omega_p^2}{\omega^2} \int \frac{\partial f_0/\partial p_\parallel}{(1 - \mathbf{v}\cdot\mathbf{k}/\omega)} d\mathbf{p} = 0. \tag{4.78}$$

For isotropic equilibrium distributions, only depending on the absolute value of $\mathbf{p}$, this can be transformed into

$$\epsilon_l(\omega, \mathbf{k}) = 1 - \frac{\omega_p^2}{\omega^2} \int \frac{\partial f_0}{\partial p} \mathcal{I}_l(p) p^2 dp = 0, \tag{4.79}$$

where we have defined the auxiliary function

$$\mathcal{I}_l(p) = 2\frac{m\gamma\omega}{kp} + \frac{m^2\gamma^2\omega^2}{k^2p^2} \ln\left[\frac{m\gamma\omega - kp}{m\gamma\omega + kp}\right] , \tag{4.80}$$

This is similar, but not identical, to the quantity introduced in eq. (4.63) for the electromagnetic waves. If we now use the Maxwell-Jüttner equilibrium distribution (4.64), we obtain

$$\epsilon_l(\omega, \mathbf{k}) = 1 + \mu\frac{\omega_p^2}{k^2c^2}\left\{1 - \frac{\mu}{K_2(\mu)}\frac{\partial^2}{\partial\mu^2}\frac{P(\mu)}{\mu}\right\} . \tag{4.81}$$

where we have used the relation

$$\gamma^2 e^{-\mu\gamma} = \frac{\partial^2}{\partial\mu^2}\left(e^{-\mu\gamma}\right) . \tag{4.82}$$

Of particular interest is the so-called *long wavelength limit*, such that $k^2c^2 \ll \omega^2$. Going back to the relativistic plasma dispersion function, $P(\mu)$, we can expand the integrant in eq. (4.67), using

$$\frac{1}{\gamma^2 - (\gamma^2 - 1)k^2c^2/\omega^2} = \sum_{n=0}^{\infty} \frac{(\gamma^2 - 1)^n}{\gamma^{2n+2}}\left(\frac{kc}{\omega}\right)^{2n} . \tag{4.83}$$

This leads to

$$P(\mu) = \sum_{n=0}^{\infty}\left(\frac{kc}{\omega}\right)^{2n}\int_1^{\infty}\frac{(\gamma^2 - 1)^{(n-1/2)}}{\gamma^{2n+2}}e^{-\mu\gamma}d\gamma , \tag{4.84}$$

where we have assumed $\sigma = 0$. Replacing this in the above dispersion relation, we obtain

$$\epsilon_l(\omega, \mathbf{k}) = 1 - \frac{\mu^2\omega_p^2}{\omega^2 K_2(\mu)}\sum_{n=0}^{\infty}\left(\frac{kc}{\omega}\right)^{2n}\frac{1}{2n+3}\int_1^{\infty}\frac{(\gamma^2 - 1)^{(n+3/2)}}{\gamma^{2n+1}}e^{-\mu\gamma}d\gamma = 0 . \tag{4.85}$$

If we further assume the weakly relativistic limit, $\mu \gg 1$, such that

$$K_2(\mu) = \sqrt{\frac{\pi}{2\mu}}\exp\left[-\mu\left(1 + \frac{1}{2\mu}\right)\right] , \tag{4.86}$$

the dispersion relation (4.85) will reduce to

$$\epsilon_l(\omega, \mathbf{k}) \simeq 1 - \frac{\omega_p^2}{\omega^2}\left(1 - \frac{5}{2\mu} + \frac{3k^2c^2}{\mu\omega^2}\right) = 0 . \tag{4.87}$$

Or, more explicitly, to

$$\omega^2 = \omega_p^2\left(1 - \frac{5}{2}\frac{T_e}{mc^2} + \frac{3T_e}{m}\frac{k^2}{\omega^2}\right) . \tag{4.88}$$

This shows a displacement of the cutoff frequency $\omega_{c.o.} \equiv \omega(k=0)$ by a factor of $5/(2\mu)$, as

$$\omega_{c.o.}^2 = \omega_p^2 \left(1 - \frac{5}{2}\frac{T_e}{mc^2}\right) . \tag{4.89}$$

On the other hand, assuming that $\omega^2 \simeq \omega_p^2$ is always true, we can see that the *Bohm-Gross dispersion relation* stays approximately valid in the weakly relativistic and long wavelength limit. Going back to the exact dispersion relation (4.81) for electrostatic waves, it is useful to consider the weakly relativistic and the ultra-relativistic limits, as we have already done for the transverse electromagnetic waves.

i) **Weakly–relativistic limit**

In this case, the electrostatic permittivity becomes

$$\epsilon_l(\omega, \mathbf{k}) = 1 + \frac{\mu}{k^2c^2}\omega_p^2 \left[1 + F(z)\right] , \tag{4.90}$$

where $z = \sqrt{\mu/2}\omega/(kc)$, and the weakly relativistic plasma dispersion function is

$$F(z) = \frac{z}{\sqrt{\pi}} \int_{-\infty}^{\infty} \frac{e^{-x^2}}{x-z} dx = \frac{z}{\sqrt{\pi}} \mathcal{P} \int_{-\infty}^{\infty} \frac{e^{-x^2}}{x-z} dx + i\sqrt{\pi} z e^{z^2} . \tag{4.91}$$

In the long wavelength limit, $z \ll 1$, we can approximately write the principal part of the integral as

$$\frac{z}{\sqrt{\pi}} \mathcal{P} \int_{-\infty}^{\infty} \frac{e^{-x^2}}{x-z} dx \simeq -\left(1 + \frac{1}{2z^2} + \frac{3}{4z^4}\right) \tag{4.92}$$

Ignoring the imaginary part of the dispersion function, to be discussed later, we are then led to

$$\epsilon_l(\omega, \mathbf{k}) \simeq 1 - \frac{\omega_p^2}{\omega^2}\left(1 + \frac{3k^2c^2}{\mu\omega^2}\right) = 0 . \tag{4.93}$$

which coincides with the non-relativistic result.

ii) **Ultra-relativistic limit**

In this case, we have $\mu \simeq 1$, and we get, for isotropic distributions

$$\epsilon_l(\omega, \mathbf{k}) \simeq 1 + 2\pi \frac{\mu\omega_p^2}{kc} \int_{-1}^{1} d(\cos\theta) \int_0^{\infty} \frac{p^2 f_0(p) \cos\theta}{(kc\cos\theta - \omega)} dp , \tag{4.94}$$

which, ignoring the imaginary part, gives

$$\epsilon_l(\omega, \mathbf{k}) = 1 + \frac{\mu\omega_p^2}{k^2c^2}\left\{1 + \frac{\omega}{2kc} \ln\left[\frac{\omega - kc}{\omega + kc}\right]\right) , \tag{4.95}$$

Let us finally consider the imaginary part of the dielectric function $\epsilon_l(\omega, \mathbf{k})$. We go back to the general expression eq. (4.81), writing it as $\epsilon_l(\omega, \mathbf{k}) = \epsilon_l' + i\epsilon"_l = 0$. We now assume that $\omega = \omega_r + i\omega_i$, with $|\omega_i| \ll \omega_r$, and use the approximation

$$\epsilon_l'(\omega) \simeq \epsilon_l'(\omega_r)\,, \quad \epsilon"_l(\omega) \simeq -\omega_i \left(\frac{\partial \epsilon_l'}{\partial \omega}\right)_{\omega_r} . \tag{4.96}$$

This then leads to

$$\epsilon"_l(\omega) = \frac{\pi}{2}\frac{\omega_p^3}{k^3c^3}\frac{\mu^2}{K_2(\mu)}\frac{\partial^2}{\partial\mu}\left\{\frac{1}{\mu}\exp\left[-\mu\frac{(kc/\omega)}{\sqrt{(kc/\omega)^2-1}}\right]\right\} . \tag{4.97}$$

Notice that, in writing this equation we used the expression of $P(\mu)$ in eq. (4.67) valid for the sub-luminal case, $\Re(\omega^2) < k^2c^2$, because superluminal waves would not be damped. We also have assumed $\omega_i \sim 0$. Taking the derivative in eq. (4.96), as approximately given by

$$\frac{\partial \epsilon_l'}{\partial \omega} \simeq \frac{\partial}{\partial \omega}\left(1 - \frac{\omega_p^2}{\omega^2}\right) \simeq 2\frac{\omega_p^2}{\omega_r^3} , \tag{4.98}$$

we get, for the imaginary part of the frequency

$$\omega_i = -\frac{\pi}{4}\frac{\omega_r^4}{k^3c^3}\frac{\mu^2}{K_2(\mu)}\frac{\partial^2}{\partial\mu}\left\{\frac{1}{\mu}\exp\left[-\frac{\mu}{\sqrt{1-\omega^2/(k^2c^2)}}\right]\right\} . \tag{4.99}$$

where $\omega_r \simeq \omega_p$. This is the expression for the electron Landau damping of electron plasma waves in a relativistic plasma. If we now take the long wavelength, $k^2c^2 \ll \omega^2$, and weakly relativistic $\mu \gg 1$ limits, we are reduced to

$$\omega_i = -\sqrt{\frac{\pi}{8}}\frac{\mu^{3/2}}{k^3c^3}\omega_r^4 \exp\left(-\frac{\mu}{2k^2c^2/\omega_p^2}\right) . \tag{4.100}$$

Noting that

$$\frac{\mu\omega_p^2}{k^2c^2} = \frac{m}{T_e}\frac{\omega_p^2}{k^2} \equiv \frac{1}{k^2\lambda_D^2} , \tag{4.101}$$

where $\lambda_D$ is the *electron Debye length*, we arrive at the expression

$$\omega_i = -\sqrt{\frac{\pi}{8}}\frac{\omega_p}{(k\lambda_D)^3}\exp\left(-\frac{1}{2k^2\lambda_D^2}\right) . \tag{4.102}$$

This coincides with the classical Landau damping, valid for non-relativistic plasmas, as first derived by Landau in 1946 [53]. The relativistic Landau damping (4.99) is represented in Figure 4.1, showing that damping strongly depends on $\mu$.

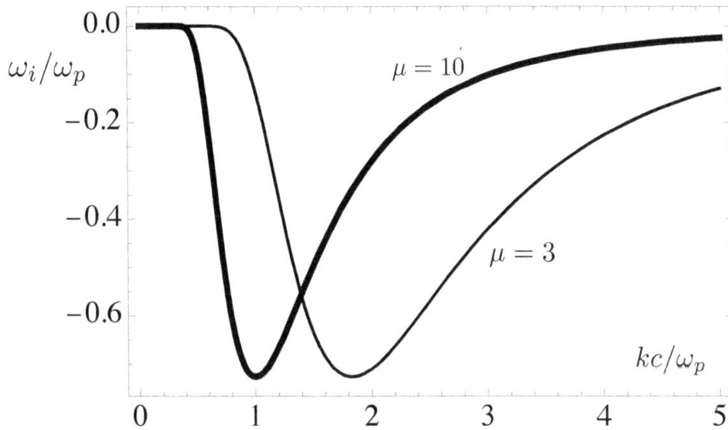


**FIGURE 4.1**
Relativistic Landau damping, $\omega_i/\omega_p$, as a function of the normalized wavenumber $kc/\omega_p$, for two different values of the temperature: $\mu = 3$ (thin curve) and $\mu = 10$ (bold).

## 4.5 Magnetized Plasma

Here, we give a general perspective of waves in a magnetized plasma, assuming a static magnetic field $\mathbf{B}_0 = B_0\mathbf{e}_z$.

### 4.5.1 Cold Plasma Description

We start with the cold plasma case, which is useful to describe the basic properties of linear wave propagation. We use the non-relativistic fluid equations, with $\gamma_\alpha = 1$ and $T_\alpha = 0$, as

$$\frac{\partial n_\alpha}{\partial t} + \nabla \cdot (n_\alpha \mathbf{v}_\alpha) = 0 \,, \quad \frac{d\mathbf{v}_\alpha}{dt} = \frac{q_\alpha}{m_\alpha} (\mathbf{E} + \mathbf{v}_\alpha \times \mathbf{B}) \,, \tag{4.103}$$

where $d/dt = \partial/\partial t + \mathbf{v}_\alpha \cdot \nabla$. Assuming wave perturbations with the form $\exp(i\mathbf{k}\cdot\mathbf{r} - i\omega t)$, and linearizing these equations with respect to perturbations, we get

$$n_\alpha = \frac{n_{0\alpha}}{\omega}(\mathbf{k} \cdot \mathbf{v}_\alpha) \,, \quad \mathbf{v}_\alpha = i\frac{q_\alpha}{m_\alpha \omega} (\mathbf{E} + \mathbf{v}_\alpha \times \mathbf{B}_0) \,, \tag{4.104}$$

where $\mathbf{B}_0$ is the static field. Following the steps of Section (4.1), this leads to a dielectric tensor of the form

$$\bar{\bar{\kappa}} = \begin{bmatrix} \kappa_1 & -i\kappa_2 & 0 \\ i\kappa_2 & \kappa_1 & 0 \\ 0 & 0 & \kappa_3 \end{bmatrix} . \tag{4.105}$$

The three independent components of this tensor are

$$\kappa_1 = 1 - \sum_\alpha \frac{\omega_{p\alpha}^2}{\omega^2 - \omega_{c\alpha}^2}, \quad \kappa_2 = -\sum_\alpha \frac{\omega_{c\alpha}}{\omega}\frac{\omega_{p\alpha}^2}{\omega^2 - \omega_{c\alpha}^2}, \quad \kappa_3 = 1 - \sum_\alpha \frac{\omega_{p\alpha}^2}{\omega^2} . \tag{4.106}$$

Here, we have used the plasma frequencies, $\omega_{p\alpha} = \sqrt{q_\alpha^2 n_{0\alpha}/(\epsilon_0 m_\alpha)}$, and the cyclotron frequencies $\omega_{c\alpha} = |q_\alpha|/(m_\alpha B_0)$ for the different particle species. Note that, in this cold plasma limit, the dielectric tensor is independent of the wavevector. Replacing this in the general dispersion relation, eq. (4.9), and assuming that the wave propagates at an angle $\theta$ with respect to the static magnetic field $\mathbf{B}_0$, we obtain

$$\begin{vmatrix} \kappa_1 - N^2\cos^2\theta & -i\kappa_2 & N^2\sin\theta\cos\theta \\ i\kappa_2 & \kappa_1 - N^2 & 0 \\ N^2\sin\theta\cos\theta & 0 & \kappa_3 - N^2\sin^2\theta \end{vmatrix} = 0 , \tag{4.107}$$

where we have used the refractive index, $N = (kc/\omega)$. This can also be written in the form

$$AN^4 - BN^2 + C = 0 , \tag{4.108}$$

with

$$\begin{aligned} A &= \kappa_1 \sin^2\theta + \kappa_3\cos^2\theta , \\ B &= \left(\kappa_1^2 - \kappa_2^2\right)\sin^2\theta + \kappa_1\kappa_1\left(1+\cos^2\theta\right) , \\ C &= \kappa_3\left(\kappa_1^2 - \kappa_2^2\right) . \end{aligned} \tag{4.109}$$

This equation can be solved in terms of $tan^2\theta$, leading to

$$\tan^2\theta = -\frac{\kappa_3\left[(N^2-\kappa_1)^2 - \kappa_2\right]}{(\kappa_1 N^2 - \kappa_1^2 + \kappa_2^2)(N^2 - \kappa_3)} . \tag{4.110}$$

This solution can be used to identify the different plasma modes, for a given angle of propagation $\theta$. The cases of parallel ($\theta = 0$) and perpendicular ($\theta = \pi/2$) propagation are particularly important, and will be discussed next.

i) **Parallel propagation**, ($\theta = 0$, $\tan^2\theta \to \infty$)

This situation corresponds to two possible conditions:

$$\kappa_3 = 0 , \quad N^2 = \kappa_1 \pm \kappa_2 , \tag{4.111}$$

which respectively define the *plasma oscillations*, with frequency

$$\omega^2 = \sum_\alpha \omega_{p\alpha}^2 \equiv \omega_p^2 \,, \tag{4.112}$$

and the two modes of propagation, defined by

$$N_\pm^2 = 1 - \sum_\alpha \frac{\omega_{p\alpha}^2}{\omega(\omega \mp \omega_{c\alpha})} \,, \tag{4.113}$$

where, by definition, the frequencies $\omega_{c\alpha}$ are all positive. In order to understand the physical nature of these two modes, it is useful to define their polarization. For this purpose, we use the determinant in eq. (4.107), noting that the electric field components of these modes are proportional to the corresponding minors. It means that

$$\frac{E_x}{(N^2 - \kappa_1)(N^2 \sin^2\theta - \kappa_3)} = \frac{E_y}{i\kappa_2(N^2 \sin^2\theta - \kappa_3)} = \frac{E_z}{(N^2 - \kappa_1)N^2 \sin\theta\cos\theta} \,. \tag{4.114}$$

This can also be stated as

$$\frac{E_y}{E_x} = \frac{i\kappa_2}{N^2 - \kappa_1} \,, \quad \frac{E_z}{E_x} = \frac{N^2 \sin\theta\cos\theta}{N^2 \sin^2\theta - \kappa^3} \,. \tag{4.115}$$

For parallel propagation ($\theta = 0$), we can see that the two transverse modes, with refractive indices $N_\pm^2$ defined by eq. (4.113), have to satisfy the relations

$$\frac{E_y}{E_x} = \frac{i\kappa_2}{N_\pm^2 - \kappa_1} = \pm i \,, \quad E_z = 0 \,. \tag{4.116}$$

This corresponds to right and left-hand circular polarization in the perpendicular plane. We conclude that, for electromagnetic waves propagating along $\mathbf{B}_0$, the field of mode $N_+$ rotates in the direction of the electron cyclotron motion, while the field of mode $N_-$ rotates in the opposite (ion) cyclotron motion. As for the plasma oscillations dictated by $\kappa_3 = 0$, they are longitudinal modes with $E_x = E_y = 0$. In a thermal plasma, they would propagate along the static field $\mathbf{B}_0$.

ii) **Perpendicular propagation**, ($\theta = \pi/2$, $\tan^2\theta \to 0$)

The plasma modes are now dictated by the two independent conditions

$$N^2 = \kappa_3 \,, \quad N^2 = \frac{1}{\kappa_1}\left(\kappa_1^2 - \kappa_2^2\right) \,, \tag{4.117}$$

The first one corresponds to the *ordinary mode*, such that

$$\omega^2 = \omega_p^2 + k^2 c^2 \,. \tag{4.118}$$

This is identical to the transverse mode that exists in isotropic plasmas. It is therefore called the *ordinary mode*. We have already discussed the thermal and relativistic corrections that are ignored in this section. Here, this mode only propagates in the perpendicular direction, and is polarized along the static magnetic field, with $E_x = E_y = 0$.

On the other hand, the second condition in eq. (4.117), corresponds to the *extraordinary mode*, which has elliptic polarization in the perpendicular plane, with a longitudinal component, as defined by

$$\frac{E_y}{E_x} = -i\frac{\kappa_1 \kappa^3}{\kappa_2^2}, \quad E_z = 0. \tag{4.119}$$

This mode has a more complex behavior and, in order to understand its main characteristics, it is useful to define the cutoff and resonance frequencies. The *cutoff frequency* is defined by the condition $k = 0$, or $N = 0$. For the ordinary mode, this is simply given by the condition $\kappa_3 = 0$, or $\omega_{c.o.} = \omega_p$. For the extraordinary mode, it is given by the conditions

$$\kappa_1 \pm \kappa_2 = 0, \tag{4.120}$$

These two cutoff conditions are actually independent of the angle of propagation $\theta$, and correspond to the case $C = 0$, as can be seen from eq. (4.108). As for the *resonance frequencies*, they correspond to the limit $k \to \infty$, or $N \to \infty$. This is satisfied for $\kappa_1 = 0$, or

$$1 - \frac{\omega_{pe}^2}{\omega^2 - \omega_{ce}^2} - \frac{\omega_{pi}^2}{\omega^2 - \omega_{ci}^2} = 0 \tag{4.121}$$

where an electron-ion plasma was assumed, $\alpha = (e, i)$. This can also be written as

$$\omega^4 - \omega_{u.h.}^2 \omega^2 + \omega_{ci}^2 \left(\omega_{ce}^2 + \omega_p^2\right) = 0, \tag{4.122}$$

where we have introduced the *upper-hybrid frequency*, $\omega_{u.h.}$, and the total cyclotron frequency $\omega_c$, defined as

$$\omega_{u.h.} = \sqrt{\omega_p^2 + \omega_c^2}, \quad \omega_c^2 = \omega_{ce}^2 + \omega_{ci}^2. \tag{4.123}$$

This quartic equation for the wave frequency shows the existence of two distinct resonances, given by the two solutions

$$\omega_I = \omega_{u.h.}, \quad \omega_{II} = \omega_{l.h.} \equiv \frac{\omega_{pi}^2 + \omega_{ci}^2}{\omega_{pi}^2 - \omega_{ce}\omega_{ci}} \omega_{ce}\omega_{ci}. \tag{4.124}$$

They are called the upper and lower-hybrid frequencies.

A final comment concerns the modes in parallel propagation. For very high frequencies ($\omega^2 \gg \omega_{ce}^2$), the two propagation modes defined by eq. (4.113) coalesce into a single one, which is identical to that of the isotropic plasma

$$N_\pm^2 \to N^2 = 1 - \frac{\omega_p^2}{\omega^2}. \tag{4.125}$$

In the opposite case of very low frequencies, such that $\omega^2 \ll (\omega_{ce}^2, \omega_{pe}^2)$, the two modes coalesce into

$$N_\pm^2 \to N^2 = 1 + \frac{c^2}{v_A^2} \,. \tag{4.126}$$

where the quantity

$$v_A = \frac{B_0}{\sqrt{\mu_0 \rho}} \,, \quad \rho = n_{0e} m_e + n_{0i} m_i \,, \tag{4.127}$$

is known as the *Alfvén velocity*. Notice that the low-frequency waves can be studied using MHD equations, as first derived by Alfvén [54]. Relativistic formulations of MHD waves can be found (see [55]).

### 4.5.2 Kinetic Description

We now turn to the kinetic description of linear waves in relativistic magnetized plasmas. For simplicity, we start with electrostatic waves and derive the corresponding dispersion relation for a generic distribution function. We then focus on the useful case of a non-relativistic plasma with a relativistic electron beam, and finally discuss the case of electromagnetic waves.

i) **Electrostatic wave dispersion**

In the electrostatic case, the wave electric field $\mathbf{E}$ is determined by the Poisson's equation

$$\nabla \cdot \mathbf{E} = \frac{1}{\epsilon_0} \sum_\alpha n_{0\alpha} q_\alpha \int f_\alpha(\mathbf{u}) d\mathbf{u} \,, \tag{4.128}$$

where $\mathbf{u} = \gamma \mathbf{v}$ is the covariant velocity and $\alpha = e, i$ represents the particle species. The distribution functions are determined by the Vlasov equation, which we write in the form

$$\frac{\partial f_\alpha}{\partial t} + \frac{\mathbf{u}}{\gamma} \cdot \frac{\partial f_\alpha}{\partial \mathbf{r}} + \frac{q_\alpha}{m_\alpha} \left( \mathbf{E} + \frac{\mathbf{u}}{\gamma} \times \mathbf{B}_0 \right) \cdot \frac{\partial f_\alpha}{\partial \mathbf{u}} = 0 \,, \tag{4.129}$$

where $\mathbf{B}_0$ is the static magnetic field. We now focus on high-frequency electrostatic waves, such that the ion contribution is negligible and can be ignored. Assuming that the electron distribution function is given by $f_e = f_0 + f_\mathbf{k} \exp(i\mathbf{k} \cdot \mathbf{r} - i\omega t)$ and using a perturbative approach, we can reduce these equations to

$$i\mathbf{k} \cdot \mathbf{E} = -\frac{e}{\epsilon_0} n_0 \int f_\mathbf{k}(\mathbf{u}) d\mathbf{u} \,, \tag{4.130}$$

and

$$\left( i\omega - i\frac{\mathbf{u}}{\gamma} \cdot \mathbf{k} \right) f_\mathbf{k} + \frac{e}{m} \left( \frac{\mathbf{u}}{\gamma} \times \mathbf{B}_0 \right) f_\mathbf{k} = -\frac{e}{m} \mathbf{E} \cdot \frac{\partial f_0}{\partial \mathbf{u}} \,. \tag{4.131}$$

We now assume the magnetic field along the $z$-axis, $\mathbf{B}_0 = B_0\mathbf{e}_z$, and use cylindrical coordinates $\mathbf{r} \equiv (r, \theta, z)$, such that we can write

$$\mathbf{k} = k_\perp \mathbf{e}_y + k_z \mathbf{e}_z \,, \quad \mathbf{u} = u_\perp (-\sin\theta \mathbf{e}_x + \cos\theta \mathbf{e}_y) + u_z \mathbf{e}_z \,. \tag{4.132}$$

We can then transform eq. (4.131) into

$$\frac{\partial f_\mathbf{k}}{\partial \theta} + \frac{i}{\omega_c} (k_\perp u_\perp \cos\theta + k_z u_z - \gamma\omega) f_\mathbf{k} = S(\theta) \,, \tag{4.133}$$

with

$$S(\theta) = \frac{\gamma}{\omega_c} \frac{eE}{mk} \left( k_\perp \cos\theta \frac{\partial f_0}{\partial u_\perp} + k_z \frac{\partial f_0}{\partial u_z} \right) \,, \tag{4.134}$$

where we have used the non-relativistic electron cyclotron frequency $\omega_c = eB_0/m$. Solutions of the homogeneous equation, when the right-hand-side of eq. (4.133) is zero, can easily be found

$$f_\pm = \exp\left[ \pm \frac{i}{\omega_c} (k_\perp u_\perp \sin\theta + k_z u_z \theta - \gamma\omega\theta) \right] \,. \tag{4.135}$$

The solution of the inhomogeneous equation can then be written as

$$f_\mathbf{k} = f_- \int_{-\infty}^{\theta} f_+(\theta') \, S(\theta') d\theta' \,. \tag{4.136}$$

After some algebra, and noting that the exponential of a sine function can be developed into a sum of Bessel functions, this solution can be written in explicit form as

$$f_\mathbf{k} = i \frac{eE}{mk} \gamma \sum_n J_n(\kappa) \left( \frac{n\omega_c}{u_\perp} \frac{\partial f_0}{\partial u_\perp} + k_z \frac{\partial f_0}{\partial u_z} \right) \frac{\exp(in\theta - i\kappa \sin\theta)}{(\gamma\omega - n\omega_c - k_z u_z)} \,, \tag{4.137}$$

where $J_n(\kappa)$ are Bessel functions of the first kind, and we have used $\kappa = k_\perp u_\perp / \omega_c$. Replacing this in Poisson's equation (4.130), and noting that $\int \exp(in\theta - i\ell\theta) d\theta = 2\pi\delta(n - \ell)$, for $n$ and $\ell$ integer, we can then derive a dispersion relation of the form

$$1 + 2\pi \frac{\omega_p^2}{k^2} \sum_n \int J_n^2(\kappa) \left[ \frac{n\omega_c}{u_\perp} \frac{\partial f_0}{\partial u_\perp} + k_z \frac{\partial f_0}{\partial z} \right] \frac{u_\perp du_\perp du_z}{(\omega - n\omega_c/\gamma - k_z u_z/\gamma)} = 0 \,. \tag{4.138}$$

This is valid for a general form of equilibrium distribution $f_0$. It is important to notice that, in the presence of a static magnetic field $B_0$, the Landau resonance previously discussed, is replaced by a *Landau-cyclotron resonance* defined by the condition $\omega = n\omega_c/\gamma + k_z u_z/\gamma$. This means that, instead of a single pole we now have an infinity of new poles, which depend on the cyclotron frequency.

ii) **Thermal plasma with a relativistic beam**

In order to understand the significance of the above result, we need to explore some specific examples. For that purpose, we focus on the useful case of a non-relativistic plasma in the presence of a relativistic electron beam. This means that, in this dispersion relation we need to replace $f_0$ by $f_0 + f_b$, where $f_0$ represents the non-relativistic background, and $f_b$ is the relativistic beam distribution. Let us start with the background, using the Maxwellian function

$$f_0 = \left(\pi\alpha^2\right)^{-3/2} \exp\left(-\frac{u_\perp^2 + u_z^2}{\alpha^2}\right), \quad \alpha^2 = \frac{2T}{m}. \tag{4.139}$$

Let us replace this distribution in eq. (4.138) and discuss the different terms associated with the integrations in velocity space. First, the integration of the first term in $u_\perp$, leads to

$$\int J_n^2(\kappa) \frac{n\omega_c}{u_\perp} \frac{\partial f_0}{\partial u_\perp} u_\perp du_\perp = -\frac{n\omega_c}{\pi^{1/2}\alpha^3} e^{-\lambda} e^{-u_z^2/\alpha^2} I_n(\lambda). \tag{4.140}$$

where we have defined the parameter

$$\lambda = \frac{k_\perp^2 \alpha^2}{2\omega_c^2} \equiv k_\perp^2 \rho_c^2, \tag{4.141}$$

and used the integral

$$\int_0^\infty J_n^2(bu_\perp) e^{-a^2 u_\perp^2} u_\perp du_\perp = \frac{1}{2a^2} e^{-b^2/2a^2} I_n(b^2/2a^2), \tag{4.142}$$

where $I_n(x)$ are Bessel functions of the second kind. If we now perform the integration in $u_z$, this leads to the following result

$$\int J_n^2(\kappa) \frac{n\omega_c}{u_\perp} \frac{(\partial f_0/\partial u_\perp)}{(\omega - n\omega_c/\gamma_e - k_z u_z/\gamma_e)} u_\perp du_\perp du_z = \frac{n\omega_c/\gamma_e}{\pi^{1/2}\alpha^3 k_z} e^{-\lambda} I_n(\lambda) Z(\alpha_n). \tag{4.143}$$

where we have used the quantities

$$\gamma_e = \sqrt{1 + \frac{5\alpha^2}{4c^2}}, \quad \alpha_n = \frac{\omega\gamma_e - n\omega_c}{\alpha k_z}, \tag{4.144}$$

and defined the new function

$$Z(\alpha_n) = i\sqrt{\pi} e^{\alpha_n^2} \left(1 + \frac{2i}{\sqrt{\pi}} \int_0^{\alpha_n} e^{t^2} dt\right). \tag{4.145}$$

It is obvious that $\gamma_e \simeq 1$, but we keep it for consistency. Similarly, for the second term in the integral of eq. (4.138), we obtain

$$\int J_n^2(\kappa) \frac{k_z(\partial f_0/\partial u_z)}{(\omega - n\omega_c/\gamma_e - k_z u_z/\gamma_e)} u_\perp du_\perp du_z = \frac{\gamma_e}{\pi\alpha^2} e^{-\lambda} I_n(\lambda) \left[\alpha_n Z(\alpha_n) + 1\right]. \tag{4.146}$$

Summing these two integrals and taking into account that

$$\sum_n e^{-\lambda} I_n(\lambda) = 1\,, \quad \frac{\omega_p^2}{\alpha^2} = \frac{1}{2\lambda_D^2}\,, \tag{4.147}$$

we finally arrive at the kinetic dispersion relation of electrostatic waves for a magnetized thermal plasma, $\mathcal{D}_0(\omega, \mathbf{k}) = 0$, as

$$\mathcal{D}_0(\omega, \mathbf{k}) \equiv 1 + k^2\lambda_D^2/\gamma_e + \alpha_0 e^{-\lambda} \sum_n I_n(\lambda) Z(\alpha_n) = 0\,, \tag{4.148}$$

with $\alpha_0 = \omega\gamma_e/\alpha k_z$. At this point, we should also note that the real part of the function $Z(\alpha_n) = X(\alpha_n) + iY(\alpha)$ can be developed asymptotically for $\alpha_n \gg 1$ as

$$-\alpha_n X(\alpha_n) = \left(1 + \frac{1}{2\alpha_n^2} + \frac{3}{4\alpha_n^4} + \frac{15}{8\alpha_n^6} + ...\right). \tag{4.149}$$

In contrast, for $\alpha_n \ll 1$, the asymptotic expansion gives

$$-X(\alpha_n) = 2\alpha_n \left(1 - \frac{2\alpha_n^2}{3} + \frac{4\alpha_n^4}{15} + ...\right). \tag{4.150}$$

We should now consider the contribution from the relativistic beam. For that purpose, we use the distribution function

$$f_b = C_b \exp\left(- \left|\epsilon - \epsilon_0\right| / T_b\right)\,, \quad \epsilon = m\gamma_z c^2\,, \quad \gamma_z = \sqrt{1 + u_z^2/c^2}\,. \tag{4.151}$$

$C_b$ is a normalization constant, $T_b$ is the beam temperature, and $\epsilon_0$ the central beam energy. Here, we have also neglected the perpendicular temperature of the beam. This means that we only need to calculate the terms associated with the derivative $\partial f_b/\partial u_z$. We then have

$$\int \frac{\partial f_b/\partial u_z}{(\omega/k_z - u_z/\gamma_z)} du_z = \frac{mC_b}{T_b} \int \frac{v_z \exp\left(- \left|\epsilon - \epsilon_0\right| / T_b\right)}{(\omega/k_z - v_z)} \left(1 + \frac{u_z^2}{c^2}\right) dv_z\,. \tag{4.152}$$

For a weak beam, with a density such that $n_b/n_0 \ll 1$, the main contribution of the beam is the introduction of a growth or damping rate $\Gamma_b(\omega, k_z)$. We can therefore neglect the contribution of the principal part of the integral and retaining only the contribution of the pole, we arrive at the following expression

$$\Gamma_b(\omega, k_z) = i2\pi \frac{\omega_{pb}^2}{k^2} \int \frac{\partial f_b/\partial u_z}{(\omega/k_z - u_z/\gamma_z)} du_z\,, \tag{4.153}$$

where the beam plasma frequency is determined by $\omega_{pb}^2 = \omega_p(n_b/n_0)$. Or, more explicitly

$$\Gamma_b(\omega, k_z) = 2\pi^2 \frac{mC_r}{T_b} \frac{\omega_{pb}^2}{k^2} \frac{\omega}{k_z} \left(1 - \frac{\omega^2}{k_z^2c^2}\right)^{-3/2} \exp\left[\frac{mc^2}{T_b}\left(1 - \frac{\omega^2}{k_z^2c^2}\right)^{-1/2} - \frac{\epsilon_0}{T_b}\right]. \tag{4.154}$$

The final dispersion relation for electrostatic waves in the presence of the beam is then given by

$$\mathcal{D}(\omega, \mathbf{k}) = \mathcal{D}_0(\omega, \mathbf{k}) - i\Gamma_b(\omega, k_z) = 0\,. \tag{4.155}$$

where the first term is given by eq. (4.119) and the second by eq. (4.153). To discuss particular cases, let us consider the real and imaginary parts of this expression. For the real part, the beam contribution is neglected, and we have

$$\mathcal{D}_r(\omega, \mathbf{k}) = 1 + k^2\lambda_D^2/\gamma_e + \alpha_0 e^{-\lambda} \sum_n I_n(\lambda) X(\alpha_n) \simeq 0\,, \tag{4.156}$$

Using the asymptotic expression (4.149), valid for $\alpha_n \gg 1$, this can be reduced to

$$k^2\lambda_D^2 - \gamma_e \sum_n I_n(\lambda) e^{-\lambda} \left( \frac{n\omega_c}{\gamma_e - n\omega_c} + \frac{\alpha_0}{2\alpha_n^3} \right) \simeq 0\,. \tag{4.157}$$

For $\lambda = k_\perp^2 \rho_c^2 \ll 1$, we can use $I_0(\lambda) = 1$ and $I_{\pm 1}(\lambda) \simeq \lambda/2$. The other therms can be neglected, and we are left with a simplified expression for the real part of the dispersion relation, given by

$$k^2\lambda_D^2 - \lambda \frac{\gamma_e \omega_c^2}{\gamma_e^2 \omega^2 - \omega_c^2} - \frac{\gamma_e}{2\alpha_0} = 0\,. \tag{4.158}$$

This can be more explicitly written as

$$k_\perp^2 \left[ 1 - \frac{\gamma_e \omega_p^2}{\gamma_e^2 \omega^2 - \omega_c^2} \right] + k_z^2 \left[ 1 - \frac{\omega_p^2}{\gamma_e \omega^2} \right] = 0\,. \tag{4.159}$$

For purely perpendicular oscillations, such that $k_z = 0$, we can recognize here the so-called *Bernstein* waves, with oscillations at the frequency $\omega^2 = \omega_p^2/\gamma_e + \omega_c^2/\gamma_e^2 \equiv \omega_{u.h}^2$. In contrast, for parallel propagation, with $k_\perp = 0$, we are reduced to plasma oscillations $\omega^2 = \omega_p^2/\gamma_e$, as expected. Of course, dispersion effects and higher-order modes, can be found in the complete dispersion condition (4.157). Let us finally consider the imaginary part of eq. (4.155). We then have

$$\mathcal{D}_i(\omega, \mathbf{k}) = \sqrt{\pi}\alpha_0 e^{-\lambda} \sum_n I_n(\lambda) e^{-\alpha_n^2} - \Gamma_b(\omega, k_z)\,, \tag{4.160}$$

Keeping only the first terms, which are the dominant ones, we get

$$\mathcal{D}_i(\omega, \mathbf{k}) = \sqrt{\pi}\alpha_0 \left( e^{-\alpha_0^2} + \frac{\lambda}{2} e^{-\alpha_1^2} \right) - \Gamma_b(\omega, k_z)\,, \tag{4.161}$$

For the beam contribution, we can also approximately write

$$\Gamma_b(\omega, k_z) = 2\pi^2 \frac{mC_r}{T_b} \frac{\omega_{pb}^2}{k^2} \frac{\omega k_z^2}{(k_z^2 - \omega^2/c^2)^{3/2}}\,. \tag{4.162}$$

The growth rate of the electrostatic instabilities generated by the beam can then be calculated using a complex frequency $\omega = \omega_r + i\gamma_i$, such that the imaginary part is dictated by the relation

$$\gamma_i = -\frac{\mathcal{D}_i(\omega_r, \mathbf{k})}{(\partial \mathcal{D}_r / \partial \omega)_{\omega_r}} . \tag{4.163}$$

As we can see from eq. (4.161), the existence of an instability implies that the beam is sufficiently strong to compensate for the eventual damping due to the thermal background.

iii) **Electromagnetic wave dispersion**

The case of weak amplitude electromagnetic waves in a relativistic magnetized plasmas was first studied by Trubnikov [56], who established the corresponding dispersion relation. A more modern account can be found in [57, 58], or in [48]. The expressions for the dielectric tensor terms can be derived using the methods explained above, and are given here for completeness. Defining the dielectric tensor as in eq. (4.164), as

$$\bar{\bar{\kappa}} = \bar{\bar{1}} + \frac{i}{\omega \epsilon_0} \bar{\bar{\sigma}}_e . \tag{4.164}$$

we can write

$$\kappa_{ij} = \delta_{ij} - \frac{\omega_p^2}{\omega^2} \frac{\mu^2}{2K_2(\mu)} \int_{-\infty}^{\infty} dp_z \int_0^{\infty} p_\perp dp_\perp \frac{e^{-\mu\gamma_e}}{\gamma_e} \sum_{n=-\infty}^{\infty} \frac{P_{ij}^n}{(\gamma_e - np_z + n\omega_c/\omega)} , \tag{4.165}$$

where $\mu = mc^2/T$, and the functions $P_{ij}^n$ are defined as

$$P_{xx}^n = \frac{n^2}{\nu_\perp^2} J_n^2(\nu_\perp p_\perp) , \quad P_{xy}^n = P_{yx}^n = \frac{ip_\perp n}{\nu_\perp} J_n(\nu_\perp p_\perp) J_n'(\nu_\perp p_\perp) , \tag{4.166}$$

$$P_{xz}^n = P_{zx}^n = -\frac{np_z}{\nu_\perp} J_n^2(\nu_\perp p_\perp) , \quad P_{yy}^n = p_\perp^2 J_n'^2(\nu_\perp p_\perp) , \tag{4.167}$$

and

$$P_{yz}^n = -P_{zy}^n = ip_z p_\perp J_n(\nu_\perp p_\perp) J_n'(\nu_\perp p_\perp) , \quad P_{zz}^n = p_z^2 J_n^2(\nu_\perp p_\perp) . \tag{4.168}$$

In these expressions $p_z$ and $p_\perp$ are normalized to $mc$, and we have used $\nu_\perp = k_\perp c/\omega_c$.

# 5

# *Nonlinear Waves*

In this chapter, we consider nonlinear wave solutions, solitons and shocks, as well as nonlinear wave-mixing and high-harmonic generation. In contrast with the previous chapter devoted to linear waves, where infinitesimal wave amplitudes were assumed, such that the dispersion relations were independent of the field amplitude, here the relativistic effects produced by the waves themselves are an essential ingredient of the physical processes, and the thermal effects are in most cases negligible.

## 5.1 Exact Wave Solutions

### 5.1.1 Electrostatic Waves

We first consider a cold plasma description, where relativistic particle motion is included. The starting point of our cold plasma theory are the so-called *Akhiezer-Polovin equations*, which are obtained from Maxwell's equations where the fields are created by the charge and current distributions, $\rho = -e(n - n_0)$ and $\mathbf{J} = -en\mathbf{v}$. The ions are assumed immobile, with equilibrium density $n_0$, and the electron mean density and velocity, $n$ and $\mathbf{v}$, are described by the mass and momentum conservation equations

$$\frac{\partial n}{\partial t} + \nabla \cdot (n\mathbf{v}) = 0\,,$$

$$\frac{\partial \mathbf{p}}{\partial t} + (\mathbf{v} \cdot \nabla)\mathbf{p} = -e(\mathbf{E} + \mathbf{v} \times \mathbf{B})\,, \tag{5.1}$$

where $\mathbf{p} = m\gamma\mathbf{v}$ and $\gamma = 1/\sqrt{1 - v^2/c^2}$. Following the pioneering work of Akhiezer and Polovin [59], we try to find exact wave solutions evolving as functions of a single space-time variable, with phase velocity $u$. For simplicity, we consider a one-dimensional problem associated with propagation along the $x$-axis. We also consider purely electrostatic waves. In this case, we consider the one-dimensional version of eqs. (5.1), which are

$$\frac{\partial n}{\partial t} + \frac{\partial}{\partial x}(nv) = 0\,, \quad \left(\frac{\partial}{\partial t} + v\frac{\partial}{\partial x}\right) p = -eE\,, \tag{5.2}$$

DOI: 10.1201/9781003514701-5

with $p = m\gamma v$. The electric field $E$ in this equation satisfies the Poisson's equation

$$\frac{\partial}{\partial x} E = \frac{e}{\epsilon_0}(n_0 - n)\,. \tag{5.3}$$

It is useful to introduce the dimensionless variable

$$\zeta = k_p(x - ut)\,, \tag{5.4}$$

where $k_p = \omega_p/c$, is a characteristic wavenumber, such that

$$\frac{\partial}{\partial x} = k_p \frac{d}{d\zeta}\,, \quad \frac{\partial}{\partial t} = -k_p u \frac{d}{d\zeta}\,, \tag{5.5}$$

We can then rewrite the one-dimensional plasma equations (5.2) as

$$\frac{dn}{d\zeta} = \frac{1}{u}\frac{d}{d\zeta}(nv)\,, \quad (v - u)\frac{dp}{d\zeta} = \frac{e}{k_p}E\,, \tag{5.6}$$

and the Poisson's equation (5.3) as

$$\frac{dE}{d\zeta} = \frac{e}{\epsilon_0 k_p}(n_0 - n)\,. \tag{5.7}$$

Now, we take a second derivative of the momentum equation, leading to

$$\frac{d}{d\zeta}(v - u)\frac{dp}{d\zeta} = \frac{e^2}{\epsilon_0 k_p}(n_0 - n)\,. \tag{5.8}$$

The first of eqs. (5.6) can easily be integrated, such that

$$n = \frac{n_0}{1 - v/u}\,. \tag{5.9}$$

Using this in the previous momentum equation, we can easily get

$$\frac{d}{d\zeta}\left(1 - \frac{v}{u}\right)\frac{dp}{d\zeta} = \frac{e^2 n_0}{\epsilon_0 k_p^2}\left(\frac{1}{1 - v/u} - 1\right)\,. \tag{5.10}$$

Using the definition of $k_p$, this can also be written as

$$\left(1 - \frac{v}{u}\right)\frac{d}{d\zeta}\left(1 - \frac{v}{u}\right)\frac{dp}{d\zeta} = mc\frac{v}{u}\,. \tag{5.11}$$

If we introduce the dimensionless momentum variable $\bar{p} = p/(mc)$, this equation becomes

$$\left(1 - \frac{v}{u}\right)\frac{d}{d\zeta}\left(1 - \frac{v}{u}\right)\frac{d\bar{p}}{d\zeta} = \frac{v}{u} = \frac{\bar{p}}{\gamma}\frac{c}{u}\,, \tag{5.12}$$

with $\gamma = \sqrt{1 + \bar{p}^2}$. In order to proceed further, it is now useful to introduce the transformation

$$\left(1 - \frac{v}{u}\right)\frac{d}{d\zeta} = \sqrt{\frac{c}{u}}\frac{d}{d\phi}\,, \tag{5.13}$$

where the new variable $\phi$ is also adimensional. This allows us to rewrite eq. (5.12) simply as

$$\frac{d^2\bar{p}}{d\phi^2} = \frac{\bar{p}}{\gamma}, \tag{5.14}$$

Integrating once, we get

$$\frac{d\bar{p}}{d\phi} = \pm\sqrt{2}\left(a - \sqrt{1+\bar{p}^2}\right)^{1/2}, \tag{5.15}$$

where $a$ is a constant of integration. We can solve this equation in terms of the elliptic integrals. For that purpose, it is convenient to write the quantity under brackets in the form

$$\left(a - \sqrt{1+\bar{p}^2}\right) = (a-1)\sin^2\theta\,. \tag{5.16}$$

This is equivalent to define a new angle, $\theta$, such that

$$\sin^2\theta = \frac{\left(a - \sqrt{1+\bar{p}^2}\right)}{a-1}\,. \tag{5.17}$$

Integration of eq. (5.15) then leads to a solution of the form

$$\phi = \sqrt{2(a+1)}E(\theta, r) - \sqrt{\frac{2}{a+1}}\,F(\theta, r) + b\,, \tag{5.18}$$

where $b$ is the second constant of integration, and

$$r = \left(\frac{a-1}{a+1}\right)^{1/2}. \tag{5.19}$$

In eq. (5.18), we have used the incomplete *elliptic integrals* of the second and first kind, respectively defined by

$$E(\theta, r) = \int_0^\theta \sqrt{1 - r^2\sin^2\theta'}\,d\theta'\,, \quad F(\theta, r) = \int_0^\theta \frac{d\theta'}{\sqrt{1 - r^2\sin^2\theta'}}\,d\theta'\,. \tag{5.20}$$

At this point, we should notice that, from the integration of eq. (5.13), we can express the variable $\zeta$ in terms of the variable $\phi$, as

$$\zeta = \frac{u}{c}\left(\phi - \frac{1}{u}\int v\,d\phi\right). \tag{5.21}$$

Using eq. (5.18), we can then write

$$\frac{c}{u}\zeta = \phi - \frac{1}{u}\frac{2r}{\sqrt{1-r^2}}\sin\theta + b\,. \tag{5.22}$$

If we restore this solution in terms of the dimensional space and time variables, $x$ and $t$, and choose the constant $b = 0$, we then get the solution, in the form

$$\omega_p \left( t - \frac{x}{u} \right) = I(\theta, r) , \tag{5.23}$$

with

$$I(\theta, r) = \frac{2}{r'} E(\theta, r) - r' F(\theta, r) - \frac{c}{u} \frac{2r}{r'} \sin\theta , \tag{5.24}$$

where $r' = \sqrt{1-r}$. These are the celebrated *Akhiezer-Polovin solutions*, valid for the nonlinear electrostatic waves in exact form. Notice that, for a fixed position in space, for instance $x = 0$, this solution determines the frequency of the nonlinear periodic oscillations. Defining the period as $t = 2\pi/\omega$, we obtain

$$\omega = 2\pi \frac{\omega_p}{I(\theta, r)} , \tag{5.25}$$

The integral in the denominator can be written in a more explicit way, as a function of the relative plasma velocity $\alpha = v/c$, such that [59, 60]

$$I(\theta, r) \to 2\sqrt{I(\alpha)} ,$$

with

$$I(\alpha) = \int_0^{\alpha_m} \left[ \left( \frac{1-\alpha^2}{1-\alpha_m^2} \right)^{1/2} - 1 \right]^{-1/2} \frac{d\alpha}{(1-\alpha^2)^{5/4}} , \tag{5.26}$$

where $\alpha_m$ is the maximum possible value of the electron velocity attained under the action of the nonlinear wave. It the wave amplitude is small, we have $\alpha \ll 1$, and eqs. (5.25)–(5.26) reduce to

$$\omega \simeq \omega_p \left( 1 - \frac{3}{16} \alpha_m^2 \right) , \quad \alpha_m = \frac{eE_0}{m\omega_p c} , \tag{5.27}$$

where $E_0$ is the maximum of the electric field amplitude, as determined by the Poisson's equation. We can see that there is a small downshift of the plasma wave with respect to the electron plasma frequency. This can be interpreted as due to a small increase in the effective electron mass, due to the plasma oscillations. This is similar to what was observed for the relativistic linear solutions studied in the previous chapter. However, here the oscillations are not simple sinusoidal, because they are determined by the elliptic integrals, and the frequency downshifts are not due to the plasma temperature (which is neglected here) but to the wave itself.

In contrast with the small amplitude approximation, for large field amplitudes such that $\alpha_m \to 1$, or $(1 - \alpha_m^2) \to 0$, instead of eq. (5.27) we get

$$\omega \simeq 2^{-3/2} \pi \omega_p \left(1 - \alpha_m^2\right)^{1/4} , \quad \left(1 - \alpha_m^2\right)^{1/4} = \frac{1}{\sqrt{2}} \left( \frac{mc}{eE_0} \right) . \tag{5.28}$$

This shows that, for large field amplitudes, the nonlinear plasma wave propagates with a frequency that is much smaller than the plasma frequency $\omega \ll \omega_p$. Of course, when this frequency approaches the ion-plasma frequency, $\omega_{pi}$, the present analysis is no longer valid and the ion motion has to be taken into account.

### 5.1.2 Electromagnetic Waves

Let us now consider the electromagnetic case, where we have to start from the full set of Maxwell's equations. Using the *Coulomb gauge*, the electromagnetic part of the field is represented by the transverse vector potential, $\mathbf{A}_\perp$, and the electrostatic part is represented by the scalar potential $V$. Poisson's equation (5.3) is then replaced by the two potential equations

$$\left(\frac{\partial^2}{\partial x^2} - \frac{\partial^2}{\partial t^2}\right)\mathbf{A} = \frac{n\mathbf{A}}{\gamma}, \quad \frac{\partial^2 V}{\partial x^2} = n - 1. \tag{5.29}$$

Here, we have normalized the space variable $x$ by $c/\omega_p$, time $t$ by $1/\omega_p$, the vector potential $\mathbf{A}$ by $mc/e$, and the electrostatic potential $V$ by $mc^2/e$. Furthermore, we have $\mathbf{p}_\perp = \mathbf{A}$, and the relativistic factor is $\gamma = \sqrt{1 + |\mathbf{A}|^2 + p^2}$, where $p$ is just the longitudinal momentum. These two equations are coupled with the plasma equations (5.2), which can be rewritten as

$$\frac{\partial n}{\partial t} + \frac{\partial}{\partial x}\left(\frac{np}{\gamma}\right) = 0, \quad \frac{\partial p}{\partial t} = \frac{\partial}{\partial x}(V - \gamma). \tag{5.30}$$

We have also used $E = -\partial V/\partial x$. They show that the scalar and vector potentials, describing the longitudinal and the transverse fields, are coupled by the relativistic factor $\gamma$.

We now focus on linearly polarized waves, such that $\mathbf{A} = a\mathbf{e}_z$. These evolution equations can then be written in terms of the new variables

$$X = a(\beta^2 - 1)^{1/2}, \quad Z = -V - (1 + a_0^2)^{1/2}, \tag{5.31}$$

where $a_0^2 \equiv |A|^2$, and $\beta = \omega/(kc)$. Written in terms of these new variables, and using $\zeta = (x - \omega t/c)$, eqs. (5.30) then take the form

$$\frac{d^2 X}{d\zeta^2} + \frac{\beta X}{\sqrt{\beta^2 - 1 + X^2 + Z^2}} = 0, \quad \frac{d^2 Z}{d\zeta^2} + \frac{\beta X}{\sqrt{\beta^2 - 1 + X^2 + Z^2}} + 1 = 0. \tag{5.32}$$

These are indeed the equations of two coupled oscillators, where $X$ describes the transverse oscillations and $Z$ the electrostatic ones. They can be derived from an Hamiltonian of the form

$$H(X, Z) = \frac{1}{2}(\dot{X}^2 + \cot Z^2) + \beta\sqrt{\beta^2 - 1 + X^2 + Z^2} + Z, \tag{5.33}$$

where we have used $\dot{X} \equiv dX/d\zeta$ and $\dot{Z} \equiv dZ/d\zeta$. It can easily be shown that this Hamiltonian is a constant of motion. Notice that $\beta = u/c$ represents the phase velocity of the nonlinear waves. For superluminal phase velocities $\beta > 1$, the Hamiltonian $H(X, Z)$ remains positive, while for the sub-luminal case this is not always true. We can see that the equations of motion (5.32) satisfy an equilibrium condition in the $(X, Z)$ plane, determined by the position

$$X_0 = 0\,, \quad Z_0 = -1\,. \tag{5.34}$$

In order to determine the stability of this fixed point, we linearize the equations around the fixed point, and obtain local equations of motion valid near the equilibrium

$$\frac{d^2}{d\zeta^2}\delta X + \delta X = 0\,, \quad \frac{d^2}{d\zeta^2}\delta Z + \left(1 - \frac{1}{\beta^2}\right)\delta Z = 0\,. \tag{5.35}$$

where $\delta X = X - X_0 \equiv X$, and $\delta Z = Z - Z_0 \equiv Z + 1$. The oscillations in the $Z$ direction represent purely longitudinal oscillations. They are recovered if we take $X = 0$ and $dX/d\zeta = 0$, and can be represented in the two-dimensional phase space $(Z, \dot{Z})$. On the other hand, if we keep the complete Hamiltonian form (5.33), we can still use this plane to represent the more complete case of coupled electrostatic and electromagnetic oscillations. For orbits crossing this plane, we can use a *Poincaré surface of section* representation of the coupled nonlinear orbits in the four-dimensional phase space. This representation shows that stochastic orbits can exist for $\beta < 1$, as demonstrated by several authors [61, 62].

## 5.2 Electron Holes

There is experimental and computational evidence that nonlinear structures that depend on the plasma temperature can locally occur in a variety of plasma conditions [63]. They are usually called *electron holes*, and are discussed here, first using a simple non-relativistic approach, and then generalizing this approach to the relativistic plasma conditions. In particular, these nonlinear structures can be excited by a beam instability, resulting for instance from two electron beams moving in opposite directions.

Here we focus on the static properties of the electron holes, leaving the beam instability processes to the next chapter. We first start with the non-relativistic Vlasov equation, coupled with the Poisson's equation, in one-dimension. They can be written as

$$\left(\frac{\partial}{\partial t} + v\frac{\partial}{\partial x} + \frac{\partial \varphi}{\partial x}\frac{\partial}{\partial v}\right) f(v, v, t) = 0\,, \tag{5.36}$$

and

$$\frac{\partial^2 \varphi}{\partial x^2} = -\rho \,, \quad \rho = n_e - n_i \,, \tag{5.37}$$

where the scale length is normalized to the Debye length, and time by the plasma frequency, as

$$x \to \frac{x}{\lambda_D} \,, \quad t \to \omega_p t \,, \tag{5.38}$$

where $\lambda_D$ and $\omega_p$ are defined with the asymptotic value of the electron density, $n_0$, far away from the electron hole. As a consequence, the electron velocity is normalized by the thermal velocity, as

$$v \to \frac{v}{v_{the}} \,, \quad v_{the} = \frac{\lambda_D}{\omega_p} = \sqrt{\frac{T_e}{m_e}} \,, \tag{5.39}$$

The total electron energy, normalized to the thermal energy $T_e$, is

$$\mathcal{E} = \frac{1}{2} v^2 - \varphi \,. \tag{5.40}$$

We now look for a static nonlinear solution, located around the origin $x = 0$. The local electron distribution function, $f(x = 0, v)$ can be written as

$$f_0(v) = f_0(\sqrt{2(\mathcal{E} + \varphi_0)}) \,, \quad \varphi_0 = \varphi(x = 0) \,. \tag{5.41}$$

For any value of the position $x$, we assume that the distribution function is a function of the plasma potential $\varphi(x)$, such that

$$f(x, v) = f_0 \left( \sqrt{v^2 - 2(\varphi - \varphi_0)} \right) \,. \tag{5.42}$$

The corresponding electron density is then calculated as

$$n_e(x) = \int f(x, v) dv = \int f_0 \left( \sqrt{v^2 - 2(\varphi - \varphi_0)} \right) dv \,. \tag{5.43}$$

As we can see, this is solely determined by the plasma potential $\varphi$, such that $n_e(x) \equiv n_e(\varphi(x))$. In order to solve this problem, we need to find the functions $\varphi(x)$ and $f_0(v)$ in explicit form. This can be done using the *integral solution method*, first proposed by [64], along the following arguments.

It is obvious that, for $\mathcal{E} < 0$, the electrons will be trapped by the local structure, while for positive energy they will escape. We should then look for two distinct electron distribution functions, one corresponding to trapped particles with energies in the interval $-\varphi < \mathcal{E} < 0$, and the other to particles with positive energy, $\mathcal{E} > 0$, that can never be trapped. Noting that $d\mathcal{E} = v dv$, and therefore

$$dv = \frac{d\mathcal{E}}{v} = \frac{d\mathcal{E}}{\sqrt{2(\mathcal{E} + \varphi)}} \,, \tag{5.44}$$

we can determine the trapped density, $n_t$, using

$$n_t = \int_0^{\infty} f_t(v)dv = \int_{-\varphi_0}^{0} f_0\left(\sqrt{2(\mathcal{E}+\varphi)}\right) \frac{d\mathcal{E}}{\sqrt{2(\mathcal{E}+\varphi)}}\,, \tag{5.45}$$

while the untrapped or free particle density is determined by $n_u = n_e - n_t$. Notice that, once we know the exact shape of the local potential $\varphi(x)$, we will be able to calculate the charge density and the electron density, as determined by the Poisson's equation (5.37) with $n_i = n_0$. We have

$$n_t \equiv n_t(\varphi) = n_0 - \frac{\partial^2 \varphi}{\partial x^2}\,. \tag{5.46}$$

On the other hand, an inversion of eq. (5.45) allows us to retrieve the electron distribution function inside the structure, as

$$f_0(v) = \int_0^{(\varphi_0 - v^2/2)} \left(\frac{dn_t}{d\varphi}\right) \frac{d\varphi}{\sqrt{(\varphi_0 - v^2/2) - \varphi}}\,. \tag{5.47}$$

An alternative approach, leading to a complete description of the electron hole, is known as the *differential solution method*, where, instead of $\varphi(x)$,we now specify the distribution function, $f_0(v)$. This can be done multiplying the Poisson's equation by the quantity $\partial\varphi/\partial x$, and then buildup a pseudo-potential, as

$$\frac{\partial \varphi}{\partial x}\frac{\partial^2 \varphi}{\partial x^2} \equiv \frac{1}{2}\frac{\partial}{\partial x}\left(\frac{\partial \varphi}{\partial x}\right)^2 = -\rho \frac{\partial \varphi}{\partial x}\,. \tag{5.48}$$

Integrating this equation with respect to $\varphi$, and assuming that $(d\varphi/dx) = 0$ for $\varphi = 0$, we get

$$\frac{1}{2}\left(\frac{\partial \varphi}{\partial x}\right)^2 = -\int_0^{\varphi} \rho(\varphi')d\varphi' \equiv -V(\varphi)\,. \tag{5.49}$$

where the new function $V(\varphi)$ can be seen as a *Sagdeev potential.* A useful model for electron holes that can be explored using both methods, was proposed by Schammel [65]. It is based on the distribution functions

$$f_t(v) = \frac{1}{\sqrt{\pi}} \exp\left(-\beta\mathcal{E} - U^2\right)\,, \quad f_u(v) = \frac{1}{\sqrt{\pi}} \exp\left[-\left(\pm\mathcal{E} + U\right)^2\right]\,, \tag{5.50}$$

where $U$ is an adjustable energy, and $\beta$ is an inverse temperature factor valid for the trapped region. For $U \ll 1$, a possible solution for the plasma potential is given by

$$\varphi(x) = \varphi_0 sech^4\left(\frac{x}{4}\right)\,. \tag{5.51}$$

A relativistic version of the above solutions, sometimes called *BGK solutions*, can also be constructed [66]. In this case, we replace eq. (5.36) by

$$\left(\frac{\partial}{\partial t} + \frac{p}{\gamma(p)}\frac{\partial}{\partial x} + \frac{\partial \varphi}{\partial x}\frac{\partial}{\partial p}\right) f(v,v,t) = 0\,, \tag{5.52}$$

where $\gamma(p) = \sqrt{1 + \theta^2 p^2}$, with $\theta = V_{the}/c$ and $\theta^2 = T_e/mc^2$. Here we use the same normalization for $x$ and $t$, and

$$p \to \frac{p}{m_e v_{the}}, \quad \varphi \to \frac{\varphi}{T_e/e}, \tag{5.53}$$

The Poisson's equation becomes

$$\frac{\partial^2 \varphi}{\partial x} = \int_{-\infty}^{\infty} f(p)dp - 1. \tag{5.54}$$

We can find solutions of these two equations of the form $f(\zeta, p)$ and $\varphi(\zeta)$, with $\zeta = (x - v_0 t)$. They correspond to electron holes moving with velocity $v_0$ along the $x$-axis. Using these new variables, the Vlasov equation (5.52) becomes

$$\left[\frac{p}{\gamma(p)} - v_0\right] \frac{\partial f}{\partial \zeta} + \frac{\partial \varphi}{\partial x} \frac{\partial f}{\partial p} = 0, \tag{5.55}$$

Following the above BGK approach, we can find a solution of this equation that is a function of the electron energy $\mathcal{E}$, as

$$f(\zeta, p) - f_0(\mathcal{E}), \tag{5.56}$$

where the (normalized) energy in the moving frame takes the form

$$\mathcal{E} = \left[\frac{1}{\theta^2}\gamma(p) - \varphi\right] - (v_0 p + \mathcal{E}_0), \quad \mathcal{E}_0 = \frac{1}{\theta^2 \gamma_0}, \tag{5.57}$$

and the reference energy $\mathcal{E}_0$ is chosen such that the energy of the trapped electrons is negative, as in the above non-relativistic case. Obviously, for untrapped electrons, the electron distribution function should be reduced to the equilibrium far away from the electron hole. In this limit, we have $\varphi \to 0$, and

$$f_0(p) = C \exp\left[-\frac{\gamma(p) - 1}{\theta^2}\right]. \tag{5.58}$$

The normalization constant $C$ is defined, as we have seen before, by

$$C = \frac{\theta \exp(-1/\theta^2)}{2K_1(1/\theta^2)}, \tag{5.59}$$

where $K_1$ is a modified Bessel function of the second-kind. In the non-relativistic limit, we get

$$C \simeq \frac{1}{\sqrt{2\pi}} \frac{1}{1 + (3/8)\theta^2} \simeq \frac{1}{\sqrt{2\pi}}, \tag{5.60}$$

and $f_0(p)$ reduces to the Maxwell-Boltzmann distribution. In order to satisfy eqs. (5.56) and (5.58) simultaneously we need to write the particle momentum as a function of the energy $\mathcal{E}$, such that $f_0(\mathcal{E}) = f_0(p(\mathcal{E}))$. Using eq. (5.57), and solving for $p$, we get

$$p = p_\pm(\mathcal{E}) = \frac{v_0}{\theta^2 \mathcal{E}_0}\left(1 + \frac{\mathcal{E}}{\mathcal{E}_0}\right) \pm \mathcal{E}_0 \sqrt{\left(1 + \frac{\mathcal{E}}{\mathcal{E}_0}\right)^2 - 1}\,. \tag{5.61}$$

Notice that, for $\mathcal{E} \to 0$, which is the boundary between trapped and free particles, we have $p(\mathcal{E}) \to \gamma_0 v_0$, where $\gamma_0 = 1/\sqrt{1 - v_0^2/c^2}$. The corresponding distribution function is

$$f(\mathcal{E} \to 0) = C \exp\left(-\frac{\gamma_0 - 1}{\theta^2}\right). \tag{5.62}$$

This result is valid for the free (or untrapped) particles, and should meet the trapped distribution at the boundary between the two regions. It is therefore natural to define, for the trapped case

$$f_t(\mathcal{E}) = C \exp\left(-\frac{\gamma_0 - 1}{\theta^2} - \beta\mathcal{E}\right), \tag{5.63}$$

where the quantity $\beta$ introduced here is a *trapping parameter*, characterizing the population in the trapped region, where $\mathcal{E} < 0$. Notice that, for $\beta = 0$ we would have a trapped population that is independent of the energy, and therefore independent of the potential $\varphi(x)$. This is called a *flat-top distribution*. On the other hand, for $\beta = 1$, we simply have a shifted Maxwell-Jüttner distribution. Making these adjustments at the boundary, $\mathcal{E} = 0$, we can completely establish the relativistic distribution function for electrostatic electron holes, as a function of the local hole potential $\varphi(x)$. We therefore, have:

i) For the free or untrapped electrons, with momenta above and below the limits determined by eq. (5.61), which means, for $p > p_+$ and $p < p_-$, the distribution function is

$$f_u(\mathcal{E}) = C \exp\left\{-\frac{1}{\theta^2}\left[\gamma(p_\pm) - 1\right]\right\}, \tag{5.64}$$

where $\gamma(p_\pm) = \sqrt{1 + \theta^2 p_\pm^2(\mathcal{E})}$.

ii) For the trapped region, corresponding to momenta in the region ($p_- \leq p \leq p_+$), the distribution function is given by eq. (5.62). A possible choice of the corresponding hole potential is [67]

$$\varphi(x) = \varphi_0 sech^4\left[\left(b\sqrt{\varphi_0}\frac{1}{15}\right)^{1/2}\right]. \tag{5.65}$$

with

$$b = \frac{1}{\sqrt{\pi}}\left(1 - \beta - 2\mathcal{E}_0^2\right)\exp(-\mathcal{E}_0^2)\,. \tag{5.66}$$

This expression generalizes the potential defined by eq. (5.51) and can be useful to compare with numerical simulation in several possible configurations. The shape of these potentials, in the relativistic and non-relativistic regimes, is illustrated in Figure 5.1.

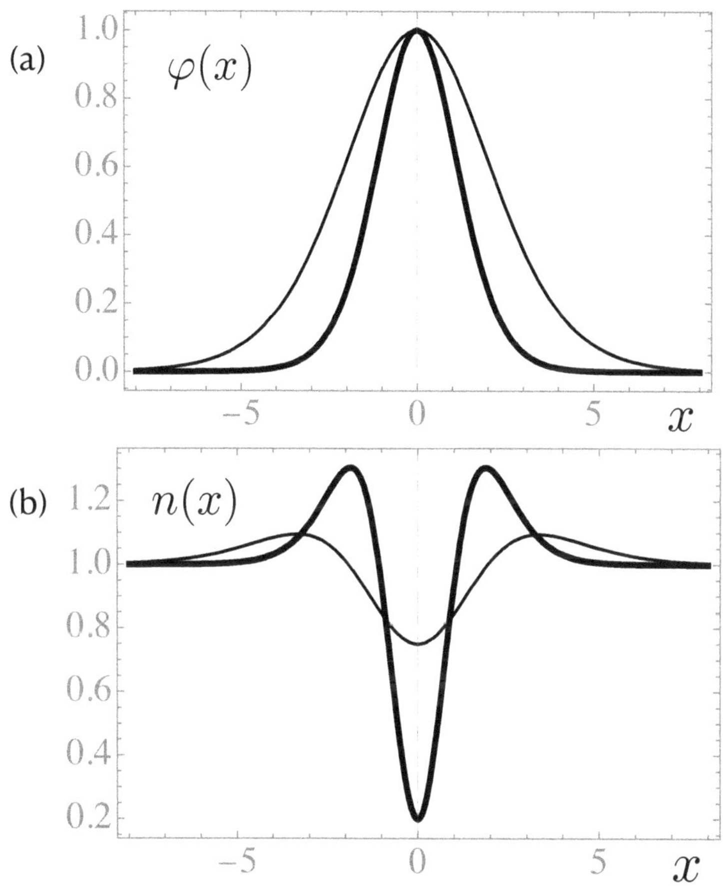


**FIGURE 5.1**
Electron Hole: (a) Electrostatic potential $\varphi(x)$ and (b) electron density $n(x)$, for non-relativistic (thin curve) and relativistic plasmas (bold), as given by eqs. (5.51) and (5.65).

## 5.3 Solitons

Solitons are particular nonlinear solutions that have been explored in several domains of physics [68]. In order to illustrate this concept in the context of relativistic plasma physics, we consider the case of electromagnetic solitons. For this purpose, we use the Coulomb gauge and write the corresponding potential equations as

$$\left(\nabla^2 - \frac{\partial^2}{\partial t^2}\right)\mathbf{A} - \nabla\left(\frac{\partial V}{\partial t}\right) = \frac{n}{\gamma}(\mathbf{P}+\mathbf{A}), \tag{5.67}$$

and

$$\nabla^2 V = n - 1, \tag{5.68}$$

where $n = n_e/n_0$ is the normalized electron density, and the ions are assumed immobile with density $n_i = n_0$. In these equations, we also use the canonical momentum $\mathbf{P} = \mathbf{p} - \mathbf{A}$, and the relativistic factor $\gamma = \sqrt{1+p^2}$, where $\mathbf{p}$ is the kinetic momentum. The obvious normalizations are, for the space-time variables

$$\mathbf{r} \to \frac{\mathbf{r}}{c/\omega_p}, \quad t \to \omega_p t, \tag{5.69}$$

and, for the scalar and vector potentials

$$V \to \frac{V}{mc/e}, \quad \mathbf{A} \to \frac{\mathbf{A}}{mc^2/e}, \tag{5.70}$$

This is to be completed with the electron fluid equations

$$\frac{\partial n}{\partial t} + \nabla \cdot (n\mathbf{v}) = 0,$$

$$\frac{\partial \mathbf{P}}{\partial t} = \nabla(V - \gamma) + \frac{1}{\gamma}(\mathbf{P}+\mathbf{A}) \times (\nabla \times \mathbf{P}), \tag{5.71}$$

where we have used $\mathbf{v} = \mathbf{p}/\gamma = (\mathbf{P}+\mathbf{A})/\gamma$. We consider propagation along the $z$-axis, such that $\partial/\partial y = \partial/\partial x = 0$. We then assume an electromagnetic perturbation of the form

$$(A_x + iA_y) = a(\zeta)\exp(-i\omega\tau), \quad A_z = 0. \tag{5.72}$$

with the variables

$$\zeta = z - ut, \quad \tau = t - uz, \tag{5.73}$$

where $u$ is the (group) velocity of the perturbation, assumed constant. We also take

$$\mathbf{P}_\perp = 0, \quad P_z = ub(\zeta). \tag{5.74}$$

Obviously, the functions $a(\zeta)$ and $b(\zeta)$ describe the transverse and longitudinal amplitude perturbations. Replacing this in eqs. (5.67)–(5.71), we obtain the following evolution equations for these two quantities.

$$\frac{d^2}{d\zeta^2}\left(\gamma - u^2 b\right) = \frac{b}{\gamma - b},$$

$$\frac{d^2 a}{d\zeta^2} = -\omega^2 a + \frac{1}{1-u^2}\frac{1}{\gamma - b}. \tag{5.75}$$

This system of equations is able to describe the evolution of nonlinear electromagnetic pulses moving with velocity $u$ along the $z$-direction. This can easily be understood when we take the limit $u \rightarrow 0$. In this limit, the above equations reduce to

$$\frac{d^2 b}{d\zeta^2} = \frac{\gamma\gamma''}{1+\gamma''}, \quad \frac{d^2 a}{d\zeta^2} = -\omega^2 a + \frac{1+\gamma''}{\gamma}, \tag{5.76}$$

where, to simplify, we have written $\gamma'' \equiv d^2\gamma/d\zeta^2$. These new equations can be easily integrated, using a solution of the form

$$a = \sinh\alpha, \quad \gamma = \cosh\alpha. \tag{5.77}$$

From here, we obtain [69]

$$a(\zeta) = \frac{2\sqrt{1-\omega^2}\cosh(\zeta\sqrt{1-\omega^2})}{\cosh^2(\zeta\sqrt{1-\omega^2}) + \omega^2 - 1}. \tag{5.78}$$

As we can see, these soliton solutions only occur for frequencies lower than $\omega_p$ which, in the present dimensionless variables, corresponds to $\omega^2 < 1$. It is also important to notice that the maximum soliton amplitude is given by

$$a_0 \equiv a(\zeta = 0) = \frac{2}{\omega^2}\sqrt{1-\omega^2}. \tag{5.79}$$

This shows that, for larger amplitudes the soliton is shorter, as illustrated in Figure 5.2, using two different values of $\omega^2$. Furthermore, for very small amplitudes, such that $a_0 \ll 1$ and $\omega^2 \sim 1$, we can approximate the above solution as

$$a(\zeta) \simeq 2\sqrt{1-\omega^2}\, sech(\zeta\sqrt{1-\omega^2}). \tag{5.80}$$

This approximate solution could also be obtained directly from the small amplitude approximation of eqs. (5.75), which reduce to a nonlinear Schrödinger equation, with $b \simeq 0$, as

$$\frac{d^2 a}{d\zeta^2} \simeq (1-\omega^2)a + a^3. \tag{5.81}$$

These solutions can be generalized to a magnetized plasma [70].

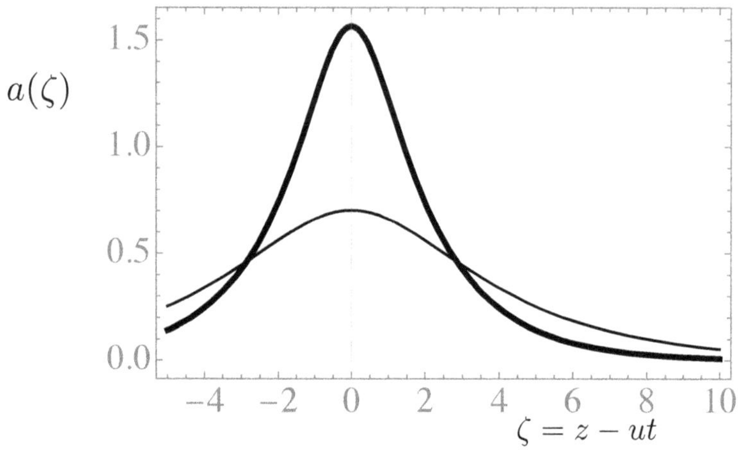


**FIGURE 5.2**
Electromagnetic solitons in a relativistic plasma, $a(\zeta)$, as described by eq. (5.78), for $\omega^2 = 0.7$ (bold curve), and $\omega^2 = 0.9$ (thin curve).

## 5.4 Nonlinear Wave-Mixing

Waves of different types and different spectra can interact nonlinearly with each other. The study of these interaction processes, first initiated in Optics [71, 72], was soon replicated in Plasmas [73, 74, 75, 76]. Here, we describe the basic three-wave-mixing process, explain the importance of phase coherence, and discuss the parametric decay and harmonic generation in plasmas. We assume that the waves under interaction satisfy the linear dispersion relations of the previous chapter. However, interactions between nonlinear waves are also possible.

### 5.4.1 Three-Wave-Mixing

The propagation of a monochromatic wave, with frequency $\omega$ and wavevector $\mathbf{k}$ in a plasma satisfies the general expression

$$\mathcal{D}(\omega, \mathbf{k}) E_{\mathbf{k}} = -\frac{i}{\omega \epsilon_0} \mathbf{e}_{\mathbf{k}}^* \cdot \mathbf{J}_{NL}(\mathbf{k}) \, , \tag{5.82}$$

where $E_{\mathbf{k}}$ is the wave amplitude and $\mathbf{J}_{NL}(\mathbf{k})$ is a nonlinear current. The linear case, where this current is absent, was discussed in Section 4.1. Here, we assume that the linear dispersion relation is still approximately valid, and therefore the frequency corresponding to the wavevector $\mathbf{k}$ is still equal to the linear mode frequency $\omega = \omega_{\mathbf{k}}$, as discussed before. In order to specify the current, we assume that it is quadratic with respect to the electric field, and can be given by

$$\mathbf{J}_{NL}(\mathbf{k}) = \int \bar{\bar{\bar{\sigma}}}(\mathbf{k}, \mathbf{k}') \cdot \mathbf{E}_{\mathbf{k}'} \mathbf{E}_{\mathbf{k}"} \, e^{i\Delta\omega t} \frac{d\mathbf{k}'}{(2\pi)^3} , \tag{5.83}$$

where the quantity $\bar{\bar{\bar{\sigma}}}(\mathbf{k}, \mathbf{k}')$ is a second-order conductivity tensor. We first assume that there is a perfect phase-matching between the three different wavevectors, but not necessarily between the three frequencies, such that

$$\mathbf{k}" = \mathbf{k} - \mathbf{k}' , \quad \omega_{\mathbf{k}"} = \omega_{\mathbf{k}} - \omega_{\mathbf{k}'} - \Delta\omega . \tag{5.84}$$

As will be seen, the strongest nonlinear wave coupling occurs when the frequency mismatch is equal to zero, $\Delta\omega = 0$. Noting that, in the presence of this nonlinear current, the linear dispersion relation is only approximately valid, we develop the quantity $\mathcal{D}(\omega, \mathbf{k})$ around the linear mode solution, $\omega = \omega_{\mathbf{k}}$, leading to

$$\mathcal{D}(\omega, \mathbf{k}) \simeq i \left( \frac{\partial D}{\partial \omega} \right)_{\mathbf{k}} \left( \frac{d}{dt} + \gamma_{\mathbf{k}} \right) , \tag{5.85}$$

where $\gamma_{\mathbf{k}}$ is a damping factor eventually associated with collisional damping and/or to Landau damping, $\mathbf{v}_{\mathbf{k}}$ is the group velocity and $d/dt$ represents the total time derivative, as defined by

$$\mathbf{v}_{\mathbf{k}} = \left. \frac{\partial \omega}{\partial \mathbf{k}} \right|_{\mathbf{k}} = - \left. \frac{\partial D / \partial \mathbf{k}}{\partial D / \partial \omega} \right|_{\mathbf{k}} , \quad \frac{d}{dt} = \frac{\partial}{\partial t} + \mathbf{v}_{\mathbf{k}} \cdot \nabla . \tag{5.86}$$

Replacing this in eq. (5.82), we obtain

$$\left( \frac{d}{dt} + \gamma_{\mathbf{k}} \right) E_{\mathbf{k}} = - \int H(\mathbf{k}, \mathbf{k}') \, E_{\mathbf{k}'} E_{\mathbf{k}"} \, e^{i\Delta\omega t} \frac{d\mathbf{k}'}{(2\pi)^3} , \tag{5.87}$$

where the nonlinear coupling factor $H(\mathbf{k}, \mathbf{k}')$ is defined as

$$H(\mathbf{k}, \mathbf{k}') = \frac{1}{\omega \epsilon_0} \left( \frac{\partial D}{\partial \omega} \right)_{\mathbf{k}} \left[ \mathbf{e}_{\mathbf{k}}^* \cdot \bar{\bar{\bar{\sigma}}}(\mathbf{k}, \mathbf{k}') \cdot \mathbf{e}_{\mathbf{k}'} \mathbf{e}_{\mathbf{k}"} \right] . \tag{5.88}$$

This equation determines the nonlinear evolution of mode $\mathbf{k}$, in the presence of other waves in the medium. This could be applied to an arbitrary spectrum, but we focus on the simple situations where only three different modes, with wavevectors $\mathbf{k}_j$ and $j = (1, 2, 3)$ are relevant to the interaction. For this purpose, we assume that

$$E_{\mathbf{k}} = (2\pi)^3 \sum_{j=1}^{3} E_j \delta(\mathbf{k} - \mathbf{k}_j) , \tag{5.89}$$

and we reduce eq. (5.87) to a system of three coupled equations, of the form

$$\begin{aligned} \left(\frac{d}{dt} + \gamma_1\right) E_1 &= -H E_2 E_3 e^{i\Delta\omega t}, \\ \left(\frac{d}{dt} + \gamma_2\right) E_2^* &= -H' E_1^* E_3 e^{-i\Delta\omega t}, \\ \left(\frac{d}{dt} + \gamma_3\right) E_3^* &= -H" E_1^* E_2 e^{-i\Delta\omega t}, \end{aligned} \tag{5.90}$$

where we have used $\Delta\omega = \omega_1 - \omega_2 - \omega_3$ and

$$H = H(k_1, k_2), \quad H' = H(-k_2, k_3), \quad H' = H(-k_3, k_2), \tag{5.91}$$

This system of equations can be simplified further, if we introduce real-field amplitudes, $a_j$, and phases $\varphi_j$, such that

$$a_j e^{i\varphi_j} = \frac{1}{2}\left(\frac{\partial D}{\partial \omega}\right)_j^{1/2} E_j, \tag{5.92}$$

The resulting evolution equations for the field mode amplitudes $a_j$ can then be written in symmetrized form, as

$$\begin{aligned} \left(\frac{d}{dt} + \gamma_1\right) a_1 &= -V a_2 a_3 \sin\theta, \\ \left(\frac{d}{dt} + \gamma_2\right) a_2 &= +V a_1 a_3 \sin\theta, \\ \left(\frac{d}{dt} + \gamma_3\right) a_3 &= +V a_1 a_2 \sin\theta, \end{aligned} \tag{5.93}$$

and the angle

$$\theta = (\varphi_1 - \varphi_2 - \varphi_3 + \Delta\omega t), \tag{5.94}$$

evolves according to the additional equation

$$\frac{d\theta}{dt} = \Delta\omega + V\left(\frac{a_1 a_3}{a_2} + \frac{a_1 a_2}{a_3} - \frac{a_2 a_3}{a_1}\right)\cos\theta. \tag{5.95}$$

The symmetrized coupling coefficient $V$ is defined by

$$V = 2H\left[\left(\frac{\partial D}{\partial \omega}\right)_1 \left(\frac{\partial D}{\partial \omega}\right)_2 \left(\frac{\partial D}{\partial \omega}\right)_3\right]^{-1/2}. \tag{5.96}$$

This nonlinear wave coupling model is well-known and was explored in a large variety of situations and models, both in optics and in plasmas.

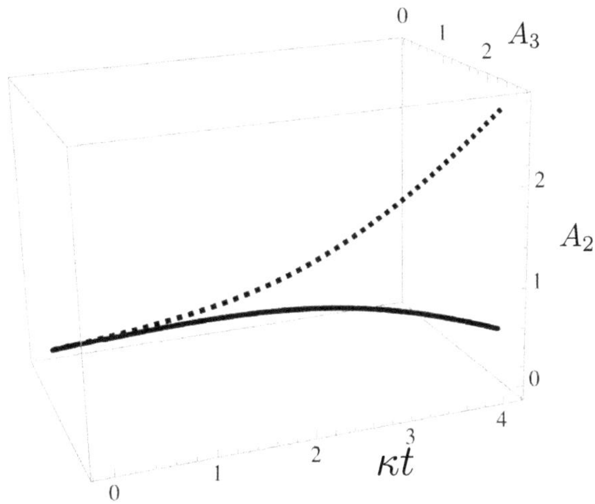


**FIGURE 5.3**
Parametric decay: Temporal evolution of the amplitudes $A_2$ and $A_3$ as a function of normalized time $\kappa t$, for $\Omega = 0$ (thin curve), and $\Omega \neq 0$ (bold curve).

Of particular interest is the case of *parametric processes*, where one of the wave amplitudes is much higher than the other two, and can be assumed constant. Let us assume $E_1 = const$. In this case, we use $E_{2,3} = A_{2,3} \exp(-\gamma_{2,3} t)$, and write eqs. (5.90) as

$$\frac{d}{dt} A_2^* = w_2 A_3 e^{-i\Omega t}, \quad \frac{d}{dt} A_3 = w_3^* A_2^* e^{i\Omega t}, \tag{5.97}$$

with $\Omega = \Delta\omega + i\Delta\gamma$ and $\Delta\gamma = \gamma_2 - \gamma_3$. The new coupling constants are proportional to the fixed amplitude of the driving field $E_1$, as $w_2 = -HE_1^*$ and $w_3^* = H'E_1$. This system of equations can be solved analytically. As an illustration, we consider the simple case of $\Omega = 0$. We therefore get

$$\frac{d^2}{dt^2} A_2^* = \kappa^2 A_2^* e, \quad \kappa^2 = w_1 w_2 * = HH' |E_1|^2, \tag{5.98}$$

Assuming initial conditions defined by $A_2(0) = 0$, and $A_3(0)$ finite, the solutions are

$$A_2(t) = \left(\frac{w_3}{w_2}\right)^{1/2} A_3(0) \sinh(\kappa t), \quad A_3(t) = A_3(0) \cosh(\kappa t). \tag{5.99}$$

This shows that the two modes $j = (2, 3)$ both grow in time, due parametric decay of the driver field $E_1$, as illustrated in Figure 5.3.

Let us consider another useful example, where mode 1 is unstable and its amplitude cannot be assumed constant. This is related with *instability saturation*: an unstable mode, with $\gamma_1 < 0$ grows in time, but at the same time decays into subharmonics that are damped by the medium. To treat this problem, we use $\omega_2 = \omega_3 = \omega_1/2$, and assume $\gamma_2 = \gamma_3 > 0$. In this case eqs. (5.93) can be reduced to

$$\left(\frac{d}{dt} + \gamma_1\right) a_1 = -V a_2^2 \sin\theta\,, \quad \left(\frac{d}{dt} + \gamma_2\right) a_2 = V a_1 a_2 \sin\theta\,, \tag{5.100}$$

and

$$\frac{d\theta}{dt} = \Delta\omega + V a_1 \left[2 - \left(\frac{a_2}{a_1}\right)^2\right] \cos\theta\,. \tag{5.101}$$

It can be shown that this new system of equations describes a complex behavior eventually leading to saturation of the unstable mode 1, but that can also lead to chaos [77]. We will return to this process later.

### 5.4.2 Random-Phase Approximation

The above description of nonlinear wave coupling is sometimes unrealistic, because it assumes that the waves are purely monochromatic and the fields behave coherently, with a well defined phase. In the case of plasma instabilities or laser-plasma interactions with ultrashort pulses, we very often have to deal with an undefined phase and a broad spectrum. It is therefore useful to describe the waves in terms of the energy, and not of the field amplitude, and to use a random-phase approximation. For that purpose, we multiply the field equation (5.87). by $E_{\mathbf{k}}^*$, and obtain

$$\left(\frac{d}{dt} + 2\gamma_{\mathbf{k}}\right) |E_{\mathbf{k}}|^2 = -2 \int H(\mathbf{k}, \mathbf{k}')\, E_{\mathbf{k}}^* E_{\mathbf{k}'} E_{\mathbf{k}"}\, e^{i\Delta\omega t} \frac{d\mathbf{k}'}{(2\pi)^3}\,, \tag{5.102}$$

The field amplitudes are complex functions that can be written as $E_{\mathbf{k}} = |E_{\mathbf{k}}| \exp(i\varphi_{\mathbf{k}})$. If the phase is random, we can make a statistical average over the field and obtain

$$< E_{\mathbf{k}} >= 0\,, \quad < E_{\mathbf{k}}^* E_{\mathbf{k}'} >= |E_{\mathbf{k}}|^2 \delta(\mathbf{k} - \mathbf{k}')\,. \tag{5.103}$$

But, if we average eq. (5.102) over a statistical ensemble, we need to calculate the average of the triple product of field amplitudes appearing inside the integral. For this purpose, we go back to eq. (5.87), and assume the approximate solution

$$E_{\mathbf{k}} \sim i \int \frac{d\mathbf{k}'}{(2\pi)^3} \left[\int H(\mathbf{k}, \mathbf{k}')(t') e^{i\Delta\omega t'} dt'\right] E_{\mathbf{k}'} E_{\mathbf{k}"}\,, \tag{5.104}$$

such that

$$\left|\int H(\mathbf{k}, \mathbf{k}')(t') e^{i\Delta\omega t'} dt'\right|^2 \simeq 2\pi\delta(\Delta\omega) \left|H(\mathbf{k}, \mathbf{k}')(t)\right|^2 \tag{5.105}$$

On the other hand, the average of products of four field amplitudes are reduced, in the random phase approximation, to

$$< E_{\mathbf{s}}^* E_{\mathbf{k-s}} E_{\mathbf{k'}} E_{\mathbf{k-k'}} >= |E_{\mathbf{s}}|^2 \, |E_{\mathbf{k-k'}}|^2 \delta(\mathbf{s} - \mathbf{k'}) \,. \tag{5.106}$$

The statistical average of eq. (5.102), we can then be written as

$$\left( \frac{d}{dt} + 2\gamma_{\mathbf{k}} \right) N_{\mathbf{k}} = \int W(\mathbf{k}, \mathbf{k'}) \left[ N_{\mathbf{k'}} N_{\mathbf{k"}} - N_{\mathbf{k}} N_{\mathbf{k"}} - N_{\mathbf{k}} N_{\mathbf{k'}} \right] \frac{d\mathbf{k'}}{(2\pi)^3} \,, \tag{5.107}$$

where we have used the photon number density $N_{\mathbf{k}}$, defined as

$$N_{\mathbf{k}} = \frac{1}{4} \left( \frac{\partial D}{\partial \omega} \right)_{\mathbf{k}} |E_{\mathbf{k}}|^2 \,. \tag{5.108}$$

The new coupling factor is

$$W(\mathbf{k}, \mathbf{k'}) = -\frac{4\pi}{\epsilon_0} |H(\mathbf{k}, \mathbf{k'})|^2 \left[ \left( \frac{\partial D}{\partial \omega} \right)_1 \left( \frac{\partial D}{\partial \omega} \right)_2 \left( \frac{\partial D}{\partial \omega} \right)_3 \right]^{-1} \,. \tag{5.109}$$

When compared with the single-frequency coupled equations (5.87), we notice that the coupling factor is now of second order with respect to the nonlinear current, which means that the nonlinear coupling in the random-phase approximation are much weaker. It is also useful to assume that the wave spectrum is only significant in three regions around $\mathbf{k} = \mathbf{k}_j$, with $j = 1, 2, 3$. After integration in $\mathbf{k'}$ around these values, we get

$$\left( \frac{dN_1}{dt} + 2\gamma_1 \right) N_1 = W_1 (N_2 N_3 - N_1 N_3 - N_1 N_2) \,, \tag{5.110}$$

where $W_1 = W(\mathbf{k}_1, \mathbf{k}_2)$, with similar equations for $N_2$ and $N_3$. As an example of application of these equations, we return to the problem of instability saturation, and assume that the first mode is unstable $\gamma_1 < 0$, while the other two are stable. Introducing a new time variable $\tau = 2|\gamma_1|t$, the coupled mode equations become

$$\frac{dN_1}{d\tau} = N_1 + C_1 [N_2 N_3 - N_1 N_3 - N_1 N_2] \,, \tag{5.111}$$

and

$$\frac{dN_{2,3}}{dt} = -\Gamma_{2,3} N_{2,3} - C_{2,3} [N_2 N_3 - N_1 N_3 - N_1 N_2] \,, \tag{5.112}$$

where $C_j = W_j / 2|\gamma_1| > 0$ and $\Gamma_{2,3} = \gamma_{2,3} / |\gamma_1| > 0$. If the modes $j = 2, 3$ are equal, we return to the above example of sub-harmonic generation. Let us first consider the equilibrium solutions, $N_j$ are constant. There are two possible equilibria, where one is trivial, $N_j = 0$, located at the origin of the three-dimensional phase-space $(N_1, N_2, N_3)$, and the other is defined by

$$N_1 = \frac{C_2 \Gamma_2 \Gamma_3}{C_1 (C_2 \Gamma_3 + C_3 \Gamma_2) - C_2 C_3} = N_2 \Gamma_2 = N_3 \Gamma_3 \frac{C_2}{C_3} \,, \tag{5.113}$$

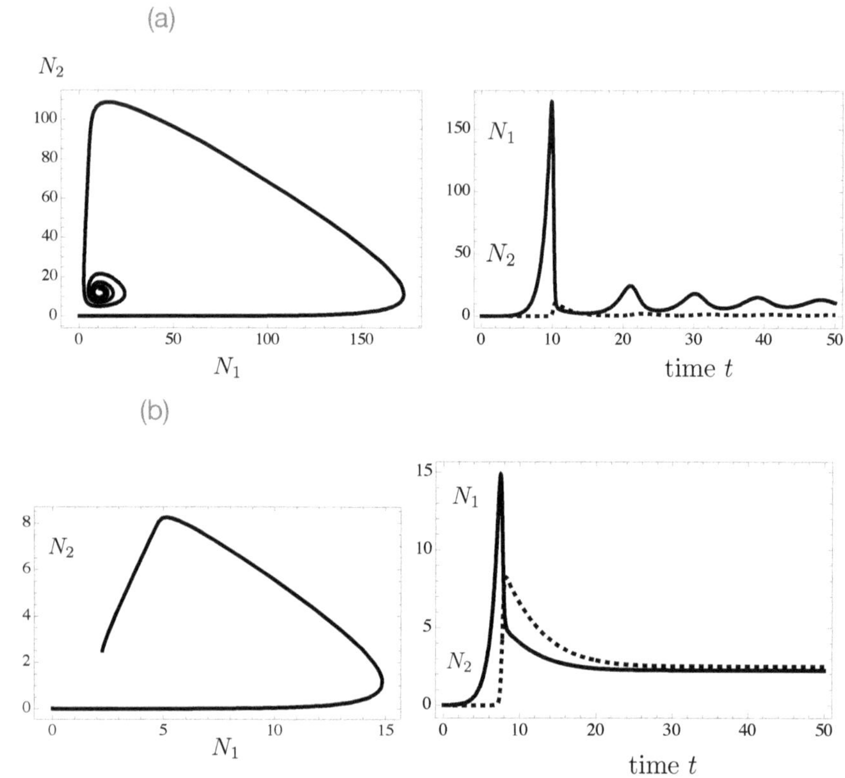


**FIGURE 5.4**
Instability saturation by subharmonic decay, for (a) weak coupling and (b) strong coupling. On the left, we represent the system trajectory on the photon number phase-space. On the right, we represent $N_1$ (bold curves) and $N_2 = N_3$ (dashed curves), as a function of time $t$.

It can easily be seen that the origin is always unstable when $\gamma_1 < 0$. As for the stable equilibrium defined by (5.113), it can considered the final state of instability saturation. Notice that it only makes sense in the first octant of phase-space, because the quantities $N_j$ are by definition positive. This implies that the denominator in eq. (5.113) has to be positive, and consequently

$$C_1(C_2\Gamma_3 + C_3\Gamma_2) \; > \; \frac{C_2 C_3}{C_1} \, . \tag{5.114}$$

This instability saturation regime is illustrated in Figure 5.4.

It is obvious that any transition to chaos, eventually occurring in the fixed-phase description (5.100)–(5.101) is absent in the random-phase approximation, where the unstable mode grows regularly until saturation is attained. More specific examples of nonlinear wave-mixing, such as those leading to stimulated scattering and to high-harmonic generation, will be examined in this and in later chapters.

## 5.5 High-Harmonic Generation

Let us turn to the important effect of harmonic generation, which can eventually lead to the formation of a very broad frequency spectrum, with a large number of harmonics. It should be noted that, in infinite and isotropic plasmas, second-harmonic generation is forbidden due to the symmetric properties of the medium. The plasma is a *centro-symmetric medium*, where the corresponding nonlinear current cannot be excited. However, if there is an axis of symmetry, due for instance to the presence of an external magnetic field, this rule is broken, and second-harmonic emission is allowed.

But this section is only focused on isotropic plasmas, which are important for intense laser-plasma interactions. In this case, there are too ways to break the centro-symmetry: i) the existence of a boundary, for incident waves in overdense plasmas and ii) the nonlinear third-order susceptibility, for waves propagating in underdense plasmas. These two cases will be examined separately. They both lead to the excitation of high-harmonics of the incident field, which can be used in the production of ultrashort pulses and to the exploration of sub-femtosecond physics.

### 5.5.1 Underdense Plasmas

Let us first focus on four-wave mixing in uniform underdense plasma. We can show that, for waves propagating with a frequency much larger than the plasma frequency, $\omega_0^2 \gg \omega_p^2$, a nonlinear cascade can take place, eventually leading to the formation of a large spectrum of high harmonics, $\omega_\nu = \nu\omega_0$, with $\nu \gg 1$. This process is based on the third-order plasma susceptibility, and is essentially different from the harmonic generation at a plasma boundary.

The four-wave mixing process can be seen as a particular case of a generic nonlinear optical process, where two waves with frequencies $\omega_0$ and $\omega_1$ are able to generate a spectral cascade of frequencies $\omega_\nu = \omega_0 + \nu(\omega_1 - \omega_0)$, where $\nu$ is an integer. This has been confirmed experimentally in nonlinear optics, and can lead to single-cycle femtosecond pulses in the visible [78]. Harmonic generation is a particular case of this mechanism, when $\omega_2 = 2\omega_1$, as in a relativistic plasma.

The main source of plasma fourth-order nonlinearity is related to the relativistic factor $\gamma$, which appears in the electron current $\mathbf{J}$. If we go back to the wave equation (4.3) for purely transverse fields ($\nabla \cdot \mathbf{E} = 0$), and write it in terms of the vector potential $\mathbf{A}$, we get

$$\left(\nabla^2 - \frac{1}{c^2}\frac{\partial^2}{\partial t}\right)\mathbf{A} = -\mu_0 \mathbf{J}\,, \tag{5.115}$$

where the electron current is

$$\mathbf{J} = -en\frac{\mathbf{u}}{c\gamma}, \quad \gamma = \sqrt{1+u^2}, \tag{5.116}$$

Here, we assume that a large amplitude wave propagates in a cold plasma, where the relativistic effects are only due to the wave. In this case, the relativistic factor only depends on the normalized covariant velocity $\mathbf{u} = \gamma\mathbf{v}/c$. To simplify we also assume that the electron density is just the equilibrium density $n \simeq n_0$, thus ignoring slow wave effects such as those associated with radiation pressure. They will be treated later, in another context. Noting the electron equations of motion in the wave field determines that $\mathbf{u} \simeq \mathbf{a} = (e\mathbf{A}/mc)$, it is convenient to rewrite the wave equation in terms of the normalized vector potential $\mathbf{a}$ and the normalzied current $\mathbf{j}$, as

$$\left(\nabla^2 - \frac{1}{c^2}\frac{\partial^2}{\partial t}\right)\mathbf{a} = -\mathbf{j}(\mathbf{a}), \tag{5.117}$$

where the normalized nonlinear current $\mathbf{j}(\mathbf{a})$ is defined as

$$\mathbf{j}(\mathbf{a}) = k_p^2 \frac{\mathbf{a}}{\sqrt{1+a^2}}, \tag{5.118}$$

with $k_p^2 = \omega_p^2/c^2$. For an incident laser pulse with frequency $\omega_0$, this nonlinear current will generate harmonics $\omega_\nu = \nu\omega_0$, for $\nu$ integer. We can therefore, write the total field as

$$\mathbf{a} = \sum_{\nu=1}^{\infty} a_\nu \mathbf{e}_0 \exp\left(i\mathbf{k}_\nu \cdot \mathbf{r} - i\omega_\nu t\right) + c.c., \tag{5.119}$$

where we assume linear polarization, with $\mathbf{e}_0 = \mathbf{a}_0/a_0$. In general, we have $\mathbf{k}_\nu \neq \nu\mathbf{k}_0$, because the harmonics, in order to propagate in the medium, have to satisfy the plasma dispersion relation. We can now expand the denominator as

$$\frac{1}{\sqrt{1+a^2}} = \frac{1}{\gamma_a}\left[1 + \sum_\ell w_\ell \left(\frac{\Sigma}{\gamma_a^2}\right)^\ell\right], \tag{5.120}$$

where $w_\ell$ are numerical coefficients such that $w_1 = -1/2$ and $|w_\ell/w_1|$ are smaller than 1. We defined the gamma factor $\gamma_a = \sqrt{1+\sum_\nu |a_\nu|^2}$, and introduced the sum

$$\Sigma = \sum_{\nu\neq\nu'} \exp\left[i\mathbf{\Delta k}_{\nu\nu'} \cdot \mathbf{r} - i\delta_{\nu\nu'} t\right] + c.c. \tag{5.121}$$

with $\mathbf{\Delta k}_{\nu\nu'} = (\mathbf{k}_\nu - \mathbf{k}_{\nu'})$, and $\delta_{\nu\nu'} = (\nu-\nu')\omega_0$. The expansion is always valid, even for very intense fields, because we always have $(|\Sigma| < \gamma_a^2)$. In the case of circular polarization, this would lead to quasi-static values of $a^2$ and $\gamma = \gamma_a$, with $S = 0$. Going back to the nonlinear current (5.118), we can see that, for circular polarization, the incident field cannot generate harmonics. This property is reminiscent of the atomic ionization-recombination model, where harmonic generation is also generally forbidden for circularly polarization [79].

Let us now retain the dominant term in this expansion. Assuming propagation in the $z$-direction, we are then able to derive a recurrence set of equations for the harmonic amplitudes $a_\nu$, such that

$$\frac{\partial}{\partial t} a_\nu = i \frac{\omega_p^2}{4\omega_\nu \gamma_a^3} \left[ (a_1^* a_2) a_{\nu-1} e^{-i\Delta k_\nu z} + (a_1 a_2^*) a_{\nu+1} e^{i\Delta k_\nu z} \right] , \tag{5.122}$$

These coupled equations are valid when the amplitudes of the first two modes $\nu = (1, 2)$ are dominant, which is often the case. Otherwise, we would need to include more terms in the summation. Here, we have used the phase-mismatch

$$\Delta k_\nu = \left( \frac{3}{4} - \frac{3}{\nu^4} \right) \frac{k_p^2}{\gamma_a k_0} . \tag{5.123}$$

This mismatch should stay small in comparison with the spectral width of the incident field, $\Delta k_0$, in order to guarantee an effective interaction between the different spectral components. Otherwise, the nonlinear cascade is broken due to phase-mixing. This means that this nonlinear mechanism for harmonic generation should stop when $\Delta k_\nu \sim \Delta k_0$. This allows us to establish the upper limit of the harmonic spectrum as

$$\nu_{max} \simeq \frac{4}{3} \frac{k_0}{k_p^2} \gamma_a \Delta k_0 . \tag{5.124}$$

In order to derive analytical solutions for the coupled mode equations, valid for the bulk of the spectrum, we neglect the phase mismatch and obtain

$$\frac{\partial}{\partial t} a_\nu = i \left[ (a_{\nu-1} e^{i\delta} + a_{\nu+1} e^{-i\delta} \right] , \tag{5.125}$$

with

$$w_\nu = \frac{\omega_p^2}{4\omega_\nu \gamma_a^3} I_0 , \quad (a_1^* a_2) = I_0 \exp(i\delta) , \tag{5.126}$$

Now, using an auxiliary coordinate $z = 2w_\nu \tau$, which is a rescaled time, and new amplitudes $A_\nu$ defined by

$$a_\nu = (-1)^{\nu/2} e^{i\nu\delta} A_\nu , \tag{5.127}$$

we can reduce eqs. (5.125) to the following expression

$$2 \frac{d}{dz} A_\nu = A_{\nu-1} - A_{\nu+1} . \tag{5.128}$$

This is formally identical to the recurrence relation valid for Bessel functions of the first kind, $J_\nu(z)$. An adequate solution for the harmonic amplitudes can then be given in terms of Bessel functions, as

$$A_\nu(t) = A_0(0) J_{\nu-1}(2w_\nu t) + A_1(0) J_\nu(z) , \tag{5.129}$$

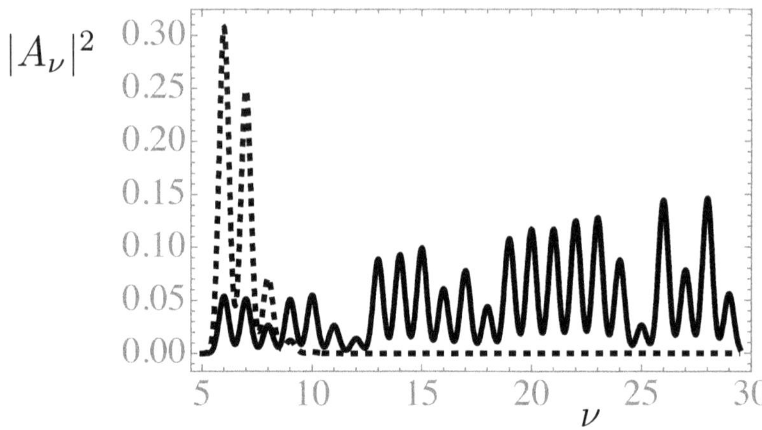


**FIGURE 5.5**
High-harmonic generation in underdense plasma: Harmonic energies $|A_\nu|^2$ (in arbitrary units), for (a) short times $w_\nu t = 50$ (dashed), and (b) long times $w_\nu t = 1000$ (in bold). The spectrum of each harmonic is artificially broadened, to simulate the width $\Delta k_0$ of the incident spectrum.

where we have assumed that only the main modes $j = 0, 1$ are initially present in the plasma, and therefore $A_j(0) = 0$ for $j \geq 2$. It is important to notice that, under quite general conditions, this solution conserves the electromagnetic energy in the plasma, for arbitrary times, given that

$$\sum_\nu |A_\nu(z)|^2 = |A_0(0)|^2 + |A_1(0)|^2 . \tag{5.130}$$

The value of $A_1(0)$ corresponds to second harmonic generation, which is injected in the medium, or created at the plasma boundary. The main features of the harmonic spectrum are illustrated in Figure 5.5.

Let us now take the asymptotic values for small arguments, $z \ll 1$. We can then use $J_\nu(z) \simeq z^\nu / 2^\nu \nu!$. If we then want to consider the spectral features of the harmonic cascade for $\nu \gg 1$, we can replace the discrete harmonic $\nu$ by a continuous variable $x = \omega/\omega_0$. Neglecting the source beams with frequencies $\omega_0$ and $2\omega_0$, which are supposed to be dominant, we could write for the spectral energy scaling of the harmonic spectrum as

$$I_\nu(z \ll 1) \propto I(\omega_0) \frac{\omega_0 (z/2)^{2\nu}}{\Gamma(\nu + 1)} . \tag{5.131}$$

In contrast, for long times, $z \gg 1$ the scaling changes, because we should now use the asymptotic expression $J_{2\nu} \simeq \sqrt{\pi z} \cos(z)$. Taking the maxima for each harmonic, we arrive at a new scaling, $I_\nu(z \to \infty) \propto \nu$. This spectrum, could extend to arbitrarily high frequencies, $\nu \gg 1$. But, it is suppressed above a cutoff value, $\omega > \omega_{c.o.} \equiv \nu_{max}\omega_0$, when the phase-mismatch becomes larger than the spectral width of the incident radiation, as given by eq. (5.124). We should however not be led to the conclusion that the cut-off frequency $\omega_{c.o.}$ is independent of the intensity $I(\omega_0)$, because self-phase modulations will tend to increase the spectral width of the incident pulse. As a result, cutoff will move to higher frequencies, almost linearly with $I(\omega_0)$.

### 5.5.2 Overdense Plasmas

Let us now consider harmonic generation at the plasma boundary. As a simple picture for intense laser interaction with solid targets, we can assume that the position of the plasma boundary layer is determined by the equality between the incident laser frequency and the local plasma density $\omega_o = \omega_p(z_0)$. Noting that the ponderomotive force has a fast component with twice the cutoff frequency, it is then plausible to assume that the plasma boundary layer will oscillate at $2\omega_0$. Because most of the incident beam is reflected, this can be described as an *oscillating mirror*, thus leading to a periodic phase modulation of the reflected radiation.

The oscillating mirror picture has been explored by several researchers [80, 81, 82]. In order describe the harmonic emission at cutoff, we return to the basic wave equation (4.3), and assume an incident field oscillating at frequency $\omega$. For an inhomogeneous plasma, this equation can then be written as

$$\nabla^2\mathbf{E} - \nabla(\nabla\cdot\mathbf{E}) = -\frac{\omega_0^2}{c^2}\epsilon(\mathbf{r})\mathbf{E}\,, \tag{5.132}$$

where the dielectric function is

$$\epsilon(\mathbf{r}) = 1 - \frac{\omega_p^2(\mathbf{r})}{\gamma(\mathbf{r})\omega_0^2}\,. \tag{5.133}$$

It should be noticed that, in general, the divergence of $\mathbf{E}$ is not negligible. From Maxwell's equations we can see that

$$\nabla\times\mathbf{H} = -i\omega_0\epsilon(\mathbf{r})\mathbf{E}\,,\quad \nabla\cdot\mathbf{E} = -\mathbf{E}\cdot\nabla\ln\epsilon(\mathbf{r})\,. \tag{5.134}$$

Given the above definition, we can easily get

$$\nabla\ln\epsilon(\mathbf{r}) = -\frac{1}{\epsilon(\mathbf{r})}\frac{\omega_p^2(\mathbf{r})}{\omega_0^2}\nabla\ln\frac{n_0}{\gamma}\,. \tag{5.135}$$

However, this simplifies when we consider normal incidence. If we assume propagation along the $z$-axis, and the plasma density depends only on $z$, we have $\omega_p^2(\mathbf{r}) \equiv \omega_p^2(z)$, and $(\nabla\cdot\mathbf{E}) = 0$. We can also forget the transverse wave profile, which is of secondary importance here, and assume $\nabla_\perp^2 = 0$. We can then reduce eq. (5.132) to

$$\frac{\partial^2}{\partial z^2}\mathbf{a} + \frac{\omega_0^2}{c^2}\epsilon(z)\mathbf{a} = 0\,, \tag{5.136}$$

where we have used the normalized field amplitude $\mathbf{a} = (eE/m\omega_0 c)$. There will be a cutoff at some position $z = z_0$, where $\omega_0^2 = \omega_p(z)^2/\gamma$. In order to determine the field around this point, we linearize the density-intensity profile around $z_0$. This then leads to the following equation

$$\frac{\partial^2 a}{\partial z^2} + \frac{\omega_0^2}{c^2}\alpha_0(z - z_0)a = 0\,, \tag{5.137}$$

where we have ignored wave polarization, and defined

$$\epsilon(z) = \alpha_0(z - z_0)\,, \quad \alpha_0 = \left(\frac{1}{\gamma n_0}\frac{\partial n_0}{\partial z}\right)_{z=z_0} . \tag{5.138}$$

The parameter $\alpha_0$ characterizes the electron density slope at cutoff, and also takes into account the local relativistic factor. In order to solve this equation, it is useful to introduce a new space variable, $\zeta$, such that

$$\zeta = (z - z_0)\left(\frac{\omega_0^2}{c^2}\alpha_0\right)^{1/3} , \tag{5.139}$$

which reduces eq. (5.140) to an equation of the form

$$\frac{\partial^2 a}{\partial \zeta^2} - \zeta a = 0\,, \tag{5.140}$$

This can be solve using the *Airy functions*, such that

$$a(\zeta) = Ai(\zeta) \equiv \frac{3}{\pi}\int_0^\infty \cos\left(\frac{t^3}{3} + \zeta t\right) dt\,. \tag{5.141}$$

Of particular importance are the asymptotic expressions of the Airy function, for positive and negative values of $\zeta$. Far away from the cutoff $\zeta = 0$, and in the low density side of the plasma profile, this solution can be approximately described as a standing wave, which is given by

$$a(\zeta) \simeq \frac{1}{\pi}|\zeta|^{-1/4}\sin\left(\frac{2}{3}|\zeta|^{3/2} + \frac{\pi}{4}\right) . \tag{5.142}$$

This standing wave, contains both the incident and the reflected wave. On the other hand, in the overdense region, the Airy function can be approximately written as

$$a(\zeta) \simeq \frac{1}{2\sqrt{\pi}}|\zeta|^{-1/4}\exp\left(-\frac{2}{3}|\zeta|^{3/2}\right) . \tag{5.143}$$

The field amplitude decays exponentially in this overdense region, as expected. It is important to notice that, for a finite wave amplitude, the cutoff position is not a constant, because all the electrons oscillate at a frequency equal to the incident frequency $\omega_0$. This means that the density profile is time-dependent, as well as the positon $z_0$, which becomes $z_0 + (\delta n_0/n_0)\sin\omega_0 t$. As a result, an oscillating phase appears in the above solutions. In the underdense region, we get a solution where $\zeta$ is replaced by $\zeta(t)$, with

$$\zeta(t) = \zeta + \delta\zeta \sin(\omega_0 t) , \quad \delta\zeta = \left( \frac{\omega_0^2 a}{c^3 \gamma} \alpha_0 \right)^{1/3} \tag{5.144}$$

This means that, far away from the cutoff, the reflected field will evolve as

$$E_r \propto \exp\left[ -i\omega_0 t + i\frac{a}{c\gamma} \sin(\omega_0 t) \right] = e^{-i\omega t} \sum_\nu J_\nu \left( \frac{\delta n_0}{n_0} \right) e^{-i\nu\omega_0 t} . \tag{5.145}$$

On the other hand, the amplitude of the reflected wave will vary as $E_r \simeq \gamma_s E_i$. This leads to a succession of intense spikes, with ultra-short durations. Due to the resulting intensity modulations, these spikes will have a typical duration of $\delta t \sim \delta t'/\gamma_s^2$, or

$$\delta t \sim \frac{\pi}{\omega_0} \frac{1}{\gamma_s^2} \tag{5.146}$$

The reflected wave amplitude cannot however increase indefinitely with $\gamma_s$, because of energy constraints. We have to keep in mind that the source of energy for the oscillating mirror is the incident laser pulse itself. This will influence the final scaling of the reflected spectrum. The resulting intensities for the harmonics $\omega_\nu = \nu\omega_0$ will then be determined by [82]

$$I_\nu \simeq \left| F_\nu \exp[i\theta_+(\nu)] - F_{-\nu} \exp[i\theta_-(\nu)] \right|^2 , \tag{5.147}$$

with $\theta_\pm(\nu) = \pm\theta_0 - n\theta_1$, and

$$F_\nu = \frac{4\sqrt{\pi}}{\nu^{4/3}} Ai \left( \frac{2}{\nu_c} \frac{\nu - \nu_c}{\nu^{1/3}} \right) , \tag{5.148}$$

where $\nu_c$ is a critical harmonic number defined by $\nu_c = 4\gamma_s^2$. From this, we can derive a simple *scaling law* for the energy of the different harmonics, as

$$I_\nu \propto \nu^{-p} , \tag{5.149}$$

where, for moderate values of $\nu_c$ we have $p = 10/3$, and for very large values of $\nu_c \gg 1$, we have $p = 8/3$. A significant cutoff of this harmonic spectrum occurs at $\nu \simeq \sqrt{8}\gamma_s^3$, where the power law is replaced by an exponential law

$$I_\nu \propto \frac{\sqrt{\nu_c}}{\nu^3} \exp\left[ -\frac{16}{3} \frac{\sqrt{2}}{\nu_c^{3/2}} (\nu - \nu_c) \right] . \tag{5.150}$$

These qualitative features are confirmed by simulations and by experiments. See Figure 5.6, for an illustration of the incident laser field near cutoff, in the linear and nonlinear regimes, and a typical harmonic spectrum.

The above models make sense for high intensity incident fields, such that $a_0 > 1$. However, harmonic generation can also take place at moderate intensities, $a_0 \ll 1$, for a neutral gas target where the plasma cutoff is meaningless, as explained by the *atomic-recombination model*, first proposed by Corkum [79]. But this is somewhat outside the area of relativistic plasmas and will not be discussed here.

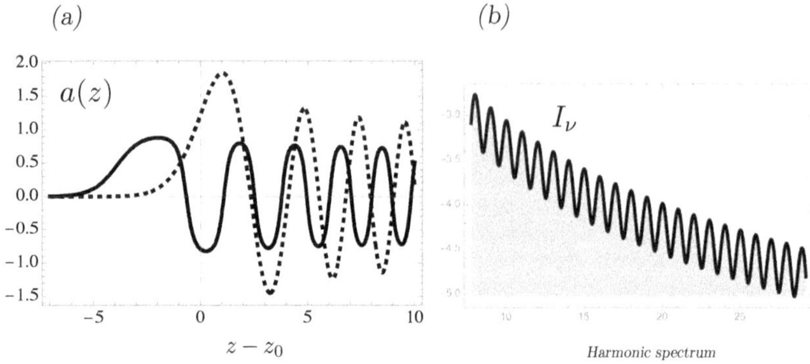


**FIGURE 5.6**
High harmonic generation at the plasma cutoff, $z - z_0 = 0$: (a) Field amplitude $a(z)$, for $\gamma_a = 1$ (dashed curve), and for $\gamma_a = \sqrt{1+a^2} = 3$ (bold curve). Displacement of the cutoff position $z = z_0$ due to relativistic self-transparency is shown; (b) Energy of the harmonic spectrum, for high harmonics $\nu < \nu_c$.

## 5.6 Relativistic Optical Guiding

Let us now discuss another important qualitative aspect of intense laser pulse propagation in a plasma. This property is the *relativistic self-focusing* and, as a consequence, the possible occurrence of *optical guiding* over distances much larger than the Rayleigh length. Our discussion is based on simple heuristic arguments, which are corroborated by numerical simulations.

We first notice that the nonlinear dispersion of a circularly polarized electromagnetic wave (representing a laser pulse) in a cold plasma, evolving as $\exp(i\mathbf{k}\cdot\mathbf{r} - i\omega t)$. In our discussion, we assume that they approximately satisfy the local dispersion relation,

$$\omega^2 = k^2c^2 + \frac{\omega_p(\mathbf{r})^2}{\gamma_a(\mathbf{r})}, \quad \gamma_a(\mathbf{r}) = \sqrt{1+a^2}, \tag{5.151}$$

where $a \equiv a(\mathbf{r})$ is the normalized wave amplitude. This is valid for very high frequencies, much larger than the electron plasma frequency $\omega \geq \omega_p(\mathbf{r})$, on a scale-length significantly larger than the local wavelength. The corresponding refractive index, $N(\omega, \mathbf{r})$ is defined as

$$N(\omega, \mathbf{r}) \equiv \frac{kc}{\omega} = \sqrt{1 - \frac{\omega_p(\mathbf{r})^2}{\omega^2\gamma_a}}. \tag{5.152}$$

In a laser pulse, the quantities $\omega_p$ and $a$ usually depend on the radial coordinate and, as a result vary across the pulse, both in the axial and radial directions, with an influence on its focusing and defocusing properties. Let us assume the wave equation, for $a \propto \exp(-i\omega t)$, written in the form

$$\left(\nabla^2 + \frac{\omega^2}{c^2}\right) a = \frac{\omega^2}{c^2}\left[1 - N(\omega, \mathbf{r})^2\right] a\,, \tag{5.153}$$

For a pulse propagating in the $z$-direction, with cylindrical symmetry, we can use

$$a(\mathbf{r}, t) = a_0(r, z, \eta) \exp(ik\eta)\,, \tag{5.154}$$

with $\eta = z - ct$. We should note that the nonlinear refractive index depends on the radial position, such that for a typical Gaussian pulse, we can use $a_0(r, z, \eta) \propto \exp(-r^2/w^2)$, where $w = w(z)$ is the beam waist. This allows us to write

$$N(\omega, \mathbf{r})^2 = 1 - \frac{\omega_p(r)^2}{\omega^2\sqrt{1 + a_0^2 \exp(-2r^2/w^2)}}\,. \tag{5.155}$$

If the plasma density stays constant, the variation of $\gamma_a$ implies the existence of a focusing effect, because the refractive index is larger at the pulse center. This focusing effect increases with the laser intensity, as illustrated in Figure 5.7 (a).

However, the plasma frequency also changes with the pulse intensity and cannot be taken as a constant. This change is due to two different reasons. First, we have laser ionization, if the medium is initially a neutral gas. This makes the value of $N(\omega, \mathbf{r})$ to increase at the pulse axis $r = 0$, which produces a defocusing effect on the pulse. Second, the laser ponderomotive force expels the electrons from the center, which contributes to the relativistic self-focusing.

These different effects, relativistic self-focusing on one side, and laser ionization on the other side, can eventually cancel each other and create a region with nearly constant refractive index, $N(\omega, \mathbf{r}) \simeq const.$ across the laser pulse cross-section, as shown in Figure 5.7 (b). Under such conditions, it is possible to observe pulse self-channeling through the medium, over distances that can be much larger than the Rayleigh length.

In order to illustrate this self-channeling, we consider a simple model, which is valid for moderate intensities, such that $a^2 \ll 1$. For that purpose, we return to the wave equation (5.153), and expand the refractive index, assuming that $\omega^2 \gg \omega_p^2$. We get

$$\left(\nabla^2 + \frac{\omega^2}{c^2}\right) a \simeq \frac{\omega_p^2}{2c^2\gamma_a} a\,, \tag{5.156}$$

Assuming that the wave amplitude $a_0$ varies on a scale much larger than the wavenumber

$$|\partial a_0/\partial z| \ll |k a_0|\,, \tag{5.157}$$

we can use $\nabla^2 = \nabla_\perp^2 + \partial^2/\partial z^2$, and obtain an equation for the slowly varying amplitude as

$$\left(\nabla^2 + \frac{\omega^2}{c^2}\right) a_0 = k^2 \frac{\omega_p^2}{2\omega^2\gamma_a} a_0\,, \tag{5.158}$$

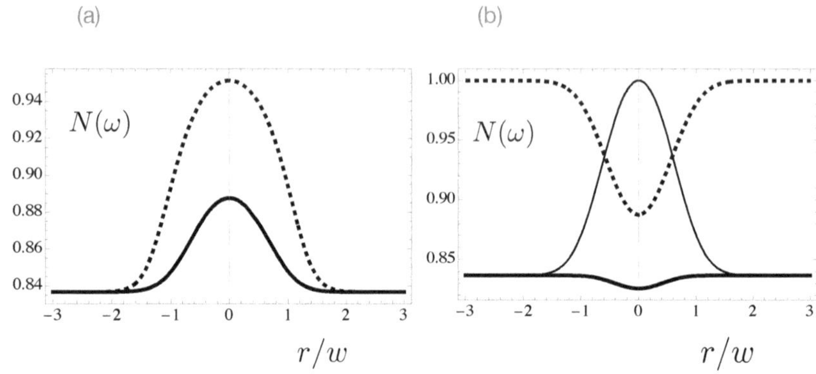


**FIGURE 5.7**
Nonlinear refractive index $N(\omega)$ as a function of the normalized laser beam radius $r/w$, for a Gaussian laser radial profile: (a) Constant Plasma density, such that $\omega_p^2/\omega^2 = 0.3$, for $a = 1$ (bold curve) and $a = 3$ (dashed curve). (b) Radially dependent plasma density, with i) $f(r) = 0.3(1 - \exp(-2r^2/w^2))$ (thin curve); ii) $f(r) = 0.3\exp(-2r^2/w^2)$ (dashed curve) and iii) $f(r) = 0.3(1 + 0.5\exp(-2r^2/w^2))$ (bold curve).

This is a *paraxial equation*. A formal solution of this equation could be represented as a superposition of orthogonal functions defined on the transverse planes, $z =$ *const.*, such as the Laguerre-Gauss or the Hermite-Gauss functions, frequently used to describe the laser modes. But here we are concerned with purely Gaussian pulses, and an approximate solution of the form [83]

$$a_0(r, z) = a_{00}\left(\frac{w_0}{w}\right)\exp\left[-(1 - i\alpha)\frac{r^2}{w^2} + i\theta\right], \tag{5.159}$$

can be used. Here, $a_{00}$ and $w_0$ are constants, $w \equiv w(z)$ is the local pulse waist which has its minimum value $w_0$ at the focus, $w_0 = w(z = 0)$, and $\alpha \equiv \alpha(z)$ and $\theta \equiv \theta(z)$. Of course, these new functions of the axial variable $z$ need to be specified. For this purpose, we expand the laser intensity as

$$a_0^2 \simeq a_{00}^2\left(1 - \frac{2r^2}{w^2}\right). \tag{5.160}$$

Defining the above functions $\alpha(z)$ and $\theta(z)$ as

$$\alpha(z) = \frac{kw}{2}\frac{dw}{dz}, \quad \frac{d\theta}{dz} = -\frac{2}{kw^2} - \frac{k}{2}\frac{\omega_p^2}{\omega^2}\left(1 - \frac{1}{2}a_{00}(r)^2\right), \tag{5.161}$$

we can derive from eq. (5.159) the following equation for the laser pulse waist

$$\frac{d^2w}{dz^2} = \frac{4}{k^2w^3}\left[1 - \frac{\omega_p^2}{2\omega^2}\frac{k^2w^3}{2w_0^2} - \frac{\omega_p^2}{2\omega^2}\frac{a_0^2}{2}k^2w^2\right]. \tag{5.162}$$

The first term on the right represents the vacuum diffraction and is characterized by the *Rayleigh length*, which can be defined as $z_R = kw_0^2/2$. The second term represents the focusing effect due to the plasma density profile (for instance, due to ionization of the background gas), and the third term is the relativistic self-focusing. Self-guided pulses exist when these terms cancel each other, and the pulse waist $w(z) \simeq w_0$ stays nearly constant over a large distance.

# 6

# *Relativistic Particle Beams*

Here we consider the important topic of charged particle beams in a plasma, relevant in many areas of physics and astrophysics. We first show that the self-magnetic fields generated by these beams impose limitations on the total beam current. We then examine the most relevant instabilities associated with beam-plasma interaction. Finally, we examine the occurrence of similar instabilities when the charged particle beams are replaced by photon beams which, using a quasi-particle description that represents electromagnetic waves in the same way as particle beams.

## 6.1 Beam Stability

A preliminary important concept is that of *Alfvén current*. To introduce this concept, we first consider a cylindrical mono-energetic electron beam, propagating with relativistic velocity $v_b \simeq c$. When we move radially, from the beam axis to the periphery, the self-generated magnetic field associated with the beam current increases with the radius. At some point, the pinch force associated with this poloidal field is not compensated by the radial electrostatic force associated with the electron repulsion. The particles then start moving backward. This defines an upper limit for the electron current, above which the electrons start moving in the opposite direction. Such a limit was first determined by Alfvén [84] and is defined as

$$I_A = 4\pi\epsilon_0 \left(\frac{mc^3}{e}\right) \frac{\beta_b}{\sqrt{1-\beta_b^2}} \simeq 1.7\beta_b\gamma_b \times 10^4 \, (Ampere) \, , \tag{6.1}$$

with $\beta_b = v_b/c$ and $\gamma_b = 1/\sqrt{1-\beta_b^2}$. But this limit can be broken, if we use a non-uniform beam, as first demonstrated by Hammer and Rostoker [85]. It is the analysis of relativistic beam stability that we discuss in this section. For this purpose, we consider an electron beam moving along the $z$-axis, in a background of stationary ions that, totally on in part, neutralize the beam charge. We start with the Poisson's equation

$$\nabla^2 V = -\frac{1}{\epsilon_0} \sum_\alpha q_\alpha n_\alpha \, , \tag{6.2}$$

DOI: 10.1201/9781003514701-6

where the index $\alpha = (e, i)$ represents the charge species. We have, $q_e = -e$ and $q_i = Ze$. Assuming axial beam symmetry, and using cylindrical coordinates such that $\mathbf{r} = (r, \theta, z)$, we can write this equation as

$$\frac{1}{r}\frac{\partial}{\partial r}\left(r\frac{\partial}{\partial r}\right)V = -\frac{1}{\epsilon_0}\sum_\alpha q_\alpha n_\alpha \,, \tag{6.3}$$

The corresponding electric field is purely radial,

$$\mathbf{E}_0 = E_r \mathbf{e}_x \,, \quad E_r(r) = -\frac{\partial V}{\partial r} \,. \tag{6.4}$$

On the other hand, the static magnetic field associated with the beam has poloidal and axial components, and can be written as $\mathbf{B} = B_\theta \mathbf{e}_\theta + B_z \mathbf{e}_z$. Using $\mathbf{B} = \nabla \times \mathbf{A}$, these field components can be determined by

$$B_\theta = -\frac{\partial A_z}{\partial r} \,, \quad B_z = \frac{1}{r}\frac{\partial}{\partial r}(r A_\theta) \,. \tag{6.5}$$

To these static self-generated fields, we should add an external magnetic field, $\mathbf{B} = B_0 \mathbf{e}_z$, which is described by the vector potential

$$\mathbf{A}_0 = \frac{r}{2} B_0 \mathbf{e}_\theta \,, \tag{6.6}$$

As for the beam field (6.5), the corresponding vector potential can be derived from the beam current, using the (static) potential equation $\nabla^2 \mathbf{A} = -\mu_0 \mathbf{J}$. This can be written explicitly as

$$\left(\frac{1}{r}\frac{\partial}{\partial r} r \frac{\partial}{\partial r}\right) A_z = -\mu_0 \sum_\alpha J_{\alpha z} \,,$$

$$\frac{\partial}{\partial r}\left(\frac{1}{r}\frac{\partial}{\partial r}\right) r A_\theta = -\mu_0 \sum_\alpha J_{\alpha\theta} \,, \tag{6.7}$$

To these field equations, we should add the charge and current densities associated with the particles and defined by the distribution functions $f_\alpha(\mathbf{p})$, as

$$n_\alpha = \int f_\alpha(\mathbf{p}) d\mathbf{p} \,, \quad \mathbf{J}_\alpha = \int \mathbf{v} f_\alpha(\mathbf{p}) d\mathbf{p} \,, \tag{6.8}$$

We assume that the ions are immobile, due to their large mass. We also assume that the ions are partially ionized, such that

$$Z n_{0i}(r) = f n_{0b}(r) \,, \tag{6.9}$$

where $Z$ is the average ionization degree and the constant $f$ is the *neutralization factor*. In order to calculate the charge and current profiles, we can specify the beam distribution function as

$$f_{0b}(\mathbf{p}, r) = \frac{N_b}{2\pi m \gamma_b} \delta(h - \gamma_b) \delta(P_z - \beta_b \gamma_b) \,, \tag{6.10}$$

where the beam density $N_b$ is constant, and the quantities $h$ and $P_z$ represent the (normalized) electron energy and the canonical momentum, defined by

$$h = \gamma(\mathbf{p}) - eV_0(r) , \quad \gamma(\mathbf{p}) = \sqrt{1 + p^2} , \tag{6.11}$$

and

$$P_z = \gamma(\mathbf{p}) \frac{v_z}{c} - eA_z(r) , \tag{6.12}$$

This beam equilibrium [85] corresponds to an electron beam with constant canonical momentum in the $z$-direction. The corresponding Poisson's equation (6.3) can be written as

$$\frac{1}{r} \frac{\partial}{\partial r} \left( r \frac{\partial}{\partial r} \right) V_0 = \frac{e}{\epsilon_0} (1 - f) n_{0b} , \tag{6.13}$$

where $Z = 1$ was assumed. Using eq. (6.10), we then get

$$n_{0b}(r) = \int f_{0b}(\mathbf{p}, r) d\mathbf{p} = N_b I_0(z) . \tag{6.14}$$

This expression is valid for $r < r_b$, where $r_b$ is the external beam radius, and $I_0(z)$ is a modified Bessel function, $I_\nu(z)$ for $\nu = 0$. The variable $z$ can be defined as

$$z = \frac{r}{\delta} \sqrt{1 - f} , \quad \delta = \frac{c}{\omega_{pb} \sqrt{\gamma_b}} . \tag{6.15}$$

Notice that the quantity $\delta$ is the collisionless *skin depth* and $\omega_{pb} = \sqrt{e^2 N_b / \epsilon_0 m}$ is the beam plasma frequency. The corresponding axial velocity profile is determined by

$$v_{0z}(r) = v_b \frac{I_0(r/\delta)}{I_0(z)} . \tag{6.16}$$

This allows us to determine the total beam current, using a radial integration, as

$$I_b = 2\pi e \int_0^{r_b} n_{0b}(r) v_{0z}(r) r dr . \tag{6.17}$$

Replacing the electron density and velocity profiles, as determined by eqs. (6.14) and (6.16), we obtain

$$I_b = 2\pi \epsilon_0 \beta_b \gamma_b \left( \frac{mc^3}{e} \right) \frac{r_b}{\delta} I_1 \left( \frac{r_b}{\delta} \right) . \tag{6.18}$$

where $I_1(r_b/\delta)$ is the modified Bessel function $I_\nu(z)$ for $\nu = 1$.

It is useful to compare this result with the Alfvén current defined in eq. (6.1). For this purpose, we need to adapt the original definition, valid for a uniform beam, to the case of a non-uniform beam. In this case, we define the current limit as the beam current for which the electron Larmor radius in the self-generated magnetic field, observed at the beam radius $r = r_b$, is

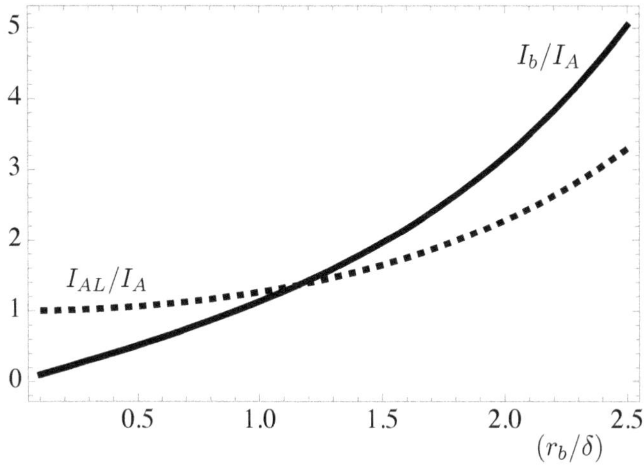


**FIGURE 6.1**
Normalized beam current $I_b/I_A$ (bold), and the Alfvén-Lawson limit $I_{AL}/I_A$ (dashed), as a function of the ratio $(r_b/\delta)$.

equal to half of the beam radius. This more general definition is known as the *Alfvén-Lawson limit* [86], and can be stated as

$$I_{AL} = \beta_b \gamma_b \left( \frac{mc^3}{e} \right) I_0 \left( \frac{r_b}{\delta} \right) = I_A I_0 \left( \frac{r_b}{\delta} \right) . \tag{6.19}$$

Comparing this with eq. (6.18), we get

$$I_b = \frac{r}{2\delta} \frac{I_1(r_b/\delta)}{I_0(r_b/\delta)} I_{AL} . \tag{6.20}$$

Using the asymptotic expressions for the modified Bessel functions $I_\nu(z)$ [87], we can then state that for a large electron beam, such that $(r_b/\delta) \gg 1$, the beam current is given by

$$I_b \simeq \frac{r}{2\delta} I_{AL} \gg I_{AL} , \tag{6.21}$$

whereas for a thin beam, such that $(r_b/\delta) \ll 1$, we obtain

$$I_b \simeq \left( \frac{r}{2\delta} \right)^2 I_{AL} \ll I_{AL} , \tag{6.22}$$

This shows that, for electron beams with radii much larger than the collisionless skin-depth $\delta$, we can produce stable beams with currents that are much larger than the Alvén-Lawson limit. This is illustrated in Figure 6.1 for different values of $(r_b/\delta)$.

## 6.2 Two-Stream Instability

We now consider the possible existence of beam instabilities, when they move in an electro-ion plasma background. For that purpose, we start from the general dispersion relation of electrostatic waves in a beam-plasma system, as given by

$$\epsilon_l(\omega, \mathbf{k}) = 1 + \sum_\alpha \chi_\alpha(\omega, \mathbf{k}) = 0\,, \tag{6.23}$$

where $\chi_\alpha$ are the susceptibilities of the different particle species, $\alpha = (e, i, b)$. For a cold plasma and high-frequency waves, we can neglect the ion response a use the electron susceptibility in its simplest form

$$\chi_e(\omega, \mathbf{k}) = -\frac{\omega_{pe}^2}{\omega^2}\,. \tag{6.24}$$

On the other hand, for a mono-kinetic relativistic electron beam, moving with velocity $\mathbf{v}_b$, we have

$$\chi_b(\omega, \mathbf{k}) = -\frac{\omega_{pb}^2}{(\omega - \mathbf{k}\cdot\mathbf{v}_b)^2}\frac{1}{\gamma_b^3}\,. \tag{6.25}$$

where $\gamma_b = 1/\sqrt{1 - v_b^2/c^2}$. By definition, we have $\omega_{pb}^2/\omega_{pe}^2 = (n_b/n_0)$, where the beam density is assumed small, $n_b \ll n_0$. In this case, the dispersion relation (6.23) can be easily solved, and gives four solutions for the frequency $\omega$. Two of them are close to the plasma frequency $\omega_{pe}$, and are given by

$$\omega = \pm\left[\omega_{pe} + \frac{\omega_{pb}}{2\gamma_b^2}\frac{1}{(\omega_{pe} \mp \mathbf{k}\cdot\mathbf{v}_b)^2}\right]\,, \tag{6.26}$$

The other two are beam modes, with frequencies close to $(\mathbf{k}\cdot\mathbf{v}_b)$, and are determined by

$$\omega = \left[(\mathbf{k}\cdot\mathbf{v}_b) \pm \frac{\omega_{pb}}{\gamma_b^3}\frac{1}{\sqrt{1 - \omega_{pe}^2/(\mathbf{k}\cdot\mathbf{v}_b)^2}}\right]\,, \tag{6.27}$$

This shows that, for $\omega_{pe} > (\mathbf{k}\cdot\mathbf{v}_b)$, one of these beam modes can be unstable, with a growth rate $\Gamma = \Im(\omega)$ given by

$$\Gamma = \frac{\omega_{pb}}{\gamma_b^3}\frac{1}{\sqrt{(\omega_{pe}/(\mathbf{k}\cdot\mathbf{v}_b))^2 - 1}}\,. \tag{6.28}$$

However, this solution is not valid in the limit of $n_b/n_0 \to 0$, because it eventually intersects the solution (6.26). It is therefore useful to examine the *double-resonance condition* associated with this intersection, and defined as

$$\omega_r = \omega_{pe} \simeq (\mathbf{k}\cdot\mathbf{v}_b)\,, \tag{6.29}$$

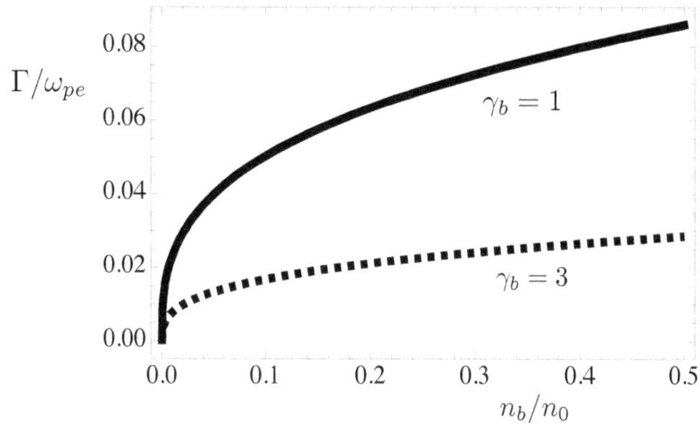


**FIGURE 6.2**
Growth rates of the beam-plasma instability, $\Gamma/\omega_{pe}$, as a function of the relative beam density $n_b/n_0$, for two different values of the beam realativistic factor, $\gamma_b = 3$ (dashed) and $\gamma_b = 1$ (bold).

where we have used $\omega = \omega_r + \epsilon$. Replacing this in eq. (6.26), we get an equation for the small deviation $\epsilon$, with respect to the plasma frequency. We have

$$\epsilon^3 \left[1 - \frac{\omega_{pe}}{(\mathbf{k} \cdot \mathbf{v}_b)}\right] \omega_{pe}\epsilon^2 - \frac{1}{2}\omega_{pe} \frac{\omega_{pb}}{\gamma_b^3} = 0\,. \tag{6.30}$$

Neglecting the second term, we are then reduced to

$$\epsilon = \left\{1, \frac{1}{2}\left(-1 \pm i\sqrt{3}\right)\right\} \left(\frac{1}{2}\omega_{pe} \frac{\omega_{pb}}{\gamma_b^3}\right)^{1/3} . \tag{6.31}$$

This result shows the existence of an unstable mode, satisfying the double condition (6.29), with a growth rate $\Gamma = \Im(\epsilon)$, determined by

$$\Gamma = \sqrt{3}\,\omega_{pe} \left(\frac{n_b}{2^4 n_0}\right)^{1/3} \frac{1}{\gamma_b} . \tag{6.32}$$

We can see that the growth rate of this *beam-plasma instability* is proportional to the cubic root of the beam density, $\Gamma \propto n_b^{1/3}$, which is a characteristic feature of beam instabilities. We also see that relativistic effects tend to stabilize the beam-plasma interaction, due to the factor $\gamma_b$. In Figure 6.2, we represent the instability growth-rates as a function of the beam density.

Let us now extend this simple analysis to include the ions. This corresponds to instabilities with a much lower frequency. To discuss this case, we return to eq. (6.23) and assume that the electrons are in Boltzmann equilibrium in the potential of the low-frequency waves, and the ion response is included but their temperature is neglected. This allows us to write the electron and ion susceptibilities as

$$\chi_e(\omega, \mathbf{k}) \simeq \frac{1}{k^2\lambda_D^2} = \frac{\omega_{pe}^2}{k^2 v_{the}^2}, \quad \chi_i(\omega, \mathbf{k}) \simeq -\frac{\omega_{pi}^2}{\omega^2}, \tag{6.33}$$

where $v_{the} = \sqrt{T_e/m_e}$ is the electron thermal velocity. The dispersion relation becomes

$$1 + \frac{1}{k^2\lambda_D^2} - \frac{\omega_{pi}^2}{\omega^2} - \frac{\omega_{pb}^2}{(\omega - \mathbf{k}\cdot\mathbf{v}_b)^2}\frac{1}{\gamma_b^2} = 0. \tag{6.34}$$

In this case, it is useful to modify the first term of the double-resonance condition (6.29), and assume that

$$\omega_r = \omega_{ac} \simeq (\mathbf{k}\cdot\mathbf{v}_b), \quad \omega_{ac} = \frac{k v_{ac}}{\sqrt{1 + k^2\lambda_D^2}}, \tag{6.35}$$

where $v_{ac} = \sqrt{T_e/m_i}$ is the *ion-acoustic velocity*. Furthermore, the role previously played by the frequency $\omega_{pb}$ is now played by $\omega_{pb}(\omega_{ac}/\omega_{pe})$. The growth rate of the instability associated with this new double-resonance condition is then given by

$$\Gamma = \sqrt{3}\,\omega_{ac}\left(\frac{\omega_{pb}}{2^2\omega_{pi}}\right)^{2/3}. \tag{6.36}$$

As above, this is valid for weak beams, such that $n_b \ll n_0(m_e/m_i)$. A general theorem for charged particle beam stability was proposed by Davidson [88] (see also the book [89]). In the following, we will describe specific cases of instabilities that are associated with beam-plasma insteractions, with relevance to the cases of two-stream instabilities, modulation and filamentation. This problem as been studied by several authors along the years [90, 91].

## 6.3 Beam-Plasma Interactions

The simple analysis of the previous section has to be seen in the frame of a more general theory of beam-plasma interactions, where electromagnetic waves propagating at an arbitrary angle with respect to the beam can eventually become unstable. In order to perform this general analysis, we start from the relativistic Vlasov equation for the electron distribution function, and assume that the ions are immobile. We also restrict our discussion to isotropic plasmas. We then use

$$\left(\frac{\partial}{\partial t} + \mathbf{v}\cdot\nabla\right) f_e = e\,(\mathbf{E} + \mathbf{v}\times\mathbf{B})\cdot\frac{\partial f_e}{\partial \mathbf{p}}, \tag{6.37}$$

where $f_e \equiv f_e(\mathbf{r}, \mathbf{p}, t)$ is the electron distribution function, $\mathbf{v} = \mathbf{p}/(m_e \gamma)$ is the electron velocity and $\gamma = [1 + (p/mc)^2]^{1/2}$ the relativistic factor. The fields $\mathbf{E}$ and $\mathbf{B}$ are described by Maxwell's equations, where the charge and and current densities are self-consistently determined by integrals over $f_e$, as seen before.

As discussed in Chapter 4, for infinitesimal perturbations $(\mathbf{E_k}, \mathbf{B_k}, f_\mathbf{k})$ evolving as $\exp(i\mathbf{k} \cdot \mathbf{r} - i\omega t)$, we get a dispersion relation of the general form

$$\mathcal{D}(\omega, \mathbf{k}) \equiv \left| \mathbf{kk} - k^2 \bar{\bar{1}} + \frac{\omega^2}{c^2} \bar{\bar{\kappa}}(\omega, \mathbf{k}) \right| = 0 \,, \tag{6.38}$$

where the dielectric tensor is

$$\bar{\bar{\kappa}}(\omega, \mathbf{k}) = \bar{\bar{1}} + \frac{e^2}{\epsilon_0 \omega} \int \frac{\mathbf{v}}{(\omega - \mathbf{k} \cdot \mathbf{v})} \frac{\partial f_0}{\partial \mathbf{p}} \cdot \left[ \left( 1 - \frac{\mathbf{k} \cdot \mathbf{v}}{\omega} \right) \bar{\bar{1}} + \frac{\mathbf{kv}}{\omega} \right] d\mathbf{p} \,, \tag{6.39}$$

and $f_0$ is the electron distribution function in equilibrium. The corresponding expression for the tensor elements is

$$\kappa_{ij}(\omega, \mathbf{k}) = \delta_{ij} + \frac{\omega_{pe}^2}{n_0 m_e} \int \left\{ \frac{p_i}{\gamma} \frac{\partial f_0}{\partial p_j} + \frac{p_i p_j}{\gamma} \frac{\mathbf{k} \cdot \partial f_0 / \partial \mathbf{p}}{(m_e \gamma \omega - \mathbf{k} \cdot \mathbf{p})} \right\} d\mathbf{p} \,, \tag{6.40}$$

In order to establish the existence of unstable modes, we need to study this general dispersion relation for electron distributions $f_0$ that can be anisotropic in momentum space. The anisotropy is due, for instance, to the presence of beams, but we still assume a cylindrical symmetry along the $z$-direction. This means that we can use distributions of the form

$$f_0(\mathbf{p}) = f_{0\perp}(\mathbf{p}_\perp) f_{0z}(p_z) \,, \quad p_\perp = \sqrt{p_x^2 + p_y^2} \,. \tag{6.41}$$

with the normalization condition

$$n_0 = \int f_0(\mathbf{p}) d\mathbf{p} = 2\pi \int f_0(p_\perp, p_z) p_\perp dp_\perp dp_z \,. \tag{6.42}$$

We also assume that the unstable modes propagate in the $yz$-plane, with wavevector $\mathbf{k} = k_x \mathbf{e}_x + k_z \mathbf{e}_z$. Replacing this in the determinant of eq. (6.38), we obtain the following expression

$$\left( \frac{\omega^2}{c^2} \kappa_{yy} - k^2 \right) \left[ \left( \frac{\omega^2}{c^2} \kappa_{xx} - k_z^2 \right) \left( \frac{\omega^2}{c^2} \kappa_{zz} - k_x^2 \right) - \left( \frac{\omega^2}{c^2} \kappa_{xz} + k_x k_z \right) \right] = 0 \,. \tag{6.43}$$

The two multiplying factors show the existence of two independent pairs of modes, where the first one satisfies the relation

$$\omega^2 = \frac{k^2 c^2}{\kappa_{yy}} \,, \tag{6.44}$$

and the second is determined by

$$\omega^2 = \left( \frac{k_x^2 \kappa_{xx} + 2k_x k_z \kappa_{xz} + k_z^2 \kappa_{zz}}{\kappa_{xx}\kappa_{zz} - \kappa_{xz}^2} \right) c^2 \,, \tag{6.45}$$

In order to understand their meaning, it is useful to consider the limit of very larger phase velocities, such that $(\omega/k) \gg v_{the}$, where the contribution of the resonances in the denominators of eq. (6.39) is negligible. In this limit, we can define the dielectric tensor by a simple expression

$$\bar{\bar{\kappa}}(\omega, \mathbf{k}) = \left(1 - \frac{\omega_{pe}^2}{\omega^2}\right) \bar{\bar{1}} \,. \tag{6.46}$$

Replacing this in the above two relations, we can reduce them to

$$\omega^2 = \omega_{pe}^2 + k^2 c^2 \,, \tag{6.47}$$

and

$$\left(\omega^2 - \omega_{pe}^2\right)\left(\omega^2 - \omega_{pe}^2 - k^2 c^2\right) = 0 \,, \tag{6.48}$$

where the first represents the electromagnetic mode in isotropic plasmas, and the second represents the product of decoupled electrostatic and electromagnetic waves. Obviously, in this limit, all the modes are stable. It means that, in order to observe instabilities, we need to consider nearly resonant interaction of these modes with the plasma particles, as one should anticipate. In the following, we consider some particular examples of beam-plasma interactions which will eventually destabilize these basic modes.

i) **Cold Beam with Return Current**

We first consider the case of a cold electron beam moving in the positive $z$-direction with density $n_b$, in a neutralizing ion background with density $n_i = n_0$. To this beam, we add a return current, moving in the opposite direction, with density $n_{pl}$. This current is due to the existence of an electric field in the plasma, produced at the beam front, which tends to establish a total current equilibrium. This return current adds to the beam current and contributes to the overall stability of the medium.

The electron population can then be represented by distribution functions of the form

$$f_0(\mathbf{p}) = f_{pl}(\mathbf{p}) + f_b(\mathbf{p}) \,, \tag{6.49}$$

with

$$f_{pl}(\mathbf{p}) = n_{pl}\delta(p_\perp)\delta(p_z + p_{pl}) \,, \quad f_b(\mathbf{p}) = n_b\delta(p_\perp)\delta(p_z - p_b) \,, \tag{6.50}$$

where $p_\alpha = m_e \gamma_\alpha v_\alpha$, for $\alpha = (pl, b)$, and the current neutralization conditions imply that $n_b = n_{pl}(v_{pl}/v_b)$. We should also keep in mind the charge neutralization condition, such that the total electron density equals the ion background density, $n_0 = n_{pl} + n_b$. Replacing this in the dispersion relation (6.43), we obtain

$$\left[\frac{\omega^2}{\omega_{pe}^2} - (1+\nu_b) - \frac{k^2c^2}{\omega_{pe}^2}\cos^2\theta_k\right]\Omega_1^2(\omega,\mathbf{k}) - \Omega_2^2(\omega,\mathbf{k}) = 0\,, \tag{6.51}$$

where we have used $\nu_b = n_b/n_{pl}$, and introduced the auxiliary functions

$$\Omega_1^2(\omega,\mathbf{k}) = \frac{\omega^2}{\omega_{pe}^2} - (1+\nu_b) + \nu_p \sum_{\pm} \frac{kv_b}{\omega_{pe}}\left[\frac{kv_b\, cos2\theta_k \pm 2\omega\cos\theta_k}{(\omega \pm kv_b\cos\theta_k)^2}\right] - \frac{k^2c^2}{\omega_{pe}^2}\sin^2\theta_k\,, \tag{6.52}$$

and

$$\Omega_2^2(\omega,\mathbf{k}) = \left[\frac{k^2c^2}{\omega_{pe}^2}\cos\theta_k + \nu_b \sum_{\pm}\left(\frac{kv_b}{kv_b\cos\theta_k \pm \omega}\right)\right]^2 \sin^2\theta_k\,. \tag{6.53}$$

We have also defined the angle $\theta_k$ between the wavevector $\mathbf{k}$ and the beam direction $\mathbf{e}_z$, and neglected the relativistic factors. In order to understand the meaning of this result, it is useful to discuss the following two limits, $\theta_k = 0$ and $\theta_k = \pi/2$. In the case of modes propagating parallel to the beam direction, $\theta_k = 0$, the above dispersion relation is reduced to

$$\left[\omega^2 - (1+\nu_b)\omega_{pe}^2 - k^2c^2\right]\left[1 - \frac{\nu_b\omega_{pe}^2}{(\omega - kv_b)^2} - \frac{\omega_{pe}^2}{(\omega + kv_b)^2}\right] = 0\,. \tag{6.54}$$

The first factor under square brackets corresponds to stable electromagnetic modes. The second factor is more interesting, and corresponds to what is called a *two-stream instability*. This dispersion relation (or factor) is formally identical to the beam-plasma configuration examined before, but where the electron plasma population now moves with a velocity $(-v_b)$ due to the return current effect. We therefore have to compare the results of the previous section with those given by

$$1 - \frac{\nu_p\omega_{pe}^2}{(\omega - kv_b)^2} - \frac{\omega_{pe}^2}{(\omega + kv_b)^2} = 0\,. \tag{6.55}$$

If we repeat the previous discussions, we arrive at an expression for the maximum growth rate (associated with the double-resonance condition $\omega_r = \omega_{pe} \simeq kv_b$), of the form

$$\Gamma_{(\theta_k=0)} \simeq \sqrt{3}\left(\frac{\nu_b}{2^4}\right)^{1/3}\omega_{pe}\,. \tag{6.56}$$

When we reintroduce the relativistic factors, this growth rate is reduce by a factor $1/\gamma_b$, as we have seen before. Let us now return to the general expression (6.51)–(6.53), and consider modes propagating in a direction perpendicular to the beam, $\theta_k = \pi/2$. In this case, we obtain

$$\left[\omega^2 - (1+\nu_b)\omega_{pe}^2\right]\left\{\omega^4 - \omega^2\left[(1+\nu_b)\omega_{pe}^2 + k^2c^2\right] - \nu_b(1+\nu_b)k^2v_b^2\omega_{pe}^2\right\} = 0\,. \tag{6.57}$$

Here, the first factor corresponds to stable plasma oscillations, while the second factor contains unstable solutions. They can easily be determined in the weak-beam limit, $\nu_b \ll 1$, where we can approximately write the above expression as

$$\omega^4 - \omega^2 \left(\omega_{pe}^2 + k^2 c^2\right) - \nu_b k^2 v_b 2\omega_{pe}^2 = 0 \,. \tag{6.58}$$

Solving for $\omega^2$, we can easily find the corresponding growth rates $\Gamma = \Im(\omega)$, as

$$\Gamma_{(\theta_k=\pi/2)} \simeq (kv_b)\sqrt{\frac{\nu_b \omega_{pe}^2}{\omega_{pe}^2 + k^2c^2}} \,. \tag{6.59}$$

This shows that, for short wavelengths such that $k^2c^2 \gg \omega_{pe}^2$, we are reduced to $\Gamma_{(\theta_k=\pi/2)} \sim (v_b/c)\sqrt{\nu_b}\omega_{pe}$. These are the so-called *filamentation instabilities*, because they tend to divide the initial electron beam into smaller beams (or filaments) with a size of order $1/k$.

## 6.4 Weibel Instability

Let us now apply the same theoretical approach to instabilities that depend on the anisotropy of the distribution function (for instance a temperature anisotropic), rather than on the existence of particle beams. For historical reasons, they are known as *Weibel instabilities* [92, 93]. For that purpose, we consider an electron distribution function of the form

$$f_{pl}(\mathbf{p}) = \frac{n_{pl}}{4p_{the}^2}\left[H(p_\perp + p_{the}) - H(p_\perp - p_{the})\right] \,, \tag{6.60}$$

where we have used Heaviside functions $H(p)$, to establish a limit to the perpendicular particle momentum. This model is usually called the *water-bag model*, because it imposes a rigid limit to the thermal spread, given by $p_{the}$. In this situation, the electromagnetic mode, determined by the first factor in the general dispersion relation (6.43), will become unstable. As we have seen, it is determined by

$$\kappa_{yy} = \frac{k^2c^2}{\omega^2} \,. \tag{6.61}$$

Calculation of the tensor term $\kappa_{yy}$, using the water-bag model, and maintaining the beam distribution $f_b(\mathbf{p})$ as in the previous section, leads to the expression

$$1 - \kappa_{yy} = \left(1 + \nu_b - \cot^2\theta_k\right)\frac{\omega_{pe}^2}{\omega^2} + \frac{(\omega^2 + \nu_b k v_b \cos\theta_k)^2 \cot^2\theta_k + k^2 v_b^2 \rho^2 \sin^2\theta_k}{(\omega^2 + \nu_b k v_b \cos\theta_k)^2 - (k v_b \rho \sin\theta_k)^2}$$
$$+ \Delta \frac{(\omega + k v_b \cos\theta_k)\cot^2\theta_k \csc\theta_k}{\rho k v_b} \,. \tag{6.62}$$

Here, we have used the auxiliary quantities $\Delta$ and $\rho$, defined by

$$\Delta = \ln \left| \frac{\omega + \nu_b k v_b \cos\theta_k - k v_b \rho \sin\theta_k}{\omega + \nu_b k v_b \cos\theta_k + k v_b \rho \sin\theta_k} \right| , \tag{6.63}$$

and

$$\rho = \frac{v_{the}}{v_b} = \frac{p_{the}}{m v_b} . \tag{6.64}$$

The dispersion relation for the electromagnetic modes becomes

$$\omega^2 = (1 + \nu_b)\, \omega_{pe}^2 + k^2 c^2 - \frac{1}{3} f(\theta_k) , \tag{6.65}$$

where

$$f(\theta_k) = \left( \frac{v_{the}^2}{v_b^2} \right) \frac{\omega_{pe}^2}{\left[ (\omega / k v_b) + \nu_b \cos\theta_k \right]^2 - (v_{the}^2 / v_b^2) \sin^2\theta_k} . \tag{6.66}$$

This expression shows that, in the limit of negligible plasma temperature, such that $p_{the} \to 0$, the additional term $f(\theta_k)$ vanishes and the electromagnetic mode is stable, as seen before. However, for a finite temperature, $p_{the} \neq 0$, this term introduces two singularities, defined by

$$\omega = -(k v_b) \nu_b \cos\theta_k \pm (k v_{the}) \sin\theta_k . \tag{6.67}$$

Instability will then occur when the minimum of the function $f(\theta_k)$, located between these two singularities, at around $\omega \simeq -k v_b \nu_b \cos\theta_k$, becomes negative. This leads to the following instability criterion

$$(1 + \nu_b) \omega_{pe}^2 + k^2 c^2 \, < \, (k v_{the})^2 \nu_b^2 \cos^2\theta_k + \frac{1}{3} \frac{\omega_{pe}^2}{\sin^2\theta_k} . \tag{6.68}$$

This shows that unstable modes will certainly exist for $\theta_k \to 0$. In this limit, all wavenumbers will be unstable. For a finite value of $\theta_k$, however, only modes satisfying

$$kc > \sqrt{\frac{1}{3 \sin^2\theta_k} - 1} , \tag{6.69}$$

are unstable. The corresponding growth rates can easily be found if we assume a weak beam, $\nu_b \ll 1$, and are

$$\Gamma_{\theta_k} \simeq \frac{\omega_{pe}}{\sqrt{3}} \frac{k v_{the}}{\sqrt{\omega_{pe}^2 + k^2 c^2}} \sqrt{1 - 3 \sin^2\theta_k} , \tag{6.70}$$

This is illustrated in Figure 6.3.

Let us now consider another mode, which is associated to the second factor in the general expression (6.43). For practical reasons, we focus on the

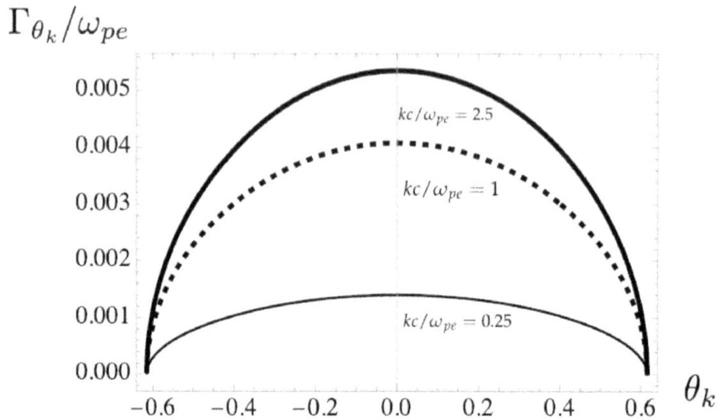


**FIGURE 6.3**
Weibel instability: growth rate $\Gamma_{\theta_k}/\omega_{pe}$ as a function of the angle with respect to the electron beam, $\theta_k$, and of the normalized wavenumber, $kc/\omega_{pe}$, in the water-bag model.

particular case of $\theta_k = \pi/2$, which corresponds to *filamentation*. From that dispersion relation, we obtain [91]

$$\left\{\omega^2 - \left[(1+\nu_b)\omega_{pe}^2 + k^2c^2\right] - \nu_b\frac{k^2v_b^2}{\omega^2}\omega_{pe}^2 - \nu_b^2\frac{k^2v_b^2}{(\omega^2 - k^2v_{the}^2)}\right\} \times$$
$$\left\{\omega^2 - \omega_{pe}^2\left[(1+\nu_b) + \frac{k^2v_{the}^2}{(\omega^2 - k^2v_{the}^2)}\right] + \nu_b^2\omega_{pe}^2\left[\frac{\omega(kv_b)}{(\omega^2 - k^2v_{the}^2)} - \frac{kv_b}{\omega}\right]^2\right\} = 0. \quad (6.71)$$

This dispersion relation has unstable solutions for $\Re(\omega) = 0$, which correspond to *purely growing modes*. Using $\omega = i\Gamma$, we are able to obtain the growth rate of these non-propagating perturbations, as

$$\Gamma_{\pi/2} = kv_b\sqrt{\nu_b}\sqrt{\frac{\omega_{pe}^2 - k^2c^2(v_b^2/v_{the}^2)}{(\omega_{pe}^2 + k^2c^2)(1 + k^2v_{the}^2/\omega_{pe}^2)}}. \quad (6.72)$$

In the cold plasma limit, $v_{the} \to 0$, this expression reduces to

$$\Gamma_{\pi/2} = kv_b\sqrt{\frac{\nu_b\omega_{pe}^2}{(\omega_{pe}^2 + k^2c^2)}}, \quad (6.73)$$

as previously derived.

## 6.5 Photon-Beam Instabilities

Let us now consider a similar kind of instabilities, but where the particle beams are replaced by photon beams, for instance a laser beam, propagation in a plasma. Here we consider the case of an electron-positron plasma, which is particularly important in astrophysics [94]. Instead of particle beam equations, we now use the wave equation for the laser field amplitude. In isotropic pair plasmas, this equation can be written in the following form

$$\left(\nabla^2 - \frac{1}{c^2}\frac{\partial^2}{\partial t^2}\right)\mathbf{a} = \frac{2\omega_p^2}{\gamma c^2}\mathbf{a}\,, \tag{6.74}$$

where $\mathbf{a} = e\mathbf{A}/mc$ is the normalized field amplitude, $\omega_p$ is the electron plasma frequency and $\gamma = \sqrt{1+a^2}$ is the relativistic gamma factor. The factor of 2 on the right is a signature of pair plasmas. Assuming the same equilibrium density for electrons and positrons, $n_{e0} = n_{p0} = n_0$, we can derive the dispersion relation

$$\omega^2 = k^2c^2 + \frac{2\omega_p^2}{\gamma^2}\,, \tag{6.75}$$

where $\omega_p^2 = e^2 n_0/\epsilon_0 m$. This is valid for circular polarization, such that $\gamma$ is constant. In the presence of density perturbations, $\tilde{n}_{e,p}(\mathbf{r},t) = n_{e,p}(\mathbf{r},t) - n_0$, we have to replace $2\omega_p^2$ in eq. (6.74) by

$$2\omega_p^2(\mathbf{r},t) = \frac{\omega_{p0}^2}{n_0}\left[n_e(\mathbf{r},t) + n_p(\mathbf{r},t)\right]\,. \tag{6.76}$$

It is now useful to introduce the concept of *photon quasi-probability* $W_{\mathbf{k}}$, sometimes called *Wigner function*, given by the field correlation function

$$W_k = \int \mathbf{a}(\mathbf{r}+\mathbf{s}/2,t)\cdot\mathbf{a}^*(\mathbf{r}-\mathbf{s}/2,t)e^{-i\mathbf{k}\cdot\mathbf{s}}\,d\mathbf{s}\,. \tag{6.77}$$

Using a procedure outlined in Chapter 11, it is possible to derive from eq. (6.74) an evolution equation for this new quantity, which is formally analogous to the kinetic equation for particles, as

$$\left(\frac{\partial}{\partial t} + \mathbf{v}_k\cdot\nabla + \mathbf{F}_k\cdot\frac{\partial}{\partial\mathbf{k}}\right)W_k = 0\,. \tag{6.78}$$

Here, the quantity $\mathbf{F}_k$ is the force acting on the photons, as defined by

$$\mathbf{F}_k = \nabla\frac{\omega_{p0}^2}{\gamma\omega_k n_0}\left[n_e(\mathbf{r},) + n_p(\mathbf{r})\right]\,. \tag{6.79}$$

The group velocity $\mathbf{v}_k = \partial\omega/\partial\mathbf{k} = c^2\mathbf{k}/\omega$ is determined by the dispersion relation (6.75), and represents the velocity of the photons, which can now be seen as quasi-particles.

The above equation describing the photon field, has to be coupled with the electron and positron fluid equations. Dropping the subscripts $(e, p)$ for the particle species, and using the symbol $s$ to define the particle charge, $s = -1$ for electrons and $s = 1$ for positrons, we can write these equations as

$$\frac{\partial n}{\partial t} + \nabla \cdot (n\mathbf{v}) = 0 \,, \tag{6.80}$$

and

$$\frac{\partial \mathbf{u}}{\partial t} = -s\left(\frac{\partial \mathbf{a}}{\partial t} + c\nabla\Phi\right) - \frac{c}{\gamma}\left[\mathbf{u}\cdot\nabla\mathbf{u} - s\mathbf{u}\times(\nabla\times\mathbf{a})\right] - \frac{\nabla P}{nmc} \,. \tag{6.81}$$

Here, we have also used the covariant velocity $\mathbf{u}$, and the scalar potential $\Phi$, defined as

$$\mathbf{u} = \gamma\frac{\mathbf{v}}{c} \,, \quad \Phi = \frac{eV}{mc^2} \tag{6.82}$$

In the Coulomb gauge, the potential is determined by Poisson's equation

$$\nabla^2\Phi = \frac{e^2}{\epsilon_0 mc^2}(n_e - n_p) \,. \tag{6.83}$$

It is now convenient to introduce the fast and slow parts of the covariant velocity, defined by a time average, such that

$$\mathbf{u} = \mathbf{u}_s + \mathbf{u}_f \,, \quad \mathbf{u}_s \equiv \langle \mathbf{u} \rangle = \frac{\omega}{2\pi}\int_0^{2\pi/\omega} \mathbf{u}(t)dt \,. \tag{6.84}$$

where $\omega$ is the characteristic frequency of the photon spectrum. From eq. (6.81), we can easily see that the fast part is simply given by $\mathbf{u}_f = -s\mathbf{a}$. As for the slow velocity, $\mathbf{u}_s$ it needs to be determined by

$$\frac{\partial \mathbf{u}_s}{\partial t} + \frac{c}{\gamma}(\mathbf{u}_s \cdot \nabla\mathbf{u}_s) = -cs\nabla\Phi - \frac{c}{2\gamma}\nabla a^2 - \frac{\nabla P}{nmc} \,. \tag{6.85}$$

For perturbations evolving with frequencies much slower than $\omega_p/\sqrt{\gamma}$, the total time derivative on the l.h.s. of this equation is negligible. Using the simple equation of state $P = nT$, and assuming a Boltzmann equilibrium in the potential $\Phi$, we get

$$\ln n = -\frac{mc^2}{T}\left(\frac{a^2}{2\gamma} + s\Phi\right) \,. \tag{6.86}$$

Defining $a^2 = a_0^2 + \delta a^2$, where $a_0$ corresponds to the equilibrium amplitude of the photon beam, we can write

$$n_e = N\exp\left(-\mu\frac{\delta a^2}{2\gamma} - \frac{eV}{T}\right) \,, \quad n_p = N\exp\left(-\mu\frac{\delta a^2}{2\gamma} - \frac{eV}{T}\right) \,, \tag{6.87}$$

where we have introduced the parameters

$$\mu = \frac{mc^2}{T} \equiv \frac{c^2}{v_{th}^2}, \quad N = n_0 \exp\left(-\mu \frac{a_0^2}{2\gamma_0}\right). \tag{6.88}$$

For small perturbations, we can expand the exponentials, and use $n_{e,p} = n_0 + \tilde{n}_{e,p}$, such that

$$\tilde{n}_e + \tilde{n}_p = -N\mu \frac{\delta a^2}{\gamma_0}. \tag{6.89}$$

Notice that this can be written in terms of the perturbed quasi-probability $\tilde{W}_k = W_k - W_{0k}$, where $W_{0k}$ defines the photon beam equilibrium. Noting that $\delta a^2$ depends on the entire beam spectrum, we get

$$\tilde{n}_e + \tilde{n}_p = -\frac{\mu N}{\gamma_0} \int \tilde{W}_k \frac{d\mathbf{k}}{(2\pi)^3}. \tag{6.90}$$

We now assume perturbations of the form $\exp(i\mathbf{q} \cdot \mathbf{r} - i\Omega t)$, corresponding to low-frequency modulations of the beam. Linearizing the photon beam equation (6.78) around equilibrium, we get the perturbation $\tilde{W}_k(\Omega, \mathbf{q})$ in terms of the density perturbations $n_{e,p}$. Using (6.90), we can then obtain a dispersion relation of the form

$$1 - \frac{\mu}{\gamma_0^2}\omega_{p0}^2 \exp\left(-\beta \frac{a_0^2}{2\gamma_0}\right) \int \frac{(\mathbf{q} \cdot \mathbf{v}_k) W_{0k}/\omega_k}{(\Omega - \mathbf{q} \cdot \mathbf{v}_k)^2 - \delta_k^2} \frac{d\mathbf{k}}{(2\pi)^3} = 0, \tag{6.91}$$

where $W_{0k}$ defines the equilibrium photon beam distribution. We also have defined $\delta_k = c^2 q^2 / 2\omega_k$. This expression is valid for large-scale modulations of the photon beam, such that $q \ll k$. It determines the dispersion and stability properties of low-frequency modulations of an intense photon beam in a relativistic pair plasma. It is useful to consider the case of a photon beam centered around some wavevector $\mathbf{k}_0$, as described by $W_{0k} = (2\pi)^3 a_0^2 \delta(\mathbf{k} - \mathbf{k}_0)$. Replacing this in eq. (6.91), we obtain

$$(\Omega - \mathbf{q} \cdot \mathbf{v}_0)^2 = \delta_0^2 + f(a_0)(\mathbf{q} \cdot \mathbf{v}_0), \tag{6.92}$$

Here, we have used $\mathbf{v}_0 \equiv \mathbf{v}_k(\mathbf{k} = \mathbf{k}_0)$, and $\delta_0 \equiv \delta_k(\mathbf{k} = \mathbf{k}_0)$. In this expression, we have also used the auxiliary function

$$f(a_0) = \frac{\mu a_0^2}{\omega_0 \gamma_0^2} \omega_{p0}^2 \exp\left(-\mu \frac{a_0^2}{2\gamma_0}\right). \tag{6.93}$$

It is obvious from eq. (6.92) that the beam is unstable for all wavevectors $\mathbf{q}$ that are anti-parallel to the beam propagation. Or, in other words, when $(\mathbf{q} \cdot \mathbf{v}_0) = -|\mathbf{q} \cdot \mathbf{v}_0|$. Let us also assume the quasi-resonant condition $\Omega = (\mathbf{q} \cdot \mathbf{v}_0) + \epsilon$. Replacing this in eq. (6.92), we obtain

$$\epsilon^2 = \delta_0^2 - f(a_0)\, |\mathbf{q} \cdot \mathbf{v}_0|. \tag{6.94}$$

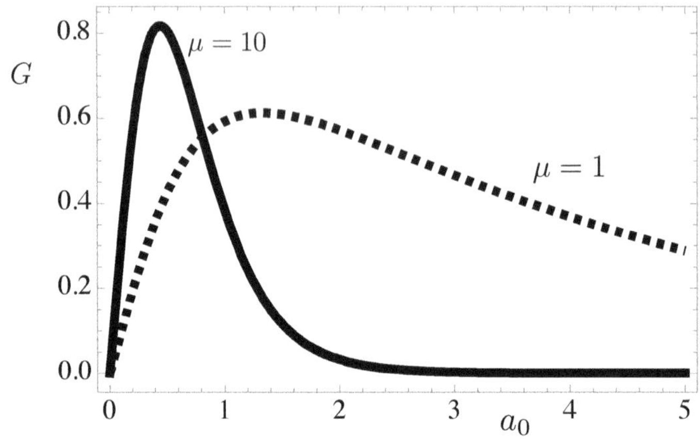


**FIGURE 6.4**
Photon-Beam Instability: growth rates in a relativistic pair plasma. We represent the normalized growth rate $G = (\Gamma/\omega_{p0})(\omega_0/|\mathbf{q}\cdot\mathbf{v}_0|)^{1/2}$, as a function of the photon beam amplitude $a_0$, for $\delta_0 = 0$, with $\mu = 10$ (bold) and $\mu = 1$ (dashed).

Instability occurs if the negative term dominates. Using $\epsilon = i\Gamma$, we can then write the instability growth rate as

$$\Gamma = \sqrt{f(a_0)\,|\mathbf{q}\cdot\mathbf{v}_0| - \delta_0^2}\,. \tag{6.95}$$

A necessary condition for instability is that the field amplitude is larger than some threshold value, $a_0 > a_{thr}$, where the threshold is determined by the equality $f(a_{thr}) = \delta_0^2/|\mathbf{q}\cdot\mathbf{v}_0|$. It can also be seen that, in the classical limit, when $\delta_0^2 \to 0$, this threshold vanishes, $a_{thr} \to 0$. In this limit, the growth rate becomes

$$\Gamma = \omega_{p0}\frac{\mu a_0}{\gamma_0}\left(\frac{|\mathbf{q}\cdot\mathbf{v}_0|}{\omega_0}\right)^{1/2}\exp\left(-\mu\frac{a_0^2}{4\gamma_0}\right)\,. \tag{6.96}$$

These results are illustrated in Figure 6.4, where the growth rates are given as a function of the beam amplitude $a_0$, for two different values of the plasma temperature.

However, the Boltzmann equilibrium defined by eq. (6.86) is no longer valid for high frequencies, when $\Omega$ becomes comparable with the relativistic plasma frequency $\omega_{p0}/\sqrt{\gamma_0}$. In this case, the density perturbations $\mathbf{u}_s$, have to be determined by the continuity equation

$$\frac{\partial n}{\partial t} + n_0 c\nabla\cdot\left(\frac{1}{\gamma}\frac{\partial \mathbf{u}_s}{\partial t}\right) = 0\,. \tag{6.97}$$

It can then be shown that modulational instabilities are still possible, but in co-propagation, when $(\mathbf{q}\cdot\mathbf{v}_0)$ becomes positive [95].

# 7

# *Wakefields*

The wakefield concept is central to the understanding of intense field interactions and of plasma acceleration processes.

In recent years, there has been an increasing interest on relativistic electron plasma waves and on their use for plasma accelerators. It is well known that electron plasma waves can propagate with relativistic phase velocities. In this regime they are decoupled from the thermal electrons of the plasma, but can strongly interact with the fast electrons and photons. In particular, they can be excited by intense laser pulses. The electrostatic fields associated with these electron plasma waves are usually called *laser wakefields*, because they follow the laser pulse along its trajectory across the plasma, with a phase velocity that equals the laser group velocity. With the advent of intense laser systems, based on the CPA concept [96], the study of laser wakefield excitation became very popular, in the context of new accelerator research.

Similarly, relativistic electron plasma waves can also be excited by beams of energetic charged particles (electrons, positrons or protons), and they are sometimes called *plasma wakefields*. This could provide a means to transfer part of the energy of these beams into higher energetic particle populations that could also be used in accelerators. Record electron acceleration of 40 $GeV$ has been obtained at SLAC in 2014, using a plasma column of 1 meter [97].

These two kinds of wakefields, *plasma wakefields* created by charged particle beams, and *laser wakefields* generated by intense laser pulses, share similar features and can be used as a basis for future plasma accelerators. The present chapter describes the theoretical models for different types of plasma and laser wakefields.

## 7.1 Plasma Wakefield

Interaction of energetic particle beams with a background plasma can occur in accelerator experiments, or in intense laser target interactions, where intense electron beams can be formed, as well as in astrophysics. Two distinctive aspects of this interaction are, first, the excitation of relativistic wakefields, and second, the acceleration of a fraction of the interacting particles to much higher energies.

DOI: 10.1201/9781003514701-7 

Let us consider a beam of energetic particles, passing through a plasma. If sufficiently short, the beam can create a wakefield, with large electrostatic field amplitude, moving with the beam, that can eventually be used to accelerate the particles. The wakefields can be seen as relativistic plasma perturbations that are resonantly excited by the particle beams. Resonant excitation occurs when the phase velocity of the wakefield oscillations is equal to the mean velocity of the particle beam.

We first discuss a *one-dimensional model*, where the particles of the beam, with charge $q_b$ and density $n_b$, propagate along the $z$-axis with velocity $v_b$. The electron mean velocity of the background plasma, $v$ is determined by the following relativistic wakefield equation [98]

$$\frac{\partial^2}{\partial \tau^2}[\gamma(\beta_b \beta - 1)] = -\omega_p^2 \left( \frac{\beta}{\beta_b - \beta} + s\frac{n_b}{n} \right) \beta_b^2 \tag{7.1}$$

where $\beta = v/c, \beta_b = v_b/c$ are the normalized plasma and beam velocities, $\gamma = (1 - \beta^2)^{-1/2}$ is the plasma relativistic factor, $s = q_b/e$ is the sign of the driving beam charge, with $s = -1$ for electrons and $s = 1$ for positrons. The time variable is defined as $\tau = t - z/v_b$, and the electron plasma density is determined by

$$n = n_0 \beta_b / (\beta_b - \beta) \, , \tag{7.2}$$

where $n_0$ is its equilibrium density. These two equations can be derived from the electron fluid equations for a cold plasma,

$$\frac{\partial n}{\partial t} + \nabla \cdot (n\mathbf{v}) = 0 \, , \quad \frac{\partial \mathbf{p}}{\partial t} + \mathbf{v} \cdot \nabla \mathbf{p} = -e\mathbf{E} \, , \tag{7.3}$$

where the electric field $\mathbf{E}$ is assumed as purely electrostatic, and determined by the Poisson's equation

$$\nabla \cdot \mathbf{E} = \frac{e}{\epsilon_0}(n_0 - n + sn_b) \, . \tag{7.4}$$

We now assume the one-dimensional case, where all vectorial variables are in the $z$-direction, and introduce the time variable $\tau = t - z/v_b$. The continuity equation can then be rewritten as

$$\frac{\partial n}{\partial \tau} = \frac{1}{v_b}\frac{\partial}{\partial \tau}(nv) \, , \quad n\left(1 - \frac{v}{v_b}\right) = n_0 \, , \tag{7.5}$$

where $n_0$ is a constant of integration defining the equilibrium density. This immediately leads to eq. (7.2). On the other hand, the momentum equation can be written in the one-dimensional version as

$$\left(1 - \frac{v}{v_b}\right) \frac{\partial p}{\partial \tau} = -eE \, , \tag{7.6}$$

which couples with the Poisson's equation

$$\frac{\partial E}{\partial \tau} = -\frac{e}{\epsilon_0}(n_0 - n + sn_b) v_b \, . \tag{7.7}$$

Taking the time derivative of eq. (7.6) and using (7.7), we then get

$$\frac{\partial}{\partial \tau}(\beta - \beta_b)\frac{\partial}{\partial \tau}(\beta\gamma) = -\omega_p^2 \left(1 - \frac{\beta_b}{\beta_b - \beta} + s\frac{n_b}{n_0}\right)\beta_b^2 , \tag{7.8}$$

where we have used $\omega_p^2 = e^2 n_0/\epsilon_0 m$. It should be noticed that

$$\frac{\partial}{\partial \tau}(\beta\gamma) \equiv \frac{\partial}{\partial \tau}\sqrt{\gamma^2 - 1} = \frac{1}{\beta}\frac{\partial \gamma}{\partial \tau} . \tag{7.9}$$

This implies that

$$\frac{\partial}{\partial \tau}(\beta - \beta_b)\frac{\partial}{\partial \tau}(\beta\gamma) = -\frac{\partial^2}{\partial \tau^2}\left[\gamma\left(\beta\beta_b -\right)\right] . \tag{7.10}$$

This then allows to rewrite eq. (7.8) in the form of our starting equation (7.1). Of course, in this one-dimensional model, the velocity, the momentum and the electric field are all purely longitudinal, and the magnetic field is assumed zero, $\mathbf{B} = 0$.

Before considering the fully relativistic solutions of eq. (7.1), it is useful to discuss the linear regime. For this purpose, we assume that $\beta \ll 1$ and $\gamma \simeq 1$. We also take $\beta_b \simeq 1$, to account for a relativistic beam. With this simplifying assumptions, we get a driven harmonic oscillator of the form

$$\left(\frac{\partial^2}{\partial \tau^2} + \omega_p^2\right)\beta = -s\frac{n_b}{n_0}\omega_p^2 , \quad \frac{n}{n_0} = \frac{1}{1 - \beta} , \tag{7.11}$$

where $n_b \equiv n_b(\tau)$ describes the shape of the driving beam. Introducing a new variable $\eta = \omega_p \tau$, we can also write this equation as

$$\left(\frac{\partial^2}{\partial \eta^2} + 1\right)\beta = s\frac{n_b}{n_0} . \tag{7.12}$$

An appropriate solution that satisfies causality, is

$$\beta(\eta) = s\int_{-\infty}^{\eta} f(\eta')\sin(\eta - \eta')d\eta' . \tag{7.13}$$

This is illustrated in Figure 7.1, where $n(\eta)$ is represented for $s = -1$ (electrons) and $s = 1$ (positrons), for a Gaussian driving field defined by

$$\frac{n_b}{n_0} = 0.3\exp\left(-\frac{\eta^2}{\delta}\right) . \tag{7.14}$$

for large beams, $\delta = 10$, and for short beams, $\delta = 1$.

This simple representation of a wakefield reveals two important features (see Figure 7.1). First, the importance of the beam width, and second the beam charge dependence. As for the beam width, we can see that the density wakes are larger for short beams, with a duration comparable to $1/\omega_p$, and

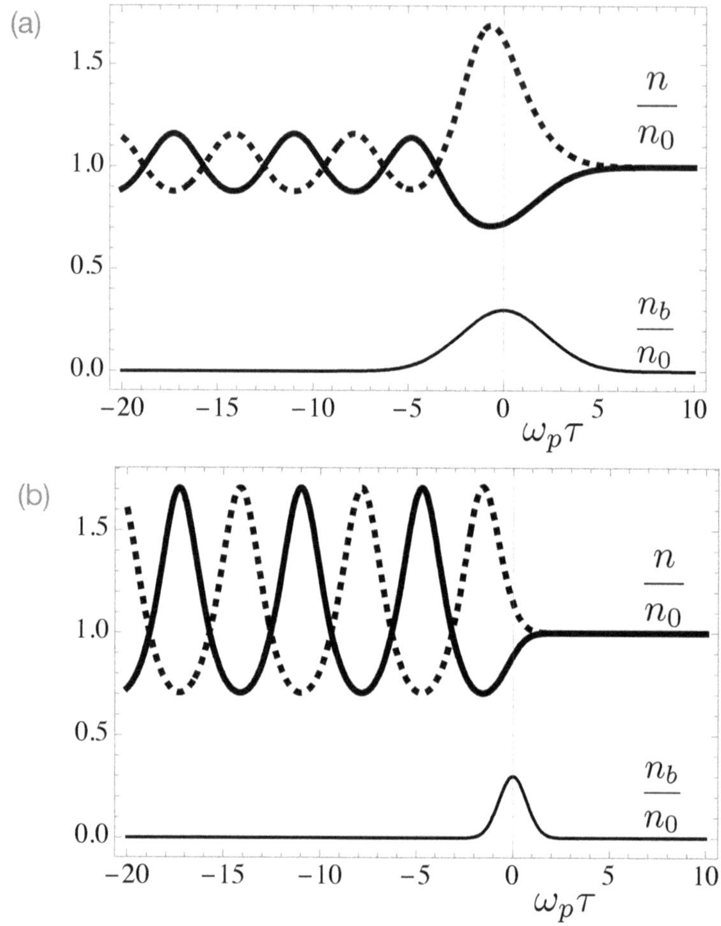


**FIGURE 7.1**
Plasma wakefields driven by an electron beam, as described by the one-dimensional model, for two different driving beams. The electron beam profiles are Gaussian, for long beams, (a) $\delta = 10$, and for short beams, (b) $\delta = 1$. The case of a positron beam is represented by the dashed curves. The value of $(n_b/n_0)_{max} = 0.3$ is assumed.

considerably drop for larger beams. In what concerns the beam charge, the effects seem to be symmetric, with the wakefield caused by a positron beam oscillating in phase opposition to the electron beam wake. However, this one-dimensional symmetry is eventually broken in three dimensions.

Let us return to eq. (7.1) and look for a more accurate solution, when the linear oscillator of eq. (7.11) is replaced by a nonlinear oscillator. For this purpose, we introduce an auxiliary function $x(\tau)$, defined by

$$x(\tau) = \sqrt{\frac{1-\beta}{1+\beta}} = \gamma(1-\beta)\,. \tag{7.15}$$

Using $\beta_b \simeq 1$, we can easily derive from eq. (7.1) the following nonlinear equation

$$\frac{\partial^2 x}{\partial \tau^2} = \omega_p^2 \left( \frac{\beta}{1-\beta} + s\frac{n_b}{n_0} \right)\,, \tag{7.16}$$

Noting that

$$\frac{2\beta}{1-\beta} = \frac{1}{x^2} - 1\,, \tag{7.17}$$

we can rewrite this equation in the form

$$\frac{d^2 x}{d\eta^2} = \frac{1}{2}\left( \frac{1}{x^2} - 1 + 2s\frac{n_b}{n_0} \right)\,. \tag{7.18}$$

In order to find analytical solutions for this nonlinear equation, it is useful to consider a driver beam with a flat profile, starting at $\eta = 0$, with a length $\ell_b$, This means that $\beta = 0$ for $\eta < 0$, and assume that $x(0) = 1$ and $x'(0) = 0$. Integration of eq. (7.18) then gives

$$\left( \frac{dx}{d\eta} \right)^2 = f(\eta)\,. \tag{7.19}$$

In order to determine the function $f(\eta)$, we take the derivative of this equation, as

$$2\frac{dx}{d\eta}\frac{d^2 x}{d\eta^2} = \frac{df}{d\eta}\,. \tag{7.20}$$

On the other hand, using eq. (7.18), we obtain

$$\frac{dx}{d\eta}\frac{d^2 x}{d\eta^2} = \frac{dx}{d\eta}\left( \frac{1}{x^2} - 1 + 2\alpha \right)\,. \tag{7.21}$$

where we have defined $\alpha = sn_b/n_0$. Comparing these two equations we are then lead to the conclusion that

$$f(\eta) = -\frac{1}{x} - (1-2\alpha)x + C\,, \tag{7.22}$$

where $C$ is a constant. In order to determine this constant, we use the boundary conditions $f(0) = 0$ and $x(0) = 1$. This leads to $C = 1 + (1-2\alpha)$. Replacing in eq. (7.19) we finally get

$$\left( \frac{dx}{d\eta} \right)^2 = 2(1-\alpha) - \frac{1}{x} - (1-2\alpha)x\,. \tag{7.23}$$

This equation allows us to describe the wakefield quantities in terms of the function $x(\eta)$, as

$$\beta = \frac{1-x^2}{1+x^2}, \quad \frac{n}{n_0} = \frac{1}{2}\left(1 + \frac{1}{x^2}\right). \tag{7.24}$$

We can also formally integrate eq. (7.23), using

$$\eta(x) = \frac{1}{\sqrt{(1-2\alpha)}} \int_1^x \left\{ \frac{x}{[(x-1)(1-2\alpha)^{-1} - x]} \right\}^{1/2} dx, \tag{7.25}$$

This can be written in terms of the *incomplete elliptical integral* of the second kind, $E(\Psi, k)$, as

$$\eta(x) = 2(1-2\alpha)^{-1/2} E(\Psi, k), \tag{7.26}$$

where we define the angle $\Psi$ and the parameter $k$, according to

$$\sin\Psi = [(x-1)(1-2\alpha)]^{1/2}, \quad k^2 = 1 - 2\alpha. \tag{7.27}$$

Based on this formal solution, and returning to the variable $\tau = \eta/\omega_p$, we can then determine the frequency of the nonlinear plasma oscillations, as

$$\omega = \omega_p \frac{\pi(1-2\alpha)^{1/2}}{2E(k)}, \tag{7.28}$$

where $E(k) \equiv E(\pi/2, k)$ is the *complete elliptical integral* of the second kind. This is illustrated in Figure 7.2. We can see that, for a low density electron beam, with $\alpha \ll 1$, we have $\omega \sim \omega_p$. However, increasing the beam density, the frequency $\omega$ decreases and tends to zero when $\alpha \to 1/2$. At the same time, the density oscillations become more nonlinear, showing the occurrence of periodic spikes. For larger beam densities, such that $\alpha > 1/2$ wakefield oscillations are suppressed.

## 7.2 Laser Wakefield

Similar wakefields can be excited when we replace the relativistic particle beams by laser beams. This similarity can physically be understood if we use the concept of *photon effective charge* [99], noting that a laser pulse is a beam of photons, and that photons behave in many respects as charged quasi-particles in a plasma. Now, the problem is electromagnetic and we need to involve both the scalar and the vector potentials, $V$ and $\mathbf{A}$. We use their normalized form as

$$\phi = \frac{eV}{mc}, \quad \mathbf{a} = \frac{e\mathbf{A}}{mc}, \tag{7.29}$$

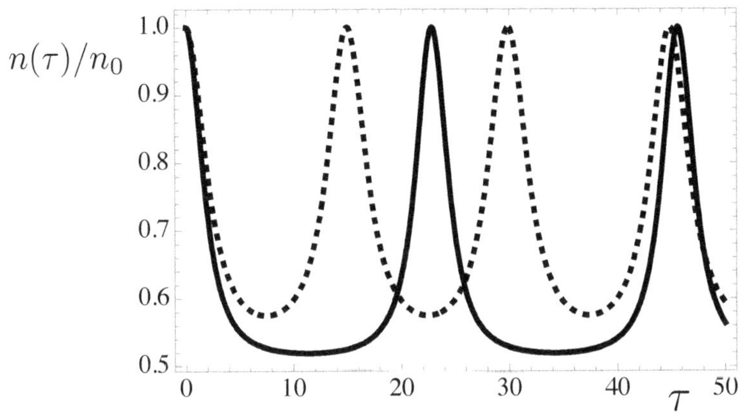


**FIGURE 7.2**
Density perturbations driven by an electron beam, $n(\tau)/n_0$, as described by eq. (7.16), for a weak beam with $n_b/n_0 = 0.1$ (dashed curves) and a strong beam with $n_b/n_0 = 0.4$ (bold curves).

In the Coulomb gauge, $\phi$ represents slow processes associated with the wakefield formation, and $\mathbf{a}$ the fast processes associated with the incident laser beam. We consider laser pulse propagation along the $z$-direction. The potential equations are then given by

$$\frac{\partial^2 \phi}{\partial z^2} = -\frac{e\rho}{\epsilon_0 mc^2}, \quad \rho = -e(n - n_0), \tag{7.30}$$

and

$$\left(\frac{\partial^2}{\partial z^2} - \frac{1}{c^2}\frac{\partial^2}{\partial t^2}\right)\mathbf{a} = -\frac{\mu_0 e}{mc}\mathbf{J}_t, \quad \mathbf{J}_t = -\frac{en}{\gamma}\mathbf{u}, \tag{7.31}$$

where $\mathbf{J}_t$ is the transverse (or divergence-free) current, and $\mathbf{u} = \gamma\mathbf{v}$, is the electron covariant velocity, with $\gamma = \sqrt{1 + u^2/c^2}$. We neglect the ions and use the relativistic electron fluid equations

$$\left(\frac{\partial}{\partial t} + \frac{\partial}{\partial z}\right)\frac{n\mathbf{u}}{\gamma} = 0, \tag{7.32}$$

and

$$\frac{\partial \mathbf{u}}{\partial t} + \frac{u_z}{\gamma}\frac{\partial \mathbf{u}}{\partial z} = \frac{\partial \mathbf{a}}{\partial t} + \frac{\partial \phi}{\partial z} + \frac{\mathbf{u}}{\gamma} \times \frac{\partial}{\partial z}(\mathbf{e}_z \times \mathbf{a}). \tag{7.33}$$

We now assume that the incident laser beam is nearly monochromatic, as described by

$$\mathbf{a} = \mathbf{a}_0 \exp(ik_0 z - i\omega_0 t), \quad \omega_0^2 = \frac{\omega_p^2}{\gamma_0} + k^2 c^2, \tag{7.34}$$

where $\gamma_0 = \sqrt{1+|a_0|^2}$, the laser is well above cutoff, $\omega_0^2 \gg \omega_p^2/\gamma$, and the field amplitude $\mathbf{a}_0$ evolves on a time-scale much slower that $1/\omega_0$. Taking a time average over the fast time scale $1/\omega_0$, and defining $< n >= n_0 + n_s$, $< \mathbf{u} >= \mathbf{u}_s$, $< \phi >= \phi_s$ and $< \mathbf{a} >= \mathbf{a}_s$ we get approximate evolution equations for the slowly varying quantities, as

$$\frac{\partial n_s}{\partial t} + n_0 \frac{\partial}{\partial z}\frac{n\mathbf{u}_s}{\gamma_0} = 0\,, \tag{7.35}$$

and

$$\frac{\partial \mathbf{u}_s}{\partial t} = \frac{\partial \mathbf{a}_s}{\partial t} + \frac{\partial \phi_s}{\partial z} - \frac{1}{2\gamma_s}\frac{\partial}{\partial z}|\mathbf{a}_s|^2\,. \tag{7.36}$$

As a first-order description of the wakefield process, we assume that the slow processes are purely electrostatic, and therefore take $\mathbf{a}_s \simeq 0$. In this case, we can reduce the above two equations to a single equation for the slow density perturbations, as

$$\left(\frac{\partial^2}{\partial t^2} + \frac{\omega_p^2}{\gamma_0}\right) n_s = \frac{c^2 n_0}{2\gamma_0^2}\frac{\partial^2}{\partial z^2}|\mathbf{a}_0|^2\,. \tag{7.37}$$

We also assume that the laser pulse envelope propagates with negligible deformation across the medium. This allows us to assume that $\mathbf{a}_0(z,t) \equiv \mathbf{a}_0(\xi)$, with $\xi = z - v_g t$ and $v_g \simeq c$. This suggests a coordinate transformation $(z,t) \to (\xi, \tau = t)$, such that in the co-moving coordinates the above equation becomes

$$\left(\frac{\partial^2}{\partial \tau^2} + \frac{\omega_p^2}{\gamma_0} + v_g^2\frac{\partial 2}{\partial \xi^2} - 2v_g\frac{\partial 2}{\partial \tau \partial \xi}\right) n_s = \frac{c^2}{2\gamma_0^2} n_0 \frac{\partial^2}{\partial \xi^2}|\mathbf{a}_0|^2\,. \tag{7.38}$$

Introducing the quasi-static approximation, $\partial/\partial\tau \simeq 0$, and the normalized space variable

$$\eta = k_p \xi\,, \quad k_p = \frac{\omega_0}{v_g\sqrt{\gamma_0}}\,, \tag{7.39}$$

we finally get the wakefield equation in the form

$$\left(\frac{\partial^2}{\partial \eta^2} + 1\right) n_s = f(\eta)\,, \quad f(\eta) = \frac{n_0}{2\gamma_0^2}\frac{\partial^2}{\partial \xi^2}|\mathbf{a}_0|^2\,, \tag{7.40}$$

An appropriate solution for this equation can be written as

$$n_s(\eta) = \int_\infty^\eta f(\eta')\sin(\eta - \eta')d\eta'\,. \tag{7.41}$$

This is formally identical to the linear solution (7.13), defined before for plasma wakefields, apart from a trivial change in the propagation direction. Integrating by parts, we get

$$n_s(\eta) = \frac{n_0}{2\gamma_0^2}k_p^2 \int_\infty^\eta \cos(\eta - \eta')\frac{\partial |\mathbf{a}_0|^2}{\partial \eta'}d\eta'\,. \tag{7.42}$$

Let us first consider the case of short laser pulses, such that the duration of the driving laser pulse is much shorter than the period of the plasma oscillation, $\Delta t \ll \sqrt{\gamma_0}/\omega_p$. During this short time interval, the cosine factor inside the integral is nearly constant, $\cos(\eta - \eta') \simeq 1$, and we get the solution

$$n_s(\eta) \simeq \frac{n_0}{2\gamma_0^2} k_p^2 |\mathbf{a}_0(\eta)|^2 . \tag{7.43}$$

Obviously, in this case, no wakefield is created, because this density perturbation is solely associated with the laser pulse itself. This case is nevertheless very useful, because it allows us to determine the photon effective charge $q_{ph}$. For this purpose, we notice that $n_s$ is an electron density perturbation, which is associated with a charge density $(-en_s)$. We can therefore state the double equality

$$-en_s = -e\frac{n_0}{2\gamma_0^2} k_p^2 |\mathbf{a}_0(\eta)|^2 \equiv q_{ph} N_0 , \tag{7.44}$$

where $N_0 \equiv N(\mathbf{k})$ is the photon number density for $\mathbf{k} = \mathbf{k}_0$. This quantity is well-known from quantum field theory, and is related with the electromagnetic energy density $W_0$ by the obvious expression $W_0 = \hbar\omega_0 N_0$. We know, by definition that the energy density associated with the field mode $\mathbf{k} = \mathbf{k}_0$ can be written in terms of the normalized field amplitude $a_0$ as

$$W_0 = \frac{\epsilon_0}{4}\left(\frac{mc}{e}\omega_0\right)^2 |a_0|^2 . \tag{7.45}$$

This leads to the following expressions for the photon number density carried by the driving laser beam, $N_0$, and for the effective photon charge in a plasma, $q_{ph}$, as

$$N_0 = \frac{\epsilon_0}{4\hbar}\left(\frac{mc}{e}\right)^2 \omega_0 |a_0|^2 , \quad q_{ph} = -e\frac{2\hbar\omega_p}{m\gamma_0 c^2}\frac{\omega_p}{\omega_0}\frac{c^2}{v_g^2} , \tag{7.46}$$

where, for convenience, we have reintroduced the factor $(c^2/v_g^2) \sim 1$. This factor is physically relevant when the laser frequency approaches the cutoff frequency $\omega_0^2 \to \omega_p^2/\gamma_0$, as this quantity diverges. It should be noticed that the photon effective charge is a direct consequence of the ponderomotive force that pushes the plasma electrons away from the laser pulse region. This means that, the photon charge is negative because the laser pulse repels the electrons due to radiative pressure.

Returning to the general wakefield solution, eq. (7.41), and integrating it a second time by parts, we obtain, for arbitrary pulse durations $\Delta t$, the following result

$$n_s(\eta) = \frac{n_0}{2\gamma_0^2} k_p^2 \left\{ |\mathbf{a}_0(\eta)|^2 - \int_\infty^\eta \sin(\eta - \eta')|\mathbf{a}_0(\eta')|^2 d\eta' \right\} . \tag{7.47}$$

In terms of the photon charge and number density, this can be rewritten as

$$n_s(\eta) = -\frac{q_{ph}}{e} \left\{ N_0(\eta) - \int_\infty^\eta \sin(\eta - \eta') N_0(\eta')|^2 d\eta' \right\} . \tag{7.48}$$

The two terms in these expressions have distinct physical meanings. The first term, as we have seen, is just the density perturbations associated with the laser pulse itself. In contrast, the second term represents the density perturbations (wakefield) that are left behind the pulse.

Let us now consider a slightly different nonlinear model. In order to describe the laser wakefield excitation, we go back to the relativistic plasma equations, where we simplify the notation: the normalized vector potential $\mathbf{a} \equiv \mathbf{a}_s$ describes the laser field propagating in a perturbed plasma, and the normalized scalar potential $\phi \equiv \phi_s$ describes the wakefield. Again, we assume that the fields only depend on the variable $\xi = z - c\beta_0 t$, where now $\beta_0 = v_g/c$ is the relative velocity of the laser pulse envelope. For simplicity, we take here $\beta_0 = 1$. The potentials are now determined by the nonlinear wave equations

$$\left( \frac{2}{c} \frac{\partial^2}{\partial \xi \partial t} - \frac{1}{c^2} \frac{\partial^2}{\partial t^2} \right) \mathbf{a} = k_p^2 \frac{n}{\gamma n_0} \mathbf{a} \,, \tag{7.49}$$

and

$$\frac{\partial^2 \phi}{\partial \xi^2} = k_p^2 \left( \frac{n}{n_0} - 1 \right) \,, \tag{7.50}$$

where the electron density $n$ is determined by

$$\frac{\partial n}{\partial t} - c \frac{\partial}{\partial \xi} [n(1 - \beta_z)] = 0 \,, \tag{7.51}$$

and the electron parallel velocity $v_z = c\beta_z$ varies according to

$$\frac{\partial}{\partial t} (\gamma \beta_z) = c \frac{\partial}{\partial \xi} [\phi - \gamma(1 - \beta_z)] \,. \tag{7.52}$$

Noting that the electron relativistic factor can be written as $\gamma = (1 + a^2)/(1 - \beta_z^2)^{1/2}$, we have here a complete set of equations describing the plasma evolution, where the finite transverse size of the laser pulse is not considered.

At this point, we reintroduce the quasi-static approximation, by neglecting the time derivative in the last two equations. This approximation is valid when the laser pulse profile described by the potential $\mathbf{a}(\xi)$ stays nearly constant during the interaction time. Equations (7.51) and (7.52) can then be easily integrated, and the result is

$$\frac{n}{n_0} = \frac{1}{(1 - \beta_z)} \,, \qquad \gamma = \frac{1 + \phi}{(1 - \beta_z)} \,. \tag{7.53}$$

Replacing this in eqs. (7.49) and (7.50), we get

$$\left( \frac{2}{c} \frac{\partial^2}{\partial \xi \partial t} - \frac{1}{c^2} \frac{\partial^2}{\partial t^2} \right) \mathbf{a} = k_p^2 \frac{\mathbf{a}}{1 + \phi} \,, \tag{7.54}$$

and

$$\frac{\partial^2 \phi}{\partial \xi^2} = \frac{k_p^2}{2} \left[ \frac{(1 + a^2)}{(1 + \phi)^2} - 1 \right] \,. \tag{7.55}$$

Let us now assume that the laser pulse is determined by a vector potential of the form

$$\mathbf{a}(\xi, t) = \frac{1}{2}\mathbf{a}_0(\xi, t)\exp(ik\xi) + c.c. \tag{7.56}$$

where $\mathbf{a}_0$ is a slowly varying amplitude describing the shape of the pulse envelope. This means that it will only change over a scale $L$, much larger than the laser wavelength, $kL \gg 1$. In the case of moderately high electrostatic fields such that $|\phi| \ll 1$, or $|eV| \ll mc^2$, we can approximate eq. (7.55) by

$$\frac{\partial^2 \phi}{\partial \xi^2} + k_p^2 \phi \simeq \frac{k_p^2}{2} a^2(\xi), \tag{7.57}$$

which can easily be integrated. For appropriate initial conditions, such that the electrostatic potential is zero before the arrival of the laser pulse, or $\phi = d\phi/d\xi = 0$ when $\xi = \rightarrow \infty$, we obtain

$$\phi(\xi) \simeq \frac{k_p}{2} \int_{\infty}^{\xi} a^2(\xi') \sin[k_p(\xi - \xi')] d\xi'. \tag{7.58}$$

Let us assume, for illustration, the simple case of a rectangular pulse with $a_0(\xi) = A_0 =$ constant inside the interval $(L > \xi \geq 0)$. This solution reduces to

$$\phi(\xi) = \frac{A_0^2}{2^3}[1 - \cos(k_p(\xi - L)], \tag{7.59}$$

which shows that the electrostatic perturbation starting at the front of the laser pulse, at $\xi = L$, remains with constant amplitude after the passage of the laser pulse, for $\xi < 0$. The above analytical models retain most of the relevant physical features of the laser produced wakefields.

It is also instructive to consider the combined effect of a laser pulse and a particle beam with charge $s$, propagating in the same direction with a given time delay $\tau_d$. The resulting combined wakefield can be described by

$$\left(\frac{\partial^2}{\partial \eta^2} + k_p^2\right) \tilde{n}(\eta) = s\omega_{pe}^2 n_b(\eta + \eta_d) + \frac{2\hbar k_p^2}{m_e \omega_0} \frac{\partial^2}{\partial \eta^2} N(\eta) \tag{7.60}$$

where the dependence in the perpendicular direction $\mathbf{r}_\perp$ is implicit, $\eta_d = c\tau_d$, $n_b(\eta)$ is the particle beam density and $N(\eta)$ is the photon number density associated with the laser pulse. The solution will then be

$$\tilde{n}(\eta) = -k_p \int^{\eta} \left[\frac{|q_{ph}|}{e} N(\eta') - s n_b(\eta' + \eta_d)\right] \sin[k_p(\eta - \eta')] d\eta' \tag{7.61}$$

Let us take the interesting example of photon and particle beams with exactly identical profiles, $N(\eta) = n_b(\eta)$. In that case, the combined wakefields can show a complete destructive interference, thus leading to the appearance of a very deep and isolated positively charged plasma bubble.

## 7.3 Wave-Breaking

An important concept in wakefield physics is that of *wave-breaking*, which sets the upper limit of the wakefield amplitude. Here, we discuss the concepts of wave-breaking, of blowout regime and the formation of bubbles (or positive cavities).

Traditionally, wave-breaking corresponds to a condition where the wave density, or velocity, becomes double-valued [100]. A simple criterium can be established for the linear and non-relativistic regime [101]. It corresponds to a wave where the density amplitude is equal to the plasma density, $\tilde{n} \simeq n_0$. This is attained when the electrostatic field in a wake is equal to a critical field, defined by

$$E_{cr} = \frac{m}{e}\omega_p v_{ph}\,, \quad v_{ph} = \frac{\omega}{k}\,. \tag{7.62}$$

We have seen that, in a plasma wakefield, the phase velocity is equal to the beam velocity, $v_{ph} = v_b$. Similarly, in a laser wakefield, it will be equal to the laser pulse group velocity, $v_{ph} = v_g = (\partial\omega/\partial k)_{\omega_0}$.

But, we also have seen that in nonlinear wakefields we have density perturbations with amplitudes that are larger than the equilibrium density, as shown in Figures (7.1)–(7.2). In the one-dimensional cold relativistic model, first explored by Akhiezer and Polovin [102] and discussed in the previous chapter, the maximum allowed field is defined by

$$E_{cr} \simeq \frac{m}{e}\omega_p c\left[2(\gamma_{ph} - 1)\right]^{1/2}, \quad \gamma_{ph} = \frac{1}{\sqrt{1 - v_{ph}^2/c^2}}\,, \tag{7.63}$$

and $v_{ph} \simeq c$. This can be $(2\gamma_{max})^{1/2}$ times larger than the critical field $E_{cr}$ defined by eq. (7.62), when $\gamma_{max} \gg 1$ is the maximum possible value of $\gamma_{ph}$. In parallel with this relativistic correction, it is also possible to introduce other corrections, due to thermal effects, leading to

$$E_{cr} \simeq \frac{m}{e}\omega_p v_{ph}\left[1 - \frac{\Theta}{2} - \frac{8}{3}\Theta^{1/4} + 2\Theta^{1/2}\right]^{1/2}, \tag{7.64}$$

where $\Theta = 2T/mv_{the}^2$. But, this is only valid in the non-relativistic limit [103].

When we take both the relativistic and thermal corrections simultaneously, a new limit can be derived, given by [104]

$$E_{cr} \simeq \frac{m}{e}\omega_p \frac{c}{\Theta^{1/4}}\left[\ln\left(2\gamma_{ph}^{1/2}\Theta^{1/4}\right)\right]^{1/2}, \tag{7.65}$$

In alternative, for the same regime, another critical field has been proposed [98], as

$$E_{cr} \simeq \frac{m}{e}\omega_p c\left[\frac{4}{9\alpha}\right]^{1/4}, \tag{7.66}$$

where $\alpha = 3T/mc^2$ is very close to $\Theta$ for relativistic wakefields $v_{ph} \simeq c$. These different results have been clarified in a unified description [105].

We now consider what is usually called the *blow-out regime*. The definition of this regime, and its relation with the wave-breaking is not very precise [106]. We prefer to refer to this regime when all the electrons are blown away from the wakefield region, and accumulate at the border, forming what is sometimes called an *ion-channel*. In two dimensions, this can be seen as a positive-bubble, as mentioned before.

This bubble regime can occur when the plasma response becomes strongly nonlinear. The main qualitative concept related with these nonlinear regimes is due to the competition between the ponderomotive force of the laser beam, which pushes the plasma electrons away from the beam path, and the attractive force of the positive ion channel remaining in the emptied region. When these two forces, seen by an electron located at a distance $r_b$, in the periphery of the laser beam, are comparable, then a single bubble can be formed. This corresponds to

$$\frac{a_0^2}{\gamma_a k_p r_b} \simeq k_p r_b \,, \tag{7.67}$$

where $a_0$ is the normalized laser field amplitude and $\gamma_a$ is the averaged electron relativistic factor due to the laser field oscillations. Noticing that this relativistic factor is of order $a_0$, we can define the approximate size of bubble as $r_b \sim 2\sqrt{a_0}/k_p$. Simulations show that the occurrence of a spontaneous bubble regime is very critical with respect to the laser spot size $w_0$, and only occur for $k_p w_0 \simeq 1$. If we then assume that $r_b \simeq w_0$, we conclude that the bubble regime is attained when $a_0 \simeq 4$.

Detailed analytical models of this regime have been proposed [106, 107]. In order to understand the physical principles of these models it is useful to discuss the energy balance inside the wakefield. The total energy has to include the field energy $W_f$ and the kinetic energy of the electrons, $W_e$, and can be defined as

$$W = W_f + W_e \,, \quad W_f = \frac{\epsilon_0}{2}\left(E^2 + c^2 B^2\right) \,, \quad W_e = m n_e (\gamma_e - 1) c^2 \,. \tag{7.68}$$

We take $\gamma_e$ as the averaged value of the electron relativistic factor, assuming that a relativistic electron beam can also be present inside the bubble, for instance a driving beam, or an injected beam. Each of these two terms satisfies a conservation law, which can be written as

$$\frac{\partial W_f}{\partial t} + \nabla \cdot \mathbf{S}_f = -\mathbf{J} \cdot \mathbf{E} \,, \tag{7.69}$$

and

$$\frac{\partial W_e}{\partial t} + \nabla \cdot \mathbf{S}_e = \mathbf{J}_e \cdot \mathbf{E} + \frac{1}{2}\frac{\partial}{\partial t}\left\langle \frac{a^2}{\gamma_e} \right\rangle n_e \,, \tag{7.70}$$

where $\mathbf{S}_f = (\mathbf{E} \times \mathbf{B})$ is the electromagnetic energy flux (the Poynting vector), and $\mathbf{S}_e = n_e < \mathbf{p} >$ is the electron energy flux. In the above expression, we also use $\mathbf{a} = e\mathbf{E}/(m\omega_0 c)$ as the normalized laser field amplitude, where $\omega_0$ is the central frequency of the driving pulse. At this point, we can write the total energy conservation as

$$\frac{\partial W}{\partial t} + \nabla \cdot \mathbf{S} = (\mathbf{J} - \mathbf{J}_e) \cdot \mathbf{E} + \frac{1}{2} \frac{\partial}{\partial t} \left\langle \frac{a^2}{\gamma_e} \right\rangle n_e \,, \tag{7.71}$$

It is also useful to consider the bubble frame, the variable $\xi = t - z/c$, and assume $\omega_0^2 \gg \omega_p^2$, such that the laser pulse group velocity is nearly equal to $c$. Furthermore, we can define the difference between the two currents in this equation as the current associated with the charge of the bubble, $\rho_b$. Introducing the quantity $\tilde{W} = W - S_z c$, which is the energy density of the plasma wake, the above conservation relation becomes

$$\frac{\partial \tilde{W}}{\partial t} + \nabla_\perp \cdot \mathbf{S}_\perp = -\rho_b E_z + \frac{1}{2} \frac{\partial}{\partial \xi} \left\langle \frac{a^2}{\gamma_e} \right\rangle n_e \,, \tag{7.72}$$

Let is also use the wakefield potential, defined as $\varphi = V - A_z c$. This allows us to rewrite the bubble energy density as $\tilde{W} = \tilde{W}_f + \tilde{W}_e$, with

$$\tilde{W}_f = \frac{1}{2} \left( (\nabla \varphi)^2 + c^2 B_z^2 \right) \,, \quad \tilde{W}_e = m n_e \left\langle (\gamma_e - 1)(1 - v_z)^2 \right\rangle \,. \tag{7.73}$$

We can then integrate eq. (7.72) in the transverse direction and obtain

$$\frac{d\Psi}{d\xi} = - \int \rho_b E_z d\mathbf{r}_\perp + \frac{1}{2} \int \frac{\partial}{\partial \xi} \left\langle \frac{a^2}{\gamma_e} \right\rangle n_e d\mathbf{r}_\perp \,, \tag{7.74}$$

where the quantity

$$\Psi(\xi) = \int \tilde{W}(\xi, \mathbf{r}_\perp \, d\mathbf{r}_\perp \,, \tag{7.75}$$

can be seen as the integrated bubble energy. Now, if we assume that the electrons in the plasma are pushed out of the bubble and accumulate at an infinitesimal layer with radius $r_b$ (the bubble radius to be determined), it is then possible establish an expression for $\Psi(\xi)$, as [107]

$$\Psi(\xi) = \frac{\pi r_b^2}{16} \left[ r_b^2 + (2r_b^2 + 8) \left( \frac{dr_b}{d\xi} \right)^2 \right] \,. \tag{7.76}$$

Noting that the longitudinal electric field $E_z$ can be written as

$$E_z = \frac{r_b}{2} \frac{dr_b}{d\xi} \,, \tag{7.77}$$

and using this in the conservation equation (7.74), we arrive at an equation for the bubble radius, of the form

$$A \frac{d^2 r_b}{d\xi^2} + B \left( \frac{dr_b}{d\xi} \right)^2 + C = \lambda_b \,, \tag{7.78}$$

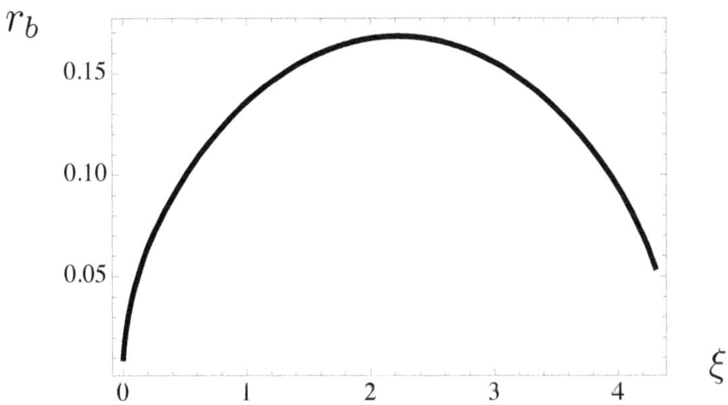


**FIGURE 7.3**
Two-dimensional model of a plasma bubble, representing the bubble radius $r_b$ as a function of the co-moving coordinate $\xi$. The driving perturbation is also shown.

where the coefficients $A, B$ and $C$ depend on $r_b$, and are defined by

$$A = r_b \left( 1 + \frac{r_b^2}{4} \right) , \quad B = 1 + \frac{r_b^2}{2} , \quad C = \frac{r_b^2}{4} . \tag{7.79}$$

The driving term in eq. (7.78) is defined as

$$\lambda_b = - \int_0^{r_b} \rho_b(r') dr' . \tag{7.80}$$

This equation can then be integrated, giving the radial structure of the two-dimensional bubble, $r_b(\xi)$, as illustrated in Figure 7.3. Numerical simulations confirm the existence of this kind of bubble structure.

## 7.4 Twisted Wakes

The wakefield concept can be extended to the case of twisted laser pulses, carrying a finite amount of orbital angular momentum. These are laser beams with an internal field structure that can be described as Laguerre-Gauss (LG) functions [108]. Their interest is that they are able to excite plasma wakes that strongly deviate from the above description. In a sense, they still preserve the same basic physical properties of a wakefield, but in a different topological scenario. To understand this new type of wakefields, it is no longer possible to use one-dimensional models, because laser pulses with orbital angular momentum can only be described in three dimensions.

First, we need to consider the properties of a twisted laser pulse propagating in a plasma. We start with the wave equation for the vector potential associated with these laser pulses, written as

$$\left( \nabla^2 - \frac{1}{c^2} \frac{\partial^2}{\partial t^2} - \frac{\omega_p^2}{\gamma c^2} \right) \mathbf{a} = 0 \,, \tag{7.81}$$

where $\mathbf{a} = e\mathbf{A}/mc$ is the normalized vector potential, and $\gamma$ the electron relativistic factor. For waves propagating along the $z$-axis with circular polarization, we write

$$\mathbf{a}(\mathbf{r}, t) = a_0 \left( \cos\varphi \, \mathbf{e}_x + \sin\varphi \, \mathbf{e}_y \right) \,, \quad \varphi = kz - \omega t \,, \tag{7.82}$$

where $a_0$ is a slowly varying amplitude. In this case, we have $\mathbf{a}^2 = a_0^2$ and, for a cold plasma, the electron relativistic factor can be written as $\gamma \simeq \gamma_a = \sqrt{1 + a_0^2}$. This means that the laser frequency and wavenumber are simply related by the dispersion relation $\omega^2 = k^2c^2 + \omega_p^2/\gamma_a$. As previously noted, this is a nonlinear relation that describes the relativistic self-transparency. Replacing this in eq. (7.81), we obtain an equation for the wave amplitude, of the form

$$\nabla_\perp^2 \left( a_0 e^{i\varphi} \right) + 2ike^{i\varphi} \left( \frac{\partial}{\partial z} + \frac{1}{v_g} \frac{\partial}{\partial t} \right) a_0 = 0 \,, \tag{7.83}$$

where $v_g = kc^2/\omega$ is the group velocity. We can also use the development

$$\nabla_\perp \left( a_0 e^{i\varphi} \right) = e^{i\varphi} \left( 1 + i \frac{a_0^2}{\gamma_a^3} \frac{\omega_p^2}{2c^2k^2} z \right) \nabla_\perp a_0 \,. \tag{7.84}$$

This expression is valid around the focal position $z = 0$, over a region of the order of the *Rayleigh length*, defined as $z_R = kw^2/2$, where $w$ is the laser beam waist. The second term inside the bracket can be ignored if

$$\frac{a_0^2}{\gamma_a^3} \frac{\omega_p^2}{2c^2k^2} z_R \equiv \frac{a_0^2}{(1 + a_0^2)^{2/3}} \frac{k_p}{4k} (k_p w) < 1 \,. \tag{7.85}$$

with $k_p = \omega_p/c$. Assuming that this inequality is satisfied in the focal region, we can then reduce eq. (7.83) to the standard *paraxial equation*, of the form

$$\nabla_\perp^2 a_0 + 2ik \left( \frac{\partial}{\partial z} + \frac{1}{v_g} \frac{\partial}{\partial t} \right) a_0 = 0 \,, \tag{7.86}$$

For a given value of the axial variable $z$, the solutions of this equation are functions of the transverse coordinates, $\mathbf{r}_\perp \equiv (r, \theta)$. These solutions can be represented in a basis of orthogonal functions of the plane, such as the Laguerre-Gauss (LG) functions or the Hermite-Gauss (HG) functions. We choose the LG functions, because they are eigenfunctions of well-defined orbital angular momentum states, while the HG functions are superpositions of different angular momentum states [109, 110]. We can then represent the solutions of eq. (7.87) as a superposition of LG modes, described as

$$\mathbf{a}(\mathbf{r}, t) = \sum_{\ell p} \mathbf{a}_{\ell p} F_{\ell p}(r, \theta) \,, \tag{7.87}$$

where the amplitudes $\mathbf{a}_{\ell p} \equiv \mathbf{a}_{\ell p}(\eta)$ depend weakly on the axial coordinate $\eta = z - v_g t$, and the LG functions, $F_{\ell p}(r, \theta)$, describe the transverse mode structure. They are defined by the expressions

$$F_{\ell p}(r, \theta) = F_{\ell p}(r) e^{i\ell\theta} \,, \quad F_{\ell, p}(r) = C_{\ell, p} X^{|\ell|/2} L_p^{|\ell|}(X) \exp\left(-\frac{X}{2}\right) \,, \tag{7.88}$$

where $L_p^\ell$ are the *associated Laguerre polynomials* of argument $X = r^2/w^2$, and the quantity $w = w(z)$ is the laser beam waist, that slowly changes in the focal region. These polynomials are defined by the formula

$$L_p^\ell(X) = \frac{e^X}{X^\ell p!} \frac{d^p}{dX^p} \left(X^{\ell+p} e^{-X}\right) \,. \tag{7.89}$$

The normalization constants $C_{\ell p}$ in eq.(7.88) are chosen to allow for orthonormality, such that the condition

$$\langle F_{\ell p} | F_{\ell' p'} \rangle \equiv \int_0^\infty r dr \int_0^{2\pi} d\theta F_{\ell p}(r, \theta) F_{\ell' p'}(r, \theta) = \delta_{pp'} \delta_{\ell\ell'} \,, \tag{7.90}$$

is satisfied. Their explicit expressions are

$$C_{\ell p}) - \frac{1}{2\sqrt{\pi}} \left[\frac{(\ell + p)!}{p!}\right]^{1/2} \,. \tag{7.91}$$

For a given LG mode, the exponential factor $\exp(i\ell\theta)$ appearing in the expression of $F_{\ell p}(r, \theta)$ adds to the phase function $\varphi = (kz - \omega t)$, allowing the field to evolve in space and time as $\exp[i(\ell\theta + kz - \omega t)]$. This implies an helical structure for the surfaces of constant phase, which deviates qualitatively from the usual plane wave solutions, as illustrated in Figure 7.4. We should however notice that the LG mode description includes the usual Gaussian laser beams as a particular case, where the absence of orbital angular momentum, corresponds to the quantum numbers ($\ell = 0, p = 0$). A striking difference between the LG beams with a finite poloidal quantum number $\ell \neq 0$ and a Gaussian beam is that the transverse electric tends to zero near the origin, when $r \to 0$. This means that the field is zero at beam axis, and maximum at a radial location of the order of the beam waist.

Such LG description can be extended to electrostatic waves in plasmas, and in particular to electron plasma waves and ion acoustic waves, with some qualitative changes [111]. The main difference is that, in contrast with electromagnetic waves, which have a spin-type of angular momentum associated

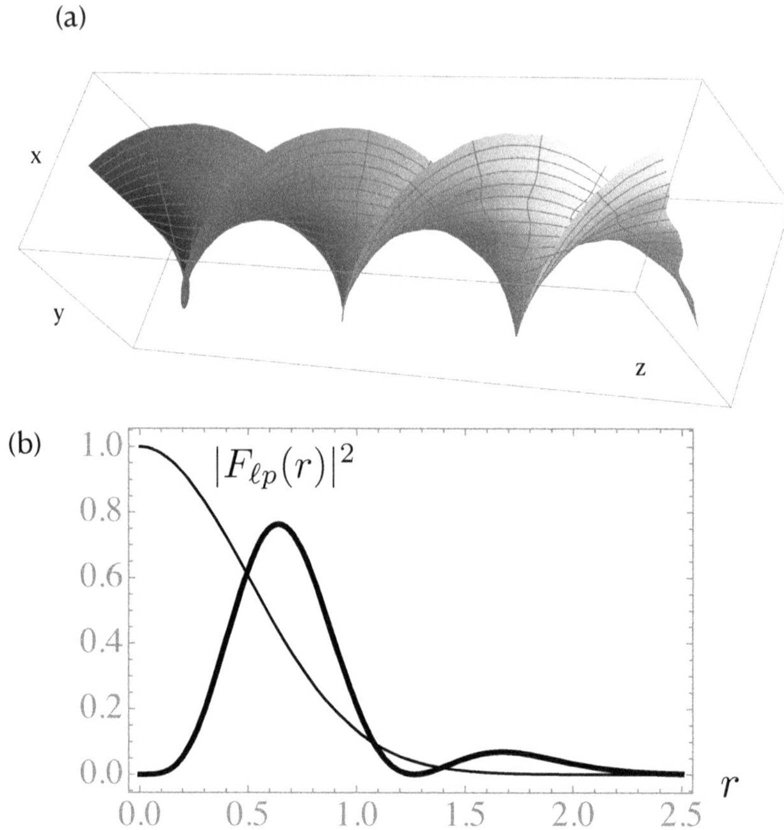


**FIGURE 7.4**
Twisted laser pulses: (a) Surfaces of constant phase: $(\ell\theta + kz - \omega t) = 0$, at a given time; ( b ) Radial intensity profile $|F_{\ell p}(r)|^2$ of LG modes, for quantum numbers $\ell = 2$ and $p = 1$ (bold curve), as compared with the Gaussian profile (thin curve).

with transverse polarization, the electrostatic waves have no spin. Therefore, the orbital angular momentum of plasmons and photons is always equal to their total angular momentum.

Let us now focus on the laser wakefield problem, and consider the electron density perturbations produced by LG laser modes. For that purpose, we consider the electron fluid equations describing the slow plasma response to these fields. The evolution equation for the electron density perturbations, $\tilde{n} = n - n_0$, associated with these laser pulses, take the form

$$\left( \frac{\partial^2}{\partial t^2} + \omega_p^2 \right) \tilde{n} = \frac{c^2}{2} n_0 \nabla \cdot \left\langle \frac{\nabla a^2}{\gamma_a} \right\rangle . \tag{7.92}$$

The notation $\langle\,\rangle$ represents an average over a time interval larger than the laser period, and shorter than the electron plasma period. Here, we have neglected the electron kinetic pressure, because we are dealing with intense laser fields, such that the electron quiver velocity is much higher than the thermal velocity.

The relativistic factor $\gamma_a = \sqrt{1+a^2}$, valid in the laser pulse region, appears in the driving term, but is neglected in the electron plasma frequency $\omega_p = (e^2 n_0/\epsilon_0 m)^{1/2}$. This is justified if we assume that the wakefield perturbations mainly occur in a region where the laser pulse is absent, $a^2 \sim 0$, staying behind the pulse. For that reason, the relativistic factor only remains in the driving force term.

Let us now use the field (7.87) to calculate the ponderomotive force appearing in eq. (7.92). To simplify the discussion of the wakefield topology, we use the notation $\nu = (\ell, p)$, and consider the useful case where the laser field is defined by a superposition of two different LG modes, identified as $\nu = 1, 2$. We therefore have

$$a^2 = \sum_{\nu=1,2} |a_j(\eta_\nu) F_\nu(r)|^2 + 2\,(\mathbf{a}_1 \cdot \mathbf{a}_2^*)\, |F_1(r)||F_2(r)| \cos\left(\Delta k z + \Delta\ell\theta - \Delta\omega t\right) , \tag{7.93}$$

with

$$\Delta k = k_1 - k_2\,, \quad \Delta\ell = \ell - \ell'\,, \quad \Delta\omega = \omega_1 - \omega_2\,. \tag{7.94}$$

We have also used the space-time variables $\eta_\nu = z - v_\nu t$, where $v_\nu$ are the group velocities of the LG modes. If the frequencies of the two modes are nearly identical, $\omega_1 \sim \omega_2$, we can use a single space-time variable $\eta_1 \sim \eta_2 = \eta$. In order to solve eq. (7.92), we could use a similar decomposition of the density perturbations $\tilde{n}$ in LG modes, as

$$\tilde{n}(\mathbf{r}, t) = \sum_{\nu'} N_{\nu'}(z, t) F_{\nu'}(r, \theta)\,. \tag{7.95}$$

This would allow us to derive an equation for each density component, $N_\nu$. An alternative, and sometimes more appropriate, approach is to use the decomposition into the principal components, according to

$$\tilde{n}(\mathbf{r}, t) = \sum_{\nu=1,2} N_\nu(z, t)\, |F_\nu(r)|^2 + N_3(z, t) F_1(r) F_2(r)\,, \tag{7.96}$$

In this case, we can derive three independent equations for the $N_{\nu=1,2}$ and $N_3$, associated with the principal components of the wakefield perturbations, such that

$$\left(\frac{\partial^2}{\partial t^2} + \omega_p^2\right) N_\nu = \frac{c^2}{2} n_0 \frac{\partial}{\partial z}\left(\frac{1}{\gamma_a}\frac{\partial}{\partial z} a_\nu^2\right)\,. \tag{7.97}$$

for $\nu = 1, 2$, and

$$\left(\frac{\partial^2}{\partial t^2} + \omega_p^2\right) N_3 = \frac{c^2}{2} n_0 \frac{\partial}{\partial z}\left[\frac{1}{\gamma_a}\frac{\partial}{\partial z} a_1 a_2^* \cos\left(\Delta k z + \Delta\ell\theta - \Delta\omega t\right)\right]\,. \tag{7.98}$$

These evolution equations can be used to discuss different wakefield configurations. Three different cases are specially interesting for practical purposes, and are considered next.

### i) Donut Wakes

This case only involves a single LG laser mode. We therefore take $a \equiv a_1$, with $a_2 = 0$. To simplify, we assume the weakly relativistic case, $\gamma_a \sim 1$, and write

$$\left( \frac{\partial^2}{\partial \eta^2} + k_p^2 \right) N_1 = \frac{\partial^2}{\partial \eta^2} V_1(\eta) \,. \tag{7.99}$$

and

$$V_1(\eta) = \frac{n_0}{2} \left| a_1(\eta) \right|^2 \,, \tag{7.100}$$

where $k_p = \omega_p / v_1, \eta = z - v_1 t$ and $v_1 = \partial \omega_1 / \partial k_1 \simeq c$. This wakefield equation can easily be integrated, leading to

$$N_1(\eta) = \delta N_1(\eta) - \int_{\eta_0}^{\eta} V_1(\eta') \sin\left[ k_p(\eta - \eta') \right] d\eta' \,, \tag{7.101}$$

where

$$\delta N_1(\eta) = V_1(\eta) - V_1(\eta_0) \cos(k_p \eta) - \left( \frac{\partial V_1}{\partial \eta} \right)_{\eta_0} \sin(k_p \eta) \,. \tag{7.102}$$

Notice that, far away from the pulse, the initial conditions can be chosen such that $V_1(\eta_0) = 0$ and $(\partial V_1 / \partial \eta)_{\eta_0} \simeq 0$. This allows us to write the wakefield solution in terms of the co-moving variable $\eta$ as

$$\tilde{n}(r, \eta) = \frac{n_0}{2} \left\{ |a_1(\eta)|^2 - \int_{\eta_0}^{\eta} |a_1(\eta')|^2 \sin\left[ k_p(\eta - \eta') \right] d\eta' \right\} \left| F_1(r) \right|^2 \,, \tag{7.103}$$

where $F_1(r) \equiv F_{p\ell}(r)$. This defines what is sometimes called a *donut wakefield*, because of the peculiar toroidal shape of the density perturbations. In this solution, the dependence on the poloidal variable $\theta$ is lost, because the ponderomotive potential in eq. (7.99) only depends on the module square of the field amplitude, therefore losing the poloidal structure. This is illustrated in Figure 7.5, where the absolute value of the density perturbations associated with the different wakefield configurations are shown.

### ii) Self-Torque Wakes

Another example of laser pulse was first considered in optics [112], and applied to plasma physics soon after. These pulses, sometimes called *self-torque pulses*, result from the superposition of two LG modes, with different values of angular momentum, $\ell_1$ and $\ell_2 \neq \ell_1$, but with the same frequency, $\omega_1 = \omega_2$. Which means that $\Delta \ell \neq 0$, and $\Delta \omega = \Delta k = 0$. Furthermore, the

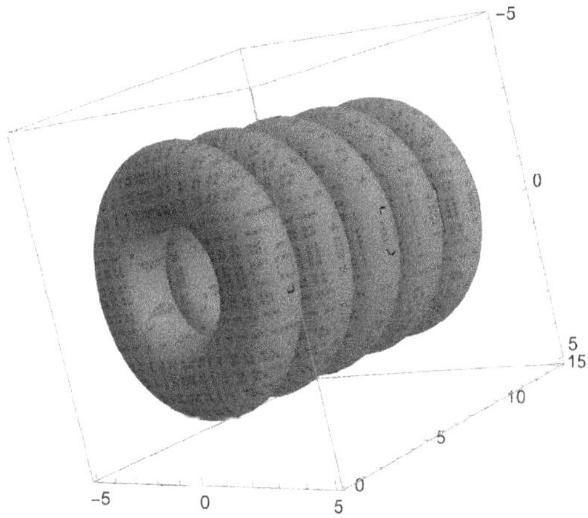


**FIGURE 7.5**
Wakefield topology - Donut shape, excited by a single LG laser mode with azimuthal index $l = 1$.

envelopes of the two LG modes can be displaced in time, which means that the angular momentum can eventually change inside the pulse, which introduces a further degree of freedom in the wake topology.

For the two Gaussian pulse envelopes, we can use.

$$a_\nu(\eta) = a_0 \exp\left[-\frac{(\eta - \eta_\nu)^2}{2\sigma^2}\right], \tag{7.104}$$

with $\nu = 1, 2$. Typically, the relative mode delay is of the order of the pulse duration, $(\eta_1 - \eta_2) \geq \sigma$. When this kind of laser pulse propagates in a plasma, these two modes produce two similar but non-overlapping wakes, changing the total wakefield structure. Instead of eqs. (7.99)–(7.100), we use $\gamma_a \sim 1$, and write

$$\left(\frac{\partial^2}{\partial \eta^2} + k_p^2\right) \tilde{n}(\mathbf{r}_\perp, \eta) = \frac{\partial^2}{\partial \eta^2} V(\mathbf{r}_\perp, \eta), \tag{7.105}$$

with the ponderomotive potential

$$V(\mathbf{r}_\perp, \eta) = \frac{n_0}{2} \left| a_1(\eta) F_1(r) + a_2(\eta) F_2(r) e^{i\Delta\ell\theta} \right|^2, \tag{7.106}$$

and $\mathbf{r}_\perp \equiv (r, \theta)$. We noticed that, when the mode delay is small and the LG modes nearly overlap, such that $(\eta_1 - \eta_2) < \sigma$, the resulting wakefield structure is reduced to a moon-shaped wake. Such a moon-shaped wake is only significant for $\Delta\ell \ll \ell_\nu$, because the radial size of the pulse depends on the azimuthal number $\ell$. This is illustrated in Figure 7.6. Notice that the corresponding wakefield potential contains an important poloidal component, eventually leading to acceleration of electrons along helical trajectories, as

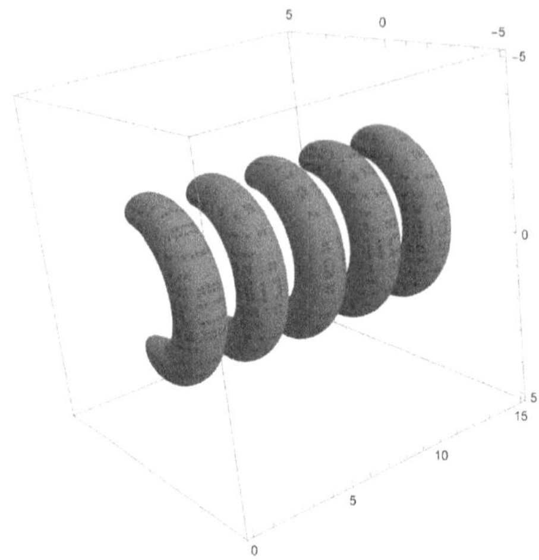


**FIGURE 7.6**
Wakefield topology - lunar-shape excited by the superposition of two different LG modes ($l_1$ and $l_2$) with the same frequency $\omega_1 = \omega_2$. No delay between the two modes is assumed, $\eta_1 = \eta_2$.

shown later. This property is even more pronounced in the case of light springs, to be discussed next.

### iii) Light-Springs

This configuration was first proposed in optics by [113], and then applied to plasma physics. It can be achieved with a superposition of two LG laser modes, using two different frequencies, $\omega_1$ and $\omega_2 \neq \omega_1$. In this case, no envelope delay is assumed between the two modes.

This particular configuration is particularly useful when applied to plasma acceleration problems, if we consider that the frequency difference is nearly equal to the plasma frequency $\Delta\omega \simeq \omega_p$. In this case, we can expect a resonant excitation of electron plasma waves with a well defined value of orbital angular momentum $\ell_p \simeq \Delta\ell = \ell_1 - \ell_2$. This means that a pure LG plasmon mode with a finite angular momentum, characterized by the quantum number $\ell_p$, can be excited by these two LG laser modes. Furthermore, this configuration can also be seen as a generalization of *beat-wave acceleration*, an historically relevant acceleration method [114]. The beat-wave process is very similar to stimulated Raman scattering, to be discussed in a later chapter.

Due to this frequency difference, the resulting density perturbations have a helical shape and can be described by eq. (7.105), where eq. (7.106) is replaced by a similar, but not identical driving potential

$$V(\mathbf{r}_\perp, \eta) = \frac{n_0}{2} |a(\eta)|^2 \left| F_1(r) + F_2(r) \exp\left(i\Delta\ell\theta + i\Delta k\eta\right)\right|^2 , \tag{7.107}$$

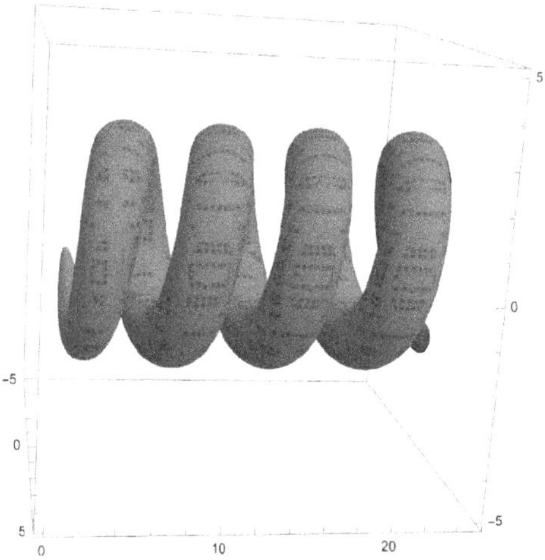


**FIGURE 7.7**
Wakefield excited by a light-spring laser pulse, resulting from the superposition of two LG modes ($l_1 = 7$ and $l_2 = 8$), with different frequencies, $\omega_1$ and $\omega_2$. Beat-wave configuration corresponds to the resonant case where $\omega_2 - \omega_1 \simeq \omega_p$.

where $\Delta k \simeq k_p$. Laser modes with the same amplitude, $a_1(\eta) = a_2(\eta)$ were assumed. Notice that the density perturbations will mainly be due to the cross-field term, proportional to $F_1(r)F_2(r)$, that oscillates at $k_p$, which is the resonant term. As a result, the plasma response can be resonant, and a pure LG mode of electron density helical structures can be excited (see Figure 7.7). The importance of this resonant term should be stressed. Notably, the similarity of the light spring configuration with the beat-wave acceleration concept, and its potential use to accelerate helical electron beams, as well as a remarkable potential use to focalize and accelerate positron beams, as shown later.

## 7.5 Magnetic Field Generation

Laser-plasma interactions in the relativistic domain usually excite quasi-static magnetic fields. There are several processes that explain the creation of these magnetic fields. We focus here on two main processes, which are: i) the inverse Faraday effect, due to the laser pulse and ii) the nonlinear effects, directly associated with an electron plasma wake. Mention should be also made to the so-called *grad n - grad T* process, which is dominant at high plasma densities, near and above the critical density and not discussed here.

### 7.5.1 Inverse Faraday Effect

Let us start with the inverse Faraday effect. The first theoretical descriptions of this effect was given in 1961 by Pitaevskii [115, 116], and the first reported experiments are described by Deschamps and collaborators in 1970 [117], who reported the generation of 20 milli-Gauss using microwaves. Much larger magnetic field generation, of several Mega-Gauss was obtained later with intense laser pulses [118]. An important ingredient of this process is the use of circularly polarized waves, which dissipate part of their angular momentum in the medium and generate quasi-static magnetic fields. However, it has also been shown that, linearly polarized laser fields can produce magnetic fields, if they have orbital angular momentum [119].

To describe this effect, we start with the conservation of angular momentum in a plasma, which can be simply written in cylindrical coordinates $\mathbf{r} \equiv (r, \theta, z)$, as

$$\frac{dM_z}{dt} = -en_e r \left[ E_\theta + \frac{1}{\gamma} \left( u_z B_r - u_r B_z \right) \right] , \tag{7.108}$$

where $\mathbf{u}$ is the slow electron covariant mean velocity, and $\gamma$ the relativistic factor. The quantity $M_z$ represents the axial component of the laser pulse angular momentum. For Gaussian-shaped beams, this quantity is only due to circular polarization, right or left, characterized by the parameter $\sigma_z = \pm 1$. In quantum terms, this can be associated with photon spin. However, for LG modes, such as considered in the previous section, there are also contributions from the orbital angular momentum, characterized by the quantum number $\ell \neq 0$. For general laser beam conditions, we can therefore define $M_z$ as the sum of two terms, one related to spin and the other to orbital angular. momentum, as

$$M_z = \frac{1}{\omega c} \left[ \ell + \sigma_z \frac{r}{2} \frac{\partial}{\partial t} \right] I(t) , \tag{7.109}$$

where $I(t)$ is the laser intensity. It should be noticed that these two terms can be of the same order, and that the spin term depends on the radial pulse shape. It is therefore equal to zero for a plane wave. This is due to a complete compensation of contributions from electrons rotating at different points of the phase plane. Along propagation, the angular momentum content of the laser pulse changes, mainly due to inverse bremsstrahlung, such that $I(t) = I(0) \exp(-\nu_{br} v_g t)$, where $\nu_{br}$ is the damping rate introduced by electron-ion collisions, when the electrons are forced to oscillate by the incident laser field. This dissipative process induces a change in the total angular momentum carried by the laser pulse.

Let us now turn to the left-hand side of eq. (7.108), which represents the plasma angular momentum. Here, only the electron motion is retained and ion motion is ignored. Assuming that the second term is small, we can see that any variation $M_z$ induces a azimuthal electric field $E_\theta$, which in turns creates a magnetic field $B_z$, as determined by

$$E_\theta \simeq -\frac{1}{enr}\frac{dM_z}{dt}\,, \qquad \frac{\partial B_z}{\partial t} = -\frac{1}{r}\frac{\partial}{\partial r}(rE_\theta)\,. \tag{7.110}$$

These are slow fields evolving on a time scale much longer than the inverse of the laser frequency. Notice that this second equation is nothing but the $z$-component of Faraday's equation. This allows us to write

$$\frac{\partial B_z}{\partial t} = \frac{1}{enr}\frac{\partial}{\partial r}\left(\frac{dM_z}{dt}\right). \tag{7.111}$$

Integration of this equation in time, then leads to an expression for the quasi-static magnetic field generated by the laser after propagation over a distance $L = v_g t$ in the plasma. The result is

$$B_z(L) = -\frac{f_{br}}{en\omega rc}\left[\ell\frac{\partial I}{\partial r} + \frac{\sigma_z}{2}\frac{\partial}{\partial r}\left(r\frac{\partial I}{\partial r}\right)\right], \tag{7.112}$$

where we have used the absorption factor $f_{br} = 1 - \exp(-\nu_{br}L)$. We can see that the magnetic field strength increases with the laser absorption. In order to derive more explicit results for the quasi-static magnetic field, we need to consider the laser intensity profile, which depends on the mode structure. For LG modes, it is given by

$$I(r) = I_0 C_{\ell p}^2 X^\ell \exp(-X) L_{\ell p}(X)^2\,, \quad X = r^2/w^2\,, \tag{7.113}$$

where $I_0$ is the maximum intensity, $C_{\ell p}$ is the LG normalization factor, and $w$ the laser beam waist. As a simple example, let us consider a circularly polarized Gaussian pulse, such that $\ell = p = 0$, and $\sigma_z = +1$. Replacing this in eq. (7.112), we obtain

$$B_z(r) = B_0\sigma_z\left(1 - \frac{r^2}{w^2}\right)\exp\left(-\frac{r^2}{w^2}\right), \quad B_0 = \frac{2f_{br}}{en\omega w^2 c}I_0\,, \tag{7.114}$$

where $B_0$ is the maximum field strength. For very intense laser pulses, this quantity can attain the Mega-Gauss level. For instance, if we take an intensity of $I_0 = 10^{23} W/m^2$, we get $B_0 \simeq 10\,MG$. This can also be written in terms of the normalized field amplitude $a_0 = eE_0/m\omega c$, as

$$B_0 \simeq f_{br}\left(\frac{en_0}{\epsilon_0\omega}\right)\left(\frac{\omega}{\omega_p}\right)^4\left(\frac{\lambda^2}{w^2}\right)a_0^2\,, \tag{7.115}$$

where the first bracket has the dimensions of a magnetic field, and all the other terms are dimensionless.

As another example, we now assume a circularly polarized laser pulse, with a finite amount of orbital angular momentum, corresponding to ($\ell = 1, p =$

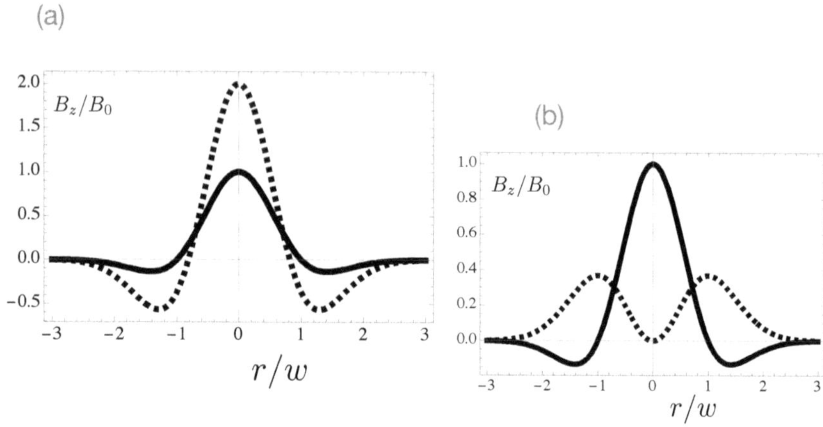


**FIGURE 7.8**
Inverse Faraday effect – Magnetic field profile, $B_z/B_0$, as a function of the normalized radius, $r/w$, generated by a twisted LG laser pulse, $(\ell = 1, p = 0)$ (black curve), as compared with a purely Gaussian pulse of the same intensity with circular polarization (dashed curve): The effect is enhanced (a), or reduced (b), according to the sign of $\sigma_z$.

0). In this case, eq. (7.113) reduces to $I(r) = I_0 X \exp(-X)$, leading to the magnetic field profile

$$B_z(r) = -B_0 \left[ \left(1 - \frac{r^2}{w^2}\right) + \sigma_z \left(1 - \frac{3r^2}{w^2} + \frac{r^4}{w^4}\right) \right] \exp\left(-\frac{r^2}{w^2}\right). \quad (7.116)$$

This shows that the contributions from the spin and orbital momentum are nearly equal in magnitude, and can almost compensate for $\sigma_z = -1$. It means that the use of twisted laser pulses can either increase or reduce the total value of the static magnetic field generated by the inverse Faraday mechanism. This is illustrated in Figure 7.8.

### 7.5.2 Magnetic Wakefield Effects

Laser wakefields are not simple electrostatic objects, as one should think from our previous discussion, because the nonlinear plasma response to these strong density perturbations also creates quasi-static magnetic fields. These magnetic wakefield effects will be described here. For simplicity, we consider $\gamma = 1$, but retain the possible effects associated with twisted wakefields. We start with the electron fluid equations, and assume a nonlinear scaling, for the plasma density as $n = n_0 + n_1 + n_2$, where $n_0$ is the equilibrium density, $n_1$ the density associated with the first-order electrostatic perturbations, and $n_2$ the density associated with possible nonlinear currents and fields. We also assume that there is no external magnetic field $B_0 = 0$. We therefore refer to $(\mathbf{v}_1, \mathbf{E}_1, \mathbf{B}_1 = 0)$ as the first-order perturbations, and $(\mathbf{v}_2, \mathbf{E}_2, \mathbf{B}_2 \neq 0)$ to nonlinear second-order effects. The nonlinear current will then be determined by

$$\mathbf{J} = -e\left(n_0\mathbf{v}_2 + n_1\mathbf{v}_1\right). \tag{7.117}$$

where the nonlinear perturbations are

$$\frac{\partial n_2}{\partial t} + n_0\nabla\cdot\mathbf{v}_2 + \nabla\cdot(n_1\mathbf{v}_1) = 0\,, \quad \frac{\partial \mathbf{v}_2}{\partial t} + \mathbf{v}_1\cdot\nabla\mathbf{v}_1 = -\frac{e}{m}\mathbf{E}_2\,, \tag{7.118}$$

We now assume a generic plasma LG mode, which also includes a Gaussian mode, described by

$$n_1 = N_\nu F_\nu(r,z)e^{i\ell\theta}\exp(ikz - i\omega t)\,, \tag{7.119}$$

where $\nu = (\ell, p)$ represent the quantum numbers of the LG mode, eventually excited by laser pulses, as seen before. From here we derive the first-order velocity and electric field, as

$$\mathbf{E}_1 = i\frac{e}{\epsilon_0 k^2}\mathbf{k}_\nu n_1, \quad \mathbf{v}_1 = \frac{\omega}{k_\nu^2}\frac{n_1}{n_0}\mathbf{k}_\nu\,, \tag{7.120}$$

where the wavenumber $\mathbf{k}_\nu$ includes the information on the transverse mode structure [120], and is defined as

$$\mathbf{k}_\nu = -\frac{i}{F_\nu}\frac{\partial F_\nu}{\partial r}\mathbf{e}_r + \frac{\ell}{r}\mathbf{e}_\theta + \left(k - \frac{i}{F_\nu}\frac{\partial F_\nu}{\partial z}\right)\mathbf{e}_z\,. \tag{7.121}$$

Although this is not a wavevector in the usual sense, it is approximately given by $\mathbf{k}_\nu \sim k\mathbf{e}_z$. Using eqs. (7.118) it is then possible to calculate the second-order quantities, and obtain

$$n_2 = -\frac{i\omega}{n_0}F_\nu^2 I_\nu(t)\,, \quad I_\nu(t) = \int^t N_\nu(t')^2 dt'\,. \tag{7.122}$$

as well as

$$\mathbf{E}_2 = i\frac{e}{\epsilon_0}\frac{\mathbf{k}_\nu}{k_\nu^2}n_2\,, \tag{7.123}$$

and

$$\mathbf{v}_2 = -\frac{\omega^2}{k^2}\left[iI_\nu(t) + \frac{\omega_p^2}{\omega^2}\int^t I_\nu(t')dt'\right]\mathbf{k}_\nu\,. \tag{7.124}$$

We are now in position to calculate the nonlinear current (7.117), which takes the form $\mathbf{J}_2 = J_2\left(\mathbf{k}_\nu/k\right)$, with

$$J_2 = \frac{e\omega}{k}\frac{F_\nu^2}{n_0}\left[i\omega I_\nu(t) + \omega_p^2\int^t I_\nu(t')dt' - N_\nu(t)^2\right]\,. \tag{7.125}$$

Notice that this current is independent of the poloidal coordinate, $J_2 \equiv J_2(r,z)$. This expression can be used to calculate the quasi-static magnetic field generated by second-order effects. Using the Ampere law, we then get, for the poloidal component of the field $\mathbf{H}_2$, the following relations

$$\frac{\partial H_\theta}{\partial z} = iJ_2 \frac{1}{F_\nu} \frac{\partial F_\nu}{\partial r}, \quad \frac{1}{r} \frac{\partial (r H_\theta)}{\partial r} = J_2 k, \tag{7.126}$$

and, for the other two components

$$\frac{\partial H_r}{\partial z} - \frac{\partial H_z}{\partial r} = -\frac{\ell}{r} J_2 . \tag{7.127}$$

It can easily be realized that the radial field component is negligible and the other two are given by

$$H_\theta(r) = -\frac{e\omega}{k r n_0} N_\nu^2 \int_0^\infty |F_\nu(r')|^2 r' dr', \quad H_z(z) = -\ell \int_0^\infty J_2 \frac{dr}{r} . \tag{7.128}$$

This shows that the poloidal field component $H_\theta$ strongly depends on the radial structure of the density perturbations, and the axial component slowly varies along the $z$-axis. Furthermore, for a Gaussian wakefield ($\ell = 0$), this axial component is absent, and in the general case, we have $|H_\theta / H_z| \simeq kw/\ell$. We can also define a second-order cyclotron frequency, such that

$$\omega_{2c} = \frac{e B_\theta}{m} \simeq \omega_p \left( \frac{N_\nu}{n_0} \right)^2 . \tag{7.129}$$

This is a signature that, at second-order, the wakefield is not simply electrostatic, and always includes a magnetic structure. This involves the existence of a nonlinear cyclotron motion. Noting that the density perturbations associated with the wakefield can be of the order, or even larger than the equilibrium density, $N_\nu \sim n_0$, we conclude that the associated magnetic fields can be significant.

# 8

# *Acceleration Processes*

In this chapter we consider the main acceleration processes that can occur in laser-plasma interaction experiments. We give special relevance to those related with the existence of wakefields and bubbles, as described before. These acceleration processes strongly depend on the laser pulse duration and intensity, as well as on the plasma density. Of particular importance are the so-called *short laser pulses*, such that their duration $\Delta t$ is much shorter than the electron plasma period $1/\omega_p$. In this regime, laser wakefield acceleration was demonstrated for the first time in three celebrated experiments [123, 124, 125]. This confirmed the theoretical predictions by Tajima and Dawson [126], which motivated intense theoretical and experimental developments (as described in several reviews [127, 128]).

Wakefield excitation is less efficient for long pulses, such that $\Delta t\ \omega_p \geq 1$, due to phase-mixing, but other mechanisms can take place. In particular, modulational instabilities can occur, which lead to the formation of electron plasma waves. Acceleration mechanisms such as *stochastic acceleration* can also occur, which eventually produces energetic electron beams with a broad energy spectrum [129]. This was first predicted by [130, 131] and confirmed by subsequent studies [132, 133]. Another interesting effect associated with long laser pulses is the *direct laser acceleration* process. This could also be linked to amplification of betatron oscillations, and subsequent emission of radiation. In this chapter, we include the description of other mechanisms, such as the *snow-plow acceleration*, which can take place at the laser pulse front, and the *surfatron acceleration*, that requires the existence of a static magnetic field. We conclude the chapter with a discussion of proton and positron acceleration.

## 8.1 Wakefield Acceleration

Most of the acceleration processes discussed in the present chapter, can be understood using single particle motion in the presence of a laser pulse. The laser is represented by the normalized field amplitude $\mathbf{a}$, and the collective plasma response, is basically represented by the electrostatic potential $\varphi$. The Coulomb gauge is used, as well as typical notations and formalism. The single particle motion is then described by the Hamiltonian, written in dimensionless

DOI: 10.1201/9781003514701-8

form as

$$H = \sqrt{1 + [\mathbf{P} + \mathbf{a}(\vec{r}, \tau)]^2} - \varphi \,. \tag{8.1}$$

The vector and scalar potentials are also normalized, according to $\mathbf{a} = e\mathbf{A}/mc$ and $\varphi = eV/mc^2$, where $e$ the electron charge, and $m$ its mass. The time variable is $\tau = ct$. The canonical momentum $\mathbf{P}$ is normalized to $mc$, as usual

$$\mathbf{P} = \gamma \frac{\mathbf{v}}{c} - \mathbf{a} \quad , \quad \gamma = \frac{1}{\sqrt{1 - v^2/c^2}} = H + \varphi \,, \tag{8.2}$$

We now assume a laser pulse with frequency $\omega$ and wavevector $\mathbf{k}$, moving in a plasma with group velocity $\mathbf{v}_g = c^2\mathbf{k}/\omega$. Using the co-moving variable $\mathbf{r}' = \mathbf{r} - \mathbf{v}_g\tau/c$, we can rewrite the particle Hamiltonian as

$$H'(\mathbf{r}', \mathbf{P}, \tau) = H - \frac{\mathbf{v}_g}{c} \cdot \mathbf{P} \,, \tag{8.3}$$

where the Hamiltonian $H$ is defined by eq. (8.1), but written as a function of the new variable $\mathbf{r}'$. Let us first focus on particle trajectories along the laser pulse direction, $\mathbf{e}_z$. such that the transverse field structure can be ignored. We consider parallel and perpendicular motion separately, using

$$\mathbf{r}' = \eta \, \mathbf{e}_z + \mathbf{r}_\perp \quad , \quad P = P \, \mathbf{e}_z + \mathbf{P}_\perp \,. \tag{8.4}$$

Ignoring the transverse field structure (to be considered in the next section), we assume that the potentials only depend on the parallel coordinate $\eta$. In this case, the perpendicular momentum is a constant of motion, which can be stated as

$$\frac{d\mathbf{P}_\perp}{d\tau} = -\frac{\partial H'}{\partial \mathbf{r}_\perp} = 0 \,, \quad \mathbf{P} = \mathbf{P}_{\perp 0} = const \,. \tag{8.5}$$

The perpendicular velocity is then given by

$$\mathbf{v}_\perp = \frac{c}{\gamma} \left( \mathbf{P}_{\perp 0} + \mathbf{a} \right) \tag{8.6}$$

We are then reduced to the study of parallel motion, and to the one-dimensional canonical equations

$$\frac{d\eta}{d\tau} = \frac{\partial H'}{\partial P} = \frac{P}{\gamma} - \frac{v_g}{c} \,, \tag{8.7}$$

and

$$\frac{dp}{d\tau} = -\frac{\partial h'}{\partial \eta} = -\frac{\partial}{\partial \eta} (\gamma - \varphi) \,. \tag{8.8}$$

Here, the relativistic gamma factor, $\gamma \equiv \gamma(\eta, p, \tau)$, can be explicitly written as

$$\gamma = \sqrt{1 + [P\mathbf{e}_z + \mathbf{P}_{\perp 0} + \mathbf{a}(\eta, \tau)]^2} \tag{8.9}$$

These are the basic equations of motion for the wakefield acceleration, as shown next.

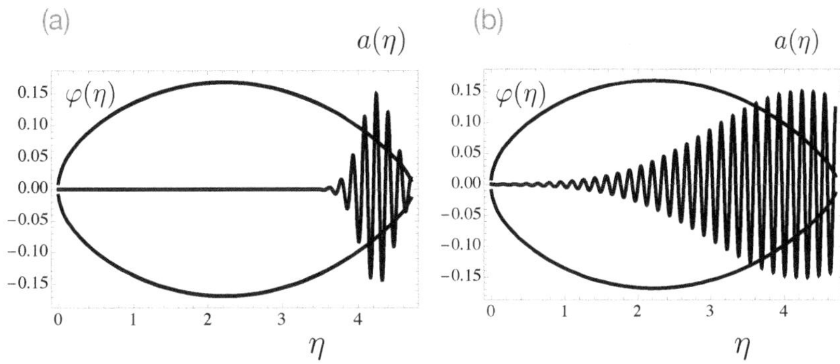


**FIGURE 8.1**
Short (a) and long (b) laser pulse wakefields: illustrative figure representing the electrostatic wakefield potential $\varphi(\eta)$, and the amplitude $a(\eta)$ of the vector potential oscillations associated with the laser pulse.

We should keep in mind that, in general, the particles are simultaneously accelerated by the transverse laser field, $\mathbf{a}(\eta, \tau)$, and by the wakefield potential $\varphi(\eta)$. This is true, even in the one-dimensional case, particularly for long pulses, where the transverse field can be present in the acceleration region. In contrast, for short pulses, we can assume $\mathbf{a} = 0$, because acceleration occurs at the rear of the wakefield cavity, and the laser pulse is mainly at the front. The short and long laser pulse conditions are illustrated in Figure 8.1.

Let us forget the transverse field structure, and describe the wakefield by a simple one-dimensional potential. We also assume circularly polarized laser field. This allows us to write

$$\varphi(\eta) = \varphi_0 \cos(k_p \eta) \quad , \quad \mathbf{a}(\eta, \tau) = a_0(\eta) \left[ \mathbf{e}_x \cos\theta + \mathbf{e}_y \sin\theta \right] , \tag{8.10}$$

where we define the angle $\theta$ as

$$\theta = k\eta - \Omega\tau/c , \quad \Omega = \omega - k v_g . \tag{8.11}$$

For an underdense plasma, such that $\omega^2 \gg \omega_p^2$, we have $\Omega \simeq \omega_p^2/\omega$. For interacting distances $\Delta\tau$ such that the dephasing between the particle and the wakefield structure is small, $(\omega_p/\omega)(\Delta\tau k_p) \ll 2\pi$, the temporal dependence of the potentials can be ignored. Two constants of motion can be defined. One is the Hamiltonian $H'(\eta, P) = H_0$, valid in the moving frame. The other is the transverse momentum, $\mathbf{P}_\perp = \mathbf{P}_{\perp 0}$. The relativistic gamma factor can then be written as

$$\gamma = \sqrt{1 + P^2 + [\mathbf{P}_{\perp 0} + \mathbf{a}(\eta)]^2} \, . \tag{8.12}$$

Of course, this will not true when the scalar and vector potentials change in the transverse plan, as considered in the next section. The parallel motion is then described by

$$\frac{d\eta}{d\tau} = \frac{P}{\sqrt{1 + P^2 + [\mathbf{P}_{\perp 0} + \mathbf{a}(\eta)]^2}} - \frac{v_g}{c}, \quad \frac{dP}{d\tau} = -\varphi_0 k_p \sin(k_p \eta) . \tag{8.13}$$

Integration of these equation leads to the following particle trajectories

$$P(\eta) = \gamma_g^2 \left\{ [H_0 + \varphi(\eta)]\beta_g \pm \sqrt{[H_0 + \varphi(\eta)]^2 - \{1 + [\mathbf{P}_{\perp 0} + \mathbf{a}(\eta)]^2\} / \gamma_g^2} \right\} , \tag{8.14}$$

where $\gamma_g = (1 - v_g^2/c^2)^{-1/2}$ is the laser pulse gamma factor. Notice that real trajectories can only exist for

$$[H_0 + \varphi(\eta)]^2 \; > \; \left\{1 + [\mathbf{P}_{\perp 0} + \mathbf{a}(\eta)]^2\right\} / \gamma_g^2 . \tag{8.15}$$

The wakefield acceleration is usually dominated by the scalar potential $\varphi(\eta)$. We can then take $\mathbf{a}(\eta) = 0$, or $\nu^2(\eta) = 1 + P_{\perp 0}^2 = const$. However, for long pulses, the canonical momentum $P(\eta)$ will vary along the longitudinal laser field structure, if the driving laser pulse coexists with the scalar potential in the acceleration region. This will not occur for short pulses, but will eventually lead to energy dispersion for long pulses. The particle trajectories for long pulses, in the acceleration region, will eventually be affected due to the presence of the laser field such that $\mathbf{a}(\eta) \neq 0$. This can lead to an energy spread of the accelerated electron spectrum. But more relevant effects appear for long pulses, and those associated with the transverse motion, as shown in the next section.

It is now useful to discuss the *separatrix*, which splits the particle phase-space $(\eta, p)$ in two different regions. This is similar to what we have already discussed in Chapter 2, and correspond to trapped and passing (or untrapped) regions. The separatrix condition $H_0 = H_{sep}$ defines a trajectory that corresponds to an unstable fixed point defined by $(a = 0, \, \varphi = \varphi_{max})$, and $\gamma = \gamma_g$. This is given by

$$H_{sep} = \varphi_{max} + \gamma_g \left( 1 - \frac{v_g^2}{c^2} \sqrt{1 + \frac{P_{\perp 0}^2 c^2}{v_g^2 \gamma_g^2}} \right) \tag{8.16}$$

The inequality$H_0 \leq H_{sep}$, defines the condition for trapped trajectories. According to this criterium, and assuming purely parallel motion, particles with initial velocity $v_0$ will stay trapped in the wakefield potential if

$$\gamma_0 \left( 1 - \frac{v_g v_0}{c^2} \right) \leq \varphi_{max} - \frac{1}{\gamma_g} . \tag{8.17}$$

The maximum acceleration attained by a trapped particle can be estimated if we consider the invariance of $H_0$. The energy difference of a trapped particle in two different positions, $\eta_1$ and $\eta_2$, will then be given by the relation

$$\gamma_2 \left[ 1 - \frac{v_g v_2}{c^2} \sqrt{1 + a^2(\eta_2)} \right] = \varphi(\eta_1) - \varphi(\eta_2) + \gamma_1 \left[ 1 - \frac{v_g v_1}{c^2} \sqrt{1 + a^2(\eta_1)} \right] \tag{8.18}$$

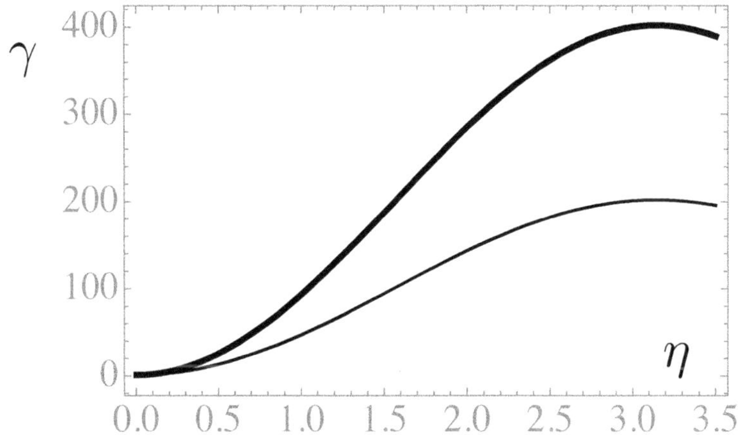


**FIGURE 8.2**
Wakefield acceleration: Evolution of the electron gamma-factor $\gamma$, as a function of $\eta$, for two different values of the laser group velocity, $v_g/c = 0.99$ (thin curve) and $v_g/c = 0.995$ (bold curve).

where $P_{\perp 0} = 0$ was assumed. Maximum acceleration is then attained when the trajectory maximizes the potential energy difference, corresponding to the values of $\varphi(\eta_1) = \varphi_{max}$ and $\varphi(\eta_2) = \varphi_{min}$. Assuming that $a(\eta_1) = 0$ and $a(\eta_2) = a_0$, the maximum relativistic factor is then given by

$$\gamma_2 \simeq \frac{\gamma_1(1 - v_g/c) + (\varphi_{max} - \varphi_{min})}{(1 - |a_0| v_g^2/c^2)} \tag{8.19}$$

This would be valid for a long laser pulse, where the maximum of the laser amplitude is assumed to coincide with the minimum of the potential well. In contrast, for a short laser pulse where $a^2(\eta) \simeq 0$ in the wakefield acceleration region, we would get

$$\Delta\gamma \equiv \gamma_2 - \gamma_1 \simeq 2\gamma_g^2(\varphi_{max} - \varphi_{min}) \tag{8.20}$$

The corresponding evolution of the particle energy $\gamma(\eta)$ along propagation is illustrated in Figure 8.2, for two values of the laser gamma-factor $\gamma_g$. Notice that this value depends essentially on the laser pulse width. We can see from this figure that the acceleration slows down and eventually stops at some point, when the maximum of the wakefield potential is attained and the accelerating electric field is reduced to zero. From that point on, the particle energy will only decrease. This means that there is an acceleration limit, which is usually described by the *dephasing length*, $L_d$. This is the distance at which the acceleration process is inverted. A simple estimate of this distance can be made, if we assume that it results from the difference between the particle velocity $v_e \simeq c$, and the wake phase velocity $v_\phi$, according to

$$L_d \simeq \frac{\lambda_p c}{(c - v_\phi)}, \quad v_\phi \simeq \left[1 - \frac{3}{2}\frac{\omega_p^2}{\omega_0^2}\right] c. \tag{8.21}$$

This can then be written as

$$L_d \simeq \frac{4\pi}{3k_p}\frac{\omega_0^2}{\omega_p^2} = \frac{4\pi}{3}\frac{\omega_0^2}{\omega_p^3}c. \tag{8.22}$$

A more detailed analysis, where the nonlinear plasma response is taken into account, would lead to a quite similar result [135]

$$L_d \simeq \frac{4}{3}\frac{\omega_0^2}{\omega_p^3}\sqrt{a_0}\, c, \tag{8.23}$$

where we now have a weak dependence on the laser amplitude, $a_0$.

## 8.2 Direct Laser Acceleration

We now consider the contribution of the transverse particle motion. For certain wakefield conditions, the transverse gradient $\partial\varphi/\partial r$ can be non-negligible when compared with the longitudinal one $\partial\varphi/\partial\eta$. This originates a new type of oscillations, called *betatron oscillations*, that take place in the perpendicular direction. These oscillations, not only modify the wakefield acceleration, but also produce strong bursts of XUV radiation [136].

Let us discuss the coupling between these transverse oscillations and the direct field acceleration due to longitudinal field gradients. Using the three dimensional equations of motion, and assuming that $\vec{P}_\perp$ is in the x-direction, we can now describe the perpendicular motion as

$$\frac{dr}{d\tau} = \frac{1}{\gamma}(P_\perp + a_0 \cos\theta) \quad , \quad \frac{dP_\perp}{d\tau} = -\frac{\partial\varphi}{\partial r}, \tag{8.24}$$

were we have assumed circular polarization. This leads to

$$\frac{d^2 P_\perp}{d\tau^2} = \frac{1}{\gamma}\frac{\partial^2\varphi}{\partial r^2}(P_\perp + a_0\cos\theta), \tag{8.25}$$

with

$$\gamma = \sqrt{\gamma_0^2 + P_\perp(P_\perp + 2a_0\cos\theta)}. \tag{8.26}$$

The quantity $\gamma_0 = (1 + p^2 + a_0^2)^{1/2}$ would be the relativistic gamma factor for a purely longitudinal motion. For particles moving along with the laser pulse and its wakefield, their motion is essentially longitudinal, and we can use the approximation

$$\gamma \simeq \gamma_0 \left( 1 + \frac{P_\perp}{\gamma_0^2} a_0 \cos\theta \right) . \tag{8.27}$$

For the same reason, we can take $\eta$ as nearly constant, which means that the transverse motion takes place in a co-moving perpendicular plane. As a result, the phase $\theta$ is nearly independent of the longitudinal variable $\eta$. Taking the simplest possible model for the transverse potential, as defined by

$$\varphi(r) \simeq \frac{1}{2} \varphi_0 r^2 , \tag{8.28}$$

which is always valid near the axis, we can transform equation (8.25) into

$$\frac{d^2 P_\perp}{d\tau^2} = -\omega_B^2 \left[ 1 + \frac{a_0^2}{2\gamma_0^2} (1 + \cos 2\theta) \right] P_\perp - \omega_B^2 a_0 \left( 1 - \frac{P_\perp^2}{\gamma_0^2} \right) \cos\theta , \tag{8.29}$$

where we have defined the *betatron frequency*, $\omega_B = \sqrt{\varphi_0/\gamma_0}$. Using $\theta = \Omega\tau/c + \theta_0$, we can rewrite this equation as

$$\frac{d^2 P_\perp}{d\tau^2} + \Omega_B^2 (1 + \epsilon \cos 2\theta) P_\perp = \epsilon_1 \cos\theta , \tag{8.30}$$

where we have used the new quantities

$$\Omega_B^2 = \frac{\omega_B^2}{\Omega^2} \left( 1 + \frac{a_0^2}{2\gamma_0^2} \right) , \quad \epsilon = \frac{a_0^2}{a_0^2 + 2\gamma_0^2} , \quad \epsilon_1 = a_0 \frac{\omega_B^2}{\Omega^2} . \tag{8.31}$$

This is a forced *Mathieu equation*, where the applied force oscillates at a frequency equal to half the modulation frequency. In the limit $\epsilon_1 \ll \epsilon$, this force could be neglected and, we would expect unstable solutions for the transverse oscillations, where $\Omega_B^2 \simeq (1 - \epsilon/2)^{-1}$. This corresponds to the first unstable region of the Mathieu equation [137]. But $\epsilon_1$ can be nearly equal, or even larger than $\epsilon$, which means that the force cannot in general be neglected. This force then leads to an exponential growth of the transverse oscillations.

Unstable solutions of eq. (8.30) are shown in Figure 8.3, for two distinct values of the force. The destabilizing effect of this term is clearly shown. This illustrates the fact that the transverse equations predict an exponential growth of the betatron oscillations, due to the coupling between the oscillations in the transverse wakefield potential and the direct laser field. We can then see that, when the betatron frequency is close to the laser frequency, $\omega_B \sim \omega$, a resonant interaction between the laser and the electrons can occur, leading to an increase in the total particle energy. This is the so-called *direct acceleration* [138]. Another important contribution of the transverse electron oscillations (transverse field) is that they increase the effective electron mass, thereby increasing the dephasing length and allowing for a longer acceleration process associated with the longitudinal wakefield. This leads to an increase of the final longitudinal momentum as well [139].

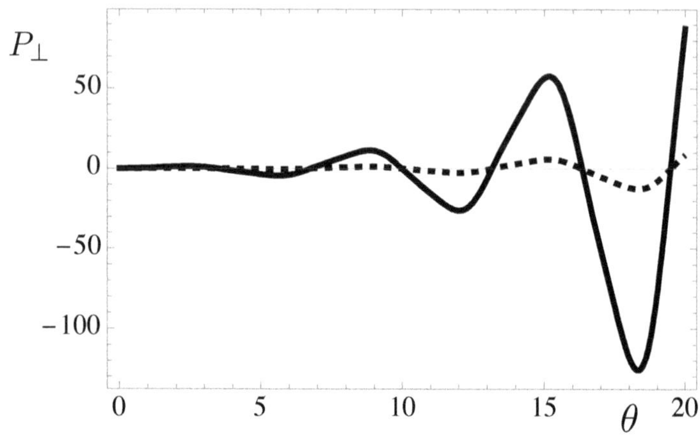


**FIGURE 8.3**
Unstable betatron oscillations: Perpendicular particle momentum, $P_\perp$, for $\Omega_B^2 = 1$ and $\epsilon = 1$. Black full line, $\epsilon_1/\epsilon = 1$; gray dashed line, $\epsilon_1/\epsilon = 0.1$.

## 8.3 Stochastic Acceleration

We now examine the problem of particle motion in the field of two counter-propagating laser fields. This could be relevant for pulse experiments where, due to backscattering or to plasma inhomogeneities, a secondary pulse propagates in the backward direction. In order to study this case, we neglect the wakefield potential $\varphi(\mathbf{r}, \tau) = 0$, and use the Hamiltonian (8.1) with the total vector potential

$$\mathbf{a}(\mathbf{r}, \tau) = \sum_{j=1,2} \mathbf{a}_j(\mathbf{r}, \tau)\,, \tag{8.32}$$

where the two potentials $\mathbf{a}_j$ oscillate in the laboratory frame with frequencies $\omega_j$, and propagate along the z-axis with wavenumbers $k_1 > 0$ and $-k_2$. Introducing, as before, the co-moving variable

$$\eta = z - \beta_1 \tau\,, \quad \beta_1 = \frac{\omega_1}{k_1 c}\,, \tag{8.33}$$

we can rewrite eq. (8.32) as

$$\mathbf{a}(\eta, \tau) = \sum_{j=1,2} \mathbf{a}_j(\eta, \tau)\,, \tag{8.34}$$

where the two terms correspond to the two counter-propagating laser beams, as described by

$$\mathbf{a}_1(\eta, \tau) = \mathbf{a}_1 \cos(\theta_1)\,, \quad \mathbf{a}_2(\eta, \tau) = \mathbf{a}_2 \cos[k_2(\theta_2)]\,, \tag{8.35}$$

with

$$\theta_1 = k_1\eta\,, \quad \theta_2 = k_2(\eta + \beta_2\tau)\,. \tag{8.36}$$

We have also defined

$$\beta_2 = \beta_1\frac{k_2}{k_1} + \frac{\omega_2}{k_2 c}\,. \tag{8.37}$$

The equation of motion in the parallel direction can then be written as

$$\frac{dP}{d\tau} = \frac{1}{\gamma}(\mathbf{P}_{\perp 0} + \mathbf{a}) \cdot \frac{\partial \mathbf{a}}{\partial \eta}\,, \tag{8.38}$$

where

$$\frac{\partial \mathbf{a}}{\partial \eta} = -\mathbf{a}_1 k_1 \sin\theta_1 - \mathbf{a}_2 k_2 \sin\theta_2\,. \tag{8.39}$$

For simplicity, we consider a purely one-dimensional motion, using $\mathbf{P}_{\perp 0} = 0$. Equation (8.38) then simplifies to

$$\begin{aligned}\frac{dP}{d\tau} &= \frac{1}{2\gamma}\sum_{j=1,2} a_j^2 k_i \sin 2\theta_j \\ &+ \frac{(\mathbf{a}_1 \cdot \mathbf{a}_2)}{2\gamma}\left[(k_1 + k_2)\sin(\theta_1 + \theta_2) + (k_1 - k_2)\sin(\theta_1 - \theta_2)\right]\,.\end{aligned} \tag{8.40}$$

This equation shows the occurrence of several nonlinear resonances in the parallel phase space $(\eta, d\eta/d\tau)$. These resonances are produced by each of the two fields, $a_{j=1,2}$ as well as by their beating, $a_1 a_2$. In the case of a single wave, $a_1 \neq 0$ and $a_2 = 0$, we would simply have

$$\frac{dP}{d\tau} = \frac{a_1^2 k_1}{2\gamma}\sin 2\theta_1\,. \tag{8.41}$$

This equation defines a single fixed point, defined by the condition $d\eta/d\tau = 0$, or equivalently, $dz/d\tau = \beta_1$. The corresponding momentum is $P = P_1$, with

$$P_1 = \gamma_1\beta_1 = \frac{\beta_1}{\sqrt{1 + a_1^2}} \quad , \quad u_1 \equiv \left(\frac{d\eta}{d\tau}\right)_1 = 0\,. \tag{8.42}$$

The corresponding *resonance half-width* can be defined in the plane $(u, \eta)$ as

$$(\Delta u)_1 = (\sqrt{1 + a_1^2} - \beta_1) - (1 - \beta_1) = \sqrt{1 + a_1^2} - 1\,. \tag{8.43}$$

For a weak laser field, such that $a_1 \ll 1$, this would simplify to $(\Delta u)_1 \simeq a_1^2/2$. We can repeat the same analysis for the other terms in the equation of motion (8.40), and obtain a similar result for the resonance associated with the second field, when $a_2 \neq 0$ and $a_1 = 0$, as

$$u_2 \equiv \left(\frac{d\eta}{d\tau}\right)_2 = -\beta_2 = -\beta_1\frac{k_2}{k_1} + \frac{\omega_2}{k_2 c} \quad , \quad (\Delta u)_2 = \sqrt{1 + a_2^2} - 1\,. \tag{8.44}$$

Finally, in the presence of both fields ($a_{1,2} \neq 0$), the beating term of eq. (8.40) introduces additional resonances, defined by

$$u_{3,4} \equiv \left(\frac{d\eta}{d\tau}\right)_{3,4} = \mp\frac{\beta_2}{1 \pm k_2/k_1} \quad , \quad (\Delta u)_{3,4} \simeq \sqrt{|\mathbf{a}_1 \cdot \mathbf{a}_2|} \tag{8.45}$$

It is well-known that an equation of motion with such properties is generally *non-integrable*, in the sense that no analytical solution of the equation can be found. Although only integrability can be demonstrated, non-integrable systems can still be identified by direct numerical analysis, through the observation of chaotic trajectories in phase-space. If a significant region of phase-space is covered by a single chaotic trajectory, we can say that we are in the presence of *large-scale stochasticity.*

The occurrence of large-scale stochasticity in phase space $(\eta, u)$, can nevertheless be predicted by qualitative methods. The simplest and most generally used is the so-called *overlapping criterion*, first proposed by Chirikov [20]. It states that large-scale stochasticity occurs when the sum of the two resonance half-widths is larger than the distance between them. Considering two of the above two dominant resonances, this criterium can be applied to our problem, leading to

$$(\Delta u)_j + (\Delta u)_{k\neq j} \ \geq \ |u_j - u_k| \, . \tag{8.46}$$

Numerical solution of the particle equations of motion can confirm (or infirm) this qualitative prediction. When this criterion is satisfied, existence of stochastic particle trajectories over large areas of the parallel phase-space can be expected. This establishes a sufficient condition for the occurrence of stochastic acceleration. Other, more elaborated stochasticity criteria [22], essentially lead to the same or similar qualitative results. This criterium could equally be applied to the discrete maps discussed in Chapter 2.

In the present case, if we assume that the incident laser amplitude is dominant, and $a_1 > a_2$, large-scale stochasticity can be predicted by the overlap of the two closest resonances, as

$$(\Delta u)_1 + (\Delta u)_3 \ \geq \ |u_1 - u_3| \, . \tag{8.47}$$

In more explicit terms, this is stated as

$$a_1 + \sqrt{|\mathbf{a}_1 \cdot \mathbf{a}_2|} \ \geq \ \left| \frac{\beta_2}{1 + k_2/k_1} \right| \, . \tag{8.48}$$

For two counter-propagating waves with the same frequency, such that $k_1 = k_2$, and $\omega_1 = \omega_2$, the right-hand side of this inequality reduces to $|\beta_1|$. And, for a standing wave, $a_1 = a_2 = a$, this overlapping criterion becomes

$$|a| \ \geq \ \frac{\omega_1}{2k_1 c} \simeq \frac{1}{2} \, . \tag{8.49}$$

These qualitative estimates agree with numerical calculations, showing the occurrence of large-scale stochasticity in the presence of two waves. On the other hand, the two resonances determined by eqs. (8.42)-(8.44), as well as the infinite number of secondary resonances that are not explicit in the above analysis of the equations of motion but can be shown to exist in phase-space, would also contribute to the creation stochastic motion on much smaller scales [22, 131].

## 8.4 Snow-Plow Acceleration

We now turn our attention to the interaction of particles with the laser front, where the ponderomotive force is dominant and the wakefield force is still not present. Snow-plow acceleration can take place when particles nearly at rest in the laboratory frame, or counter-propagating with the laser pulse, are reflected by the laser pulse front [140]. This effect can also be called *photon mirror* effect [141], because the laser creates a moving mirror of radiation where charged particles can be reflected, in the same way as photons are reflected by a moving mirror. Here, the moving mirror of the classical Einstein model is replaced by the incoming laser pulse (a mirror of photons reflecting charged particles).

To simplify the discussion, we ignore the transverse field structure. Particle reflection by the laser front occurs when its trajectory in phase-space $(\eta, P)$ starts from the far positive region $\eta \to \infty$ at $\tau \to -\infty$, and returns asymptotically to that same region when $\tau \to \infty$. This situation is associated with the following condition for trajectory reversal in phase-space, when at some position $\eta = \eta_r$ the derivative of the canonical momentum diverges

$$\left(\frac{dP}{d\eta}\right)_{\eta_r} \to \pm\infty \,. \tag{8.50}$$

It means that, when this occurs, the particle momentum changes sign. Taking the derivative of eq. (8.14) we get

$$\frac{dP}{d\eta} = \frac{P}{(H_0 + \varphi)}\frac{d\varphi}{d\eta} \pm \frac{\gamma_g^2 (H_0 + \varphi)}{\sqrt{1 - f(\eta)}}\frac{df}{d\eta}\,, \tag{8.51}$$

with

$$f(\eta) = \frac{1 + [\mathbf{P}_{\perp 0} + \mathbf{a}(\eta)]^2}{\gamma_g^2 (H_0 + \varphi)^2}\,. \tag{8.52}$$

We see that condition (8.50) can be satisfied in two situations. First when $H_0 = -\varphi(\eta_r)$. Second when $f(\eta_r) = 1$. The first situation cannot occur for electrons, because by definition $H_0 > 0$, and the electrostatic potential is attractive, $\varphi(\eta) > 0$, due to the ion background. It could however occur for positrons, as discussed later. The second situation is more interesting because it applies to electrons, and corresponds to

$$(H_0 + \varphi)^2 = (1 - \beta_g^2) \left\{1 + [\mathbf{P}_{\perp 0} + \mathbf{a}(\eta_r)]^2\right\} . \tag{8.53}$$

Let us call $H_r$ the value of the Hamiltonian satisfying this equality, at position $\eta = \eta_r$. We can write it explicitly as

$$H_r = -\varphi(\eta_r) + \frac{1}{\gamma_g}\sqrt{1 + P_{\perp 0}^2 + a^2(\eta_r) + 2\mathbf{P}_{\perp 0} \cdot \mathbf{a}(\eta_r)} . \tag{8.54}$$

Particle reflection will occur if its initial energy is smaller or equal to this limit. This defines the photon mirror effect. Snow-plow acceleration can then take place, because the particles are pushed away by the front of the incoming laser pulse. At the pulse front region, the electrostatic potential is negligible and the ponderomotive force associated with the laser field dominates. Taking the maximum possible value for this field, and assuming purely one-dimensional motion, $P_{\perp 0} = 0$, the condition for snow-plow acceleration can then be stated as

$$H_0 \leq H_r \simeq \frac{1}{\gamma_g}\sqrt{1 + a_0^2} . \tag{8.55}$$

As a result, the energy gain of a particle reflected by the laser pulse front can be quite substantial, and is determined by [141]

$$\Delta\gamma \equiv \gamma_f - \gamma_i = \beta_g \left[\sqrt{\gamma_f^2 - 1} - \sqrt{\gamma_i^2 - 1}\right] , \tag{8.56}$$

where $\gamma_i$ and $\gamma_f$ are the initial and final values of the relativistic gamma factor. If initially we have a beam of energetic particles, such that $\gamma_i^2 \gg 1$, but the laser field amplitude is still compatible with the snow-plow criterion (8.55) and the energy gain upon reflection is given by $\Delta\gamma \simeq \beta_g(\gamma_i + \gamma_f)$. Solving for $\gamma_f$, we then get

$$\gamma_f = \gamma_i \frac{1 + \beta_g}{1 - \beta_g} \simeq 4\gamma_i\gamma_g^2 . \tag{8.57}$$

This shows a double Doppler shift of the particle energy, similar to the Einstein formula for photons reflected by a relativistic mirror. But now we are talking about electrons and not photons. Large energy gains can is still be observed for low energy particles or at rest in the lab frame, with $\gamma_i \simeq 1$. In this case, eq. (8.56) leads to

$$\Delta\gamma = \frac{1}{1 - \beta_g} \simeq 2\gamma_g^2 \quad , \quad \gamma_f \simeq 2\gamma_i\gamma_g^2 . \tag{8.58}$$

This shows that snow-plow acceleration can be very effective, with an energy enhancement factor of order $\gamma_g^2 \gg 1$. It should however be noticed that this particle reflection process takes place over a significant time interval in the laboratory frame, and for this reason, it is difficult to achieve experimentally.

Let us now make a short comment on the influence of perpendicular motion. For this purpose we use the equation of motion

$$\frac{dP_\perp}{d\tau} = \frac{1}{\gamma}\left[\mathbf{P}_\perp - \mathbf{a}(\eta, r_\perp, \tau)\right]^2 \cdot \frac{\partial \mathbf{a}}{\partial \mathbf{r}_\perp} \,. \tag{8.59}$$

Assuming a Gaussian profile in the perpendicular direction, defined by

$$\mathbf{a}(\eta, \mathbf{r}_\perp, \tau) = \mathbf{a}(\tau) \exp\left[-\frac{(\eta - b r_\perp)^2}{2\sigma^2} - \frac{r_\perp^2}{2\sigma_\perp^2}\right] , \tag{8.60}$$

where $b$ is the bending parameter of the pulse front, we obtain

$$\frac{\partial \mathbf{a}}{\partial r_\perp} = -\left[\frac{1}{\sigma_\perp^2} r_\perp + \frac{b}{\sigma^2}(\eta - b r_\perp)\right] \mathbf{a} \,. \tag{8.61}$$

For a flat Gaussian beam, such that $b = 0$, this expression shows that for particles with no transverse momentum, $P_{\perp 0} = 0$, the value of $P_\perp(t)$ grows in time with the impart parameter (distance to the laser beam axis), $r_\perp / \sigma_\perp^2$. The perpendicular motion is therefore negligible if $r_\perp / \sigma_\perp \ll 1$. For large impact parameters and long interaction times, the transverse acceleration has to be taken into account, and eventually leads to particle side-scattering. But, for bended beam fronts, such that $b \neq 0$, a focusing effect eventually occurs, with a significant reduction of side-scattering.

## 8.5 Surfatron Acceleration

Another important configuration for particle acceleration is that of magnetized particles, when they interact with a wave propagating in a direction perpendicular to the static magnetic field $\mathbf{B}_0$. In this case, the particle trajectory is similar to that of a surfer in ocean waves, where gravity is replaced by the magnetic field [142]. For this reason this configuration is often called the *surfatron*. This term is also sometimes used for microwave discharges produced by surface waves [143]. But these two surfatron configurations have nothing in common.

Although historically relevant, surfatron acceleration is much less interesting for laboratory applications than wakefield acceleration, because it implies large transverse plasma dimensions. However, the surfatron process can be relevant in space and astrophysical scenarios [144], where it could eventually provide an alternative mechanism for the formation of cosmic rays, because it is a regular acceleration process, much faster in principle than the Fermi mechanism, which is based on diffusion [145].

Let us then consider a particle of charge $q$ and mass $m$ moving in a static magnetic field $\vec{B}_0 = B_0 \vec{e}_z$ , under the action of an electrostatic wave propagating along the $y$-axis. The wave electric field can be described as $\mathbf{E} = E_0 \mathbf{e}_y \sin(ky - \omega t)$. It is useful to write the particle equations of motion in the wave frame. The normalized velocity of this reference frame is $\beta_0 = \omega/kc$, and the new space and time coordinates are $x' = x$, $z' = z$ and

$$y' = \gamma_0 \left( y - \frac{\omega t}{k} \right) , \quad t' = \gamma_0 \left( t - \frac{\omega y}{kc^2} \right) . \tag{8.62}$$

Here, $\gamma_0 = (1 - \beta_0^2)^{-1/2}$ is the relativistic gamma factor of the moving frame. In this new frame, the particle motion is described by the equations

$$\frac{d\mathbf{r}'}{d\tau} = \mathbf{u}' , \quad \frac{d\mathbf{u}'}{d\tau} = \frac{q}{m} (\mathbf{E}' + \mathbf{u}' \times \mathbf{B}') . \tag{8.63}$$

The electric and magnetic fields seen by the particle in this moving frame are determined by a Lorentz transformation, and are given by

$$E'_x = \gamma_0 \frac{\omega}{k} B_0 , \quad E'_y = E_0 \sin \left( \frac{ky'}{\gamma_0} \right) , \quad E'_z = 0 , \tag{8.64}$$

and

$$B'_x = 0 , \quad B'_y = 0 , \quad B'_z = \gamma_0 B_0 . \tag{8.65}$$

We can see that the electric field $\mathbf{E}'$ is reduced to a static field, because the wave is static in its reference frame, and the magnetic field $\mathbf{B}'$ is amplified by the relativistic factor $\gamma_0$. The equations of motion (8.63) can then be written in explicit form as

$$\begin{aligned} \frac{du'_x}{d\tau} &= \omega_c \gamma_0 \left[ u'_y + \left( \frac{\omega}{k} \right) \gamma' \right] , \\ \frac{du'_y}{d\tau} &= \left( \frac{qE_0}{m} \right) \gamma' \sin \left( \frac{ky'}{\gamma_0} \right) - \omega_c \gamma_0 u'_x , \\ \frac{du'_z}{d\tau} &= 0 , \end{aligned} \tag{8.66}$$

where we have used the cyclotron frequency, defined in the rest frame $\omega_c = qB_0/m$, and the particle relativistic factor $\gamma'$. Elimination of $u'_y$ then leads to the following equation of motion

$$\frac{d^2 u'_x}{d\tau^2} + \omega_c^2 \gamma_0^2 \, u'_x = \left( \frac{qE_0}{m} \right) \omega_c \gamma_0 \gamma' \sin \left( \frac{ky'}{\gamma_0} \right) . \tag{8.67}$$

This corresponds to a nonlinear forced oscillator, and can be solved approximately in the following way. We notice that the force acting on the particle in the parallel wave direction is

$$F'_y = q \left[ E_0 \gamma' \sin \left( \frac{ky'}{\gamma_0} \right) - \gamma_0 u'_x B_0 \right] . \tag{8.68}$$

The first term of this force can be called the *trapping* term and the second the *detrapping* one. If the first term is dominant, a particle that is initially trapped by the wave potential will never be detrapped. This occurs for

$$E_0 \; > \; \gamma_0 |u'_x| \frac{B_0}{\gamma'} \, . \tag{8.69}$$

Noting that $u'_x = \gamma' v'_x$, and that the maximum possible value for $v'_x$ is $c$, we can also write this condition as $E_0 > \gamma_0 c B_0$. Let us then assume that the trapping condition is largely satisfied, and furthermore, that the particle is deeply trapped inside the wave potential. We can then write $u'_y \simeq 0$ and $\gamma' \simeq 1$, and obtain an approximate solution for eq. (8.67), such that

$$u'_x \simeq \omega_c \gamma_0 \frac{\omega}{k} \tau \, . \tag{8.70}$$

This expression shows unlimited acceleration in the transverse direction, which can be called a surfing effect. Going back to the rest frame and using $v_y \simeq \omega/k$, we obtain

$$v_x = \frac{\omega_c (\omega/k) t}{\gamma_0 \sqrt{1 + \omega_c^2 (\omega/kc)^2 t^2}} \, . \tag{8.71}$$

Integration in time leads to

$$x(t) = \frac{c}{\gamma_0 a} \left( \sqrt{1 + a^2 t^2} - 1 \right) , \tag{8.72}$$

where we have assumed $x(0) = 0$ and used $a = \omega_c \omega / kc$. This expression gives the distance travelled by the particle in the direction perpendicular to the wave propagation. Using these results, we can now calculate the energy increase in the perpendicular direction, as

$$\gamma(x) = \gamma_0 \left( 1 + \frac{a}{c} \gamma_0 x \right) , \tag{8.73}$$

and, using $t = yk/\omega$, the energy increase in the longitudinal direction is

$$\gamma(y) = \gamma_0 \left( 1 + \frac{\omega_c^2}{c^2} y^2 \right)^{1/2} \, . \tag{8.74}$$

These results are valid of $u'_y \simeq 0$. If we now include the evolution of $u'_y = dy'/d\tau$, the motion of a particle very close to the bottom of the potential well ($y' \simeq 0$), will satisfy the equation

$$\frac{d^2 y'}{d\tau^2} + \omega_B^2 \frac{\gamma'}{\gamma_0} y' = -\omega_c^2 \gamma_0^2 \frac{\omega}{k} \tau \, , \tag{8.75}$$

where the particle bounce frequency $\omega_B = (qE_0 k/m)^{1/2}$ was used. A solution of this equation, valid for $\gamma' \simeq 1$, can be written as

$$y'(\tau) = a_0 \sin(\omega_B \tau + \phi) - \frac{\omega \omega_c^2}{2k} \gamma_0^2 \tau^2 \, , \tag{8.76}$$

where $a_0$ and $\phi$ are arbitrary amplitude and phase, respectively. This shows that, if the particle starts moving near the bottom of the potential well, it will be delayed with respect to the wave, due to the increase of its relativistic mass. But, the above two equations are only valid for $ky'/\gamma_0 \ll 1$.

An improved analytical description of the particle motion can be obtained using the *multiple time scales method*, where the temporal variable $\tau$ is decomposed into different variables, $\tau_j$, with $j = 0, 1, 2, ...$, such that

$$\frac{d}{d\tau} = \frac{\partial}{\partial\tau_0} + \epsilon\frac{\partial}{\partial\tau_1} + \epsilon^2\frac{\partial}{\partial\tau_2} + \dots , \tag{8.77}$$

where $\epsilon \ll 1$ is an ordering parameter. We then seek for solutions of the form

$$f = f_0 + \epsilon f_1 + \epsilon^2 f_2 + \dots \tag{8.78}$$

valid for all variables appearing in the equations of motion, $f \equiv (x', y', u'_x, u'_y)$. The resulting solutions, to the lowest order, can be written as

$$u'_x(\tau) = \frac{\omega}{k}\left(\frac{\omega_c\gamma_0}{\Omega_0}\right)^3 \sin(\Omega_0\tau) + \omega_c\gamma_0\frac{\omega}{k}\left[1 - \left(\frac{\omega_c\gamma_0}{\Omega_0}\right)^2\right]\tau , \tag{8.79}$$

and

$$u'_y = \frac{\omega}{k}\left(\frac{\omega_c\gamma_0}{\Omega_0}\right)^2 [\cos(\Omega_0\tau) - 1] , \tag{8.80}$$

where we have used a new frequency, defined as

$$\Omega_0 = \left(\frac{\omega_B^2}{\gamma_0} + \omega_c^2\gamma_0^2\right)^{1/2} . \tag{8.81}$$

Notice that this new solution reduces to the previous one in the limit

$$\left(\frac{\omega_c\gamma_0}{\Omega_0}\right)^2 \ll 1 , \tag{8.82}$$

which can also be stated as $\omega_B^2 \gg \omega_c^2\gamma_0^3$. In this limit, eq. (8.79) tends to eq. (8.70), and eq. (8.80) reduces to $u'_y \simeq 0$, as above. This shows that, for waves with a large phase velocity, such that $\gamma_0 \gg 1$, this condition will be difficult to satisfy, and the multiple time scales solution shows a drastic reduction in the acceleration efficiency. The corrected expression for the surfatron energy growth, will be given by

$$\gamma(x) = \gamma_0\left[1 + \frac{ax}{c}\gamma_0(1 - \delta)\right] , \quad \delta = \left(\frac{\omega_c\gamma_0}{\Omega_0}\right)^2 , \tag{8.83}$$

where $x$ is still determined by eq. (8.72), but with $a$ multiplied by the factor $(1 - \delta)$. This means that the acceleration is less efficient than indicated by the simple estimate of eq. (8.73), and decays with $(1 - \delta)$.

## 8.6 Transverse Force

The shape of the accelerated particle beam strongly depends on the transverse structure of the laser pulses and of the corresponding wakefield structures. A simple bubble structure is naturally able to accelerate unidimensional electron beams. But we also have seen that any radial motion will eventually lead to betatron oscillations. For the same reason, a donut bubble will accelerate a cylindrical hollow electron beam. And a twisted wakefield will eventually accelerate helical electron beams. These three distinct configurations are illustrated in Figure 8.4.

In order to discuss the equilibrium and stability of these beam structures, we need to consider the accelerating electrostatic fields that result from the different wakefield configurations. The longitudinal field $E_z(\mathbf{r})_\perp$ is directly linked by the Poisson's equation to the density wake perturbations, and satisfies a driving equation of the form

$$\left( \frac{\partial^2}{\partial t^2} + \omega_p^2 \right) E_z = - \frac{e n_0 c^2}{2\epsilon_0} \left\langle \frac{\nabla a^2}{\gamma_a} \right\rangle . \tag{8.84}$$

This is formally analogous to eq. (7.92). For the generic twisted wakes discussed in Section 7.4, we can use eq. (7.93) to describe the laser field, and decompose $E_z$ into principal field components, as

$$E_z(\mathbf{r}, t) = \sum_{\nu} E_{z\nu} \left| F_\nu(r) \right|^2 + E_{z3} F_1(r, \theta) F_2(r, \theta) \, , \tag{8.85}$$

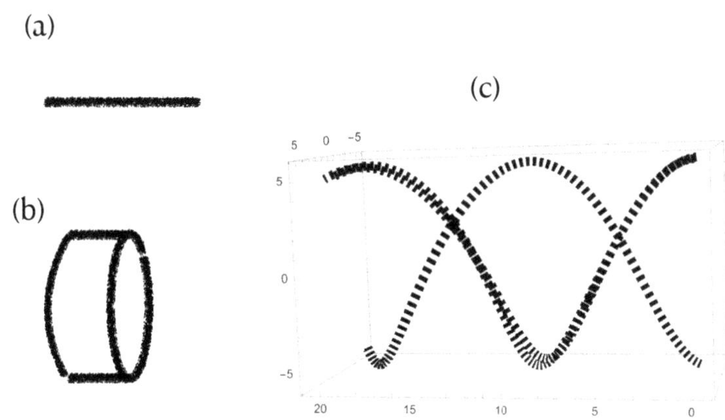


**FIGURE 8.4**
Electron beam configurations: (a) Nearly one-dimensional beam created by bubble wakes; (b) Cylindrical hollow electron beam created by donut wakes; and (c) Helical electron beams created by light spring wakes.

similar to eq. (7.96). The electric field amplitudes are determined now by three wakefield equations of the form

$$\left( \frac{\partial^2}{\partial \eta^2} + k_p^2 \right) \tilde{E}_{z\nu} = - \frac{\partial}{\partial \eta} \phi_\nu(\mathbf{r}_\perp, \eta) \,. \tag{8.86}$$

For the first two components of eq. (8.85), where $\nu = 1, 2$, we use the potential functions

$$\phi_\nu(\eta) = \frac{e n_0}{2\epsilon_0 \gamma_a} a_\nu^2 \,, \tag{8.87}$$

while, for the third one, $\nu = 3$, we have to use

$$\phi(\eta) = \frac{e n_0}{2\epsilon_0 \gamma_a} a_1 a_2^* \cos(\eta + \Delta\ell\theta) \,. \tag{8.88}$$

This can be solved in the same way as for one-dimensional density perturbations, and we get

$$E_{z\nu}(\eta) = \int_{\eta_0}^{\eta} \phi_\nu(\eta') \cos(\eta - \eta') d\eta' \,. \tag{8.89}$$

These solutions contain, as particular cases, two relevant situations. First, the single *bubble wakes*, where $a_1 \neq 0$, and $a_2 = 0$ (or vice-versa), and the laser field is represented by a simple Gaussian beam, with $\ell = p = 0$. Second, the *donut wakes* where we still have a single laser mode, but the laser beam is twisted and contains a finite amount of orbital angular momentum, such that the field $a_1$, is described by a LG mode with $\ell_1 \neq 0$. Finally, the solution of $\nu = 3$ is more interesting, and can be written explicitly as

$$E_{z3}(\eta, \theta) \simeq - \frac{e n_0}{2\epsilon_0 \gamma_0} (a_1 a_2^*) \int_{\eta_0}^{\eta} \cos(k_p \eta' + \Delta\ell\theta) \cos\left[k_p(\eta - \eta')\right] d\eta' \,. \tag{8.90}$$

Here, we have used an averaged value of the relativistic gamma factor inside the pulse, $\gamma_0$. We can see that, in the case of a self-torque pulse, the resonant part of the potential in $\eta$ is absent, and we are reduced to

$$E_{z3}(\eta, \theta) \simeq - \frac{e n_0}{2\epsilon_0 \gamma_0} (a_1 a_2^*) \cos(\Delta\ell\theta) \int_{\eta_0}^{\eta} \cos\left[k_p(\eta - \eta')\right] d\eta' \,. \tag{8.91}$$

which, apart from the dependence in the azimuthal angle $\theta$, is formally identical to the fields with $\nu = 1, 2$. But this azimuthal dependence contains a new physical meaning, because it is related with twisted plasma wakes, and therefore is able to generate helical acceleration. This is particularly relevant to the *light spring* configuration, when the mode frequency difference $\Delta\omega = \omega_1 - \omega_2$ is approximately equal to the plasma frequency, $\omega_p$. Finally, this is also relevant to *self-torque beams*, where resonant effects are absent, $\Delta\omega \simeq 0$, but transverse particle acceleration is still possible, although less effective than in the light spring case.

In order to fully understand the process of particle beam formation, and to elucidate transverse acceleration, it is convenient to use the *Panofsky-Wenzel theorem* [146]. This theorem describes, in general terms, the relation between transverse and longitudinal forces acting on charged particle beams. Assuming that a thin cylindrical electron beam is formed inside the wake, and that the theorem still applies in more complex geometries such as those relevant to these wakefields, we can state that

$$\nabla_\perp E_z = \frac{\partial}{\partial \eta} W_\perp \,, \quad W_\perp = \mathbf{E}_\perp + c\,(\mathbf{e}_z \times \mathbf{B}_\perp) \tag{8.92}$$

where $\mathbf{B}_\perp$ is the magnetic field associated with the electron beam. This allows us to calculate a poloidal field component $E_\theta$ such that

$$\frac{1}{r}\frac{\partial}{\partial \theta} E_z = \frac{\partial}{\partial \eta}(E_\theta + cB_r) \,. \tag{8.93}$$

The poloidal force is then determined by

$$E_\theta + cB_r = \frac{\Delta \ell}{r} \int E_z(\eta) d\eta \,, \tag{8.94}$$

where the axial force associated with $E_z$ depends on the longitudinal shape of the wavefield. This means that, for the light spring case, where have two modes of the same amplitude $a_1 = a_2 = a_0$, we can use eqs. (8.87) and (8.90) and write $E_z(\eta)$ as

$$E_z(\eta) = -E_0 \sin(\eta + \Delta\ell\theta) \,, \quad E_0 = \frac{e n_0 k_p}{2\epsilon_0 \gamma_a} a_0^2 \,. \tag{8.95}$$

which leads to

$$(E_\theta + cB_r) = E_0 \frac{\Delta \ell}{r} \sin(\eta + \Delta\ell\theta) \,. \tag{8.96}$$

For very weak beams, we can neglect the self-generated magnetic field $B_r$ and get an estimate of the electric field component $E_\theta$. This poloidal field implies the existence of electron acceleration in the transverse direction, and eventually produces helical beams. The ratio between the axial and poloidal electron momentum will be approximately given by

$$\frac{p_\theta}{p_z} = \frac{\Delta \ell}{k_p r_0} \,, \quad r_0 \simeq w_0 \sqrt{\frac{\ell_1 + \ell_2}{2}} \,, \tag{8.97}$$

where $r_0$ is the radius of maximum intensity of the light spring and $w_0$ is the laser beam waist at focus. The resulting helical beam trajectories are illustrated in Figure 8.4.

## 8.7 Other Acceleration Processes

Intense laser-plasma interactions are particularly fitted for electron acceleration, according to the various processes discussed in previous sections. This can lead to the formation of intense and sometimes nearly mono-energetic electron beams. But, energetic beams of protons, ions, neutrons and other particles can also be produced as well. In general these other energetic particles result from an energy cascade, where the electrons receive energy from the laser field and then couple to protons and ions. The case of positrons is somewhat different, because they have the same mass as the electron and can receive directly energy from the laser field. However, the wakefield configurations are usually not very favorable for positron acceleration because they correspond to positively charged bubbles that are attractive to electrons but repulsive to positrons. A short discussion of ion and positron acceleration processes is given here.

### 8.7.1 Ion Acceleration

Production of nearly mono-energetic proton and ion beams was shown by simulations [147] and experiments [148]. Such processes can be described by a simple analytical model. These mechanisms are particularly important for different areas, such as *inertial fusion* and *proton therapy*.

In order to discuss the ion beam formation, we consider an incident laser pulse with intensity $I_0$, frequency $\omega_0$ and duration $\Delta t$, interacting with a thin foil. For very intense pulses, the critical layer where the laser frequency equals the electron plasma frequency, can be accelerated. Such acceleration is due the ponderomotive force, and creates a shock front that is pushed forward by the incident radiation, as illustrated in Figure 8.5. The motion of this shock front can be easily estimated if we assume that the radiation pressure is balanced at the critical layer by the ion pressure [149, 150]. The electron pressure can be ignored, due to the small electron mass.

In order to establish this pressure balance, we use the shock front frame, moving in the forward laser direction with velocity $v_f$. The laser intensity and frequency in this frame are determined by Lorentz transformations, as

$$I_0' = I_0 \, \frac{(1-\beta_f)}{(1+\beta_f)} , \quad \omega_0' = \omega_0 \, \gamma_f (1-\beta_f) , \tag{8.98}$$

where $\beta_f = v_f/c$, and $\gamma_f = (1-\beta_f^2)^{-1/2}$. Assuming total reflection, the radiation pressure acting on the shock front is given by $P_{rad} = 2I_0'/c$. On the other hand, the particle pressure can be calculated by assuming that the plasma ions, initially at rest in the lab frame, and therefore, initially counter-propagate with velocity $-v_f$ in the front frame, are reflected from the shock front with velocity $+v_f$. This elastic reflection is assumed in order to satisfy

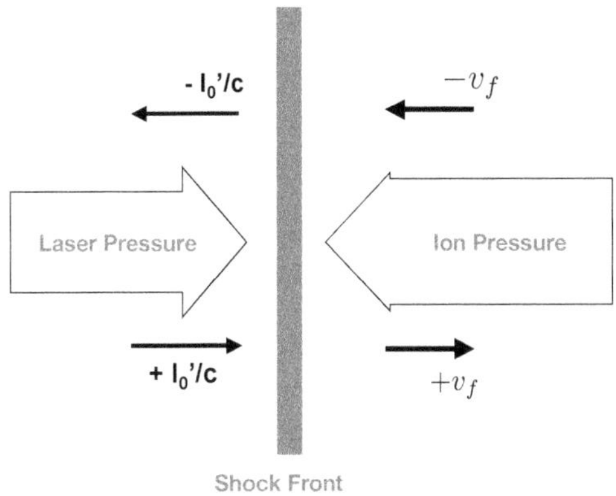


**FIGURE 8.5**
Ion acceleration at the shock front: the laser ponderomotive force is exactly balanced by the ion pressure. Quantities are shown in the moving frame.

the conservation of the particle number density. Noting that the single ion momentum is $m_i \gamma_f v_f$, where $m_i$ is the ion mass, we can then estimate the ion pressure as

$$P_i = 2n_0' m_i \gamma_f v_f^2 = 2n_0 m_i \gamma_f^2 v_f^2 \,, \tag{8.99}$$

where $n_0$ is the density of the overdense plasma region, assumed constant. The pressure balance equation is simply given by $P_{rad} = P_i$, and leads to a relation between the laser intensity and the beam velocity, as

$$B\frac{(1-\beta_f)}{(1+\beta_f)} = \gamma_f^2 \beta_f^2 \,, \quad B = \left( \frac{I_0}{n_0 m_i c^3} \right)^{1/2} \tag{8.100}$$

This can be rewritten as

$$(B^2 - 1)\beta_f^2 - 2B^2 \beta_f + B^2 = 0 \,. \tag{8.101}$$

The shock front velocity is then determined by

$$v_f = \frac{cB}{(1+B)} \,, \quad \gamma_f^2 = \frac{(1+B^2)}{(1+2B)} \,. \tag{8.102}$$

Using a Lorentz transformation, we can then retrieve the ion beam velocity in the laboratory frame

$$v_i = \frac{2v_f}{(1+\beta_f^2)} \,. \tag{8.103}$$

We notice that, for $B \ll 1$, we are in the non-relativistic regime where $\beta_f \simeq B$, and $v_i \simeq 2cB$. In the opposite case of $B \gg 1$, we are in the strongly relativistic regime, where $\gamma_f \sim \sqrt{B/2}$. Finally, the ion kinetic energy in the laboratory frame can be determined by

$$\epsilon_i = 2m_i c^2 \left[ \frac{B^2}{(1+2B)} \right] . \tag{8.104}$$

This simple treatment can be generalized to inhomogeneous plasma and to time-dependent models, without changing the basic ion acceleration mechanisms [150, 151].

Another interesting acceleration process is that of a *Coulomb explosion*, which can occur when a laser pulse interacts with an atomic cluster [152]. Micron-scale atomic clusters can be field ionized by short laser pulses, before the ions are able to respond. In a first step, the electrons are pushed away by the laser, and we are left with a positively charged cluster only made of ions, which then starts to expand due to the Coulomb repulsion between the nearby unshielded ions. This expansion (or explosion, depending on the assumed time-scale) can easily be described by a non-relativistic electrostatic model. Assuming that we have an ion cluster with charge $Ze$ and total number of ions $N_i$, the initial cluster density is determined by

$$n(r, t=0) = n_0 H(r_0 - r) , \quad n_0 = \frac{3N_i}{4\pi r_0^3} , \tag{8.105}$$

where $r$ is the radial coordinate, $r_0$ is the initial cluster radius and $H(r)$ is the Heaviside function. During expansion, the density will evolve according to

$$\frac{\partial n}{\partial t} + \frac{Ze^2}{\epsilon_0 \nu_{coll}} n^2 = 0 , \tag{8.106}$$

where $\nu_{coll}$ is the ion-ion collision frequency. This can easily be integrated, leading to

$$n(t) = \frac{n_0}{(1 + t/t_e)} , \quad t_e = \frac{\epsilon_0 \nu_{coll}}{Ze^2 n_0} , \tag{8.107}$$

whre $t_e$ is the expansion time. The corresponding evolution of the cluster radius $R(t)$, and expansion velocity $v_e(t) = v_e(r = R(t))$, are

$$R(t) = r_0 \left( 1 + \frac{t}{t_e} \right) , \quad v_e(t) = \frac{Ze^2}{\epsilon_0 \nu_{coll}} \frac{N}{4\pi R(t)^2} . \tag{8.108}$$

These results are illustrated in Figure 8.6, and qualitatively agree with particle-in-cell simulations [153]. For targets with a complex structure, such as micro-structured targets and foam targets, the two physical mechanisms described above, of shock acceleration and Coulomb explosion, can eventually take place simultaneously.

### 8.7.2 Positrons

We have seen that the trajectories of electrons and positrons are perfectly symmetrical in one-dimension. However, due to the intrinsic positive charge of the

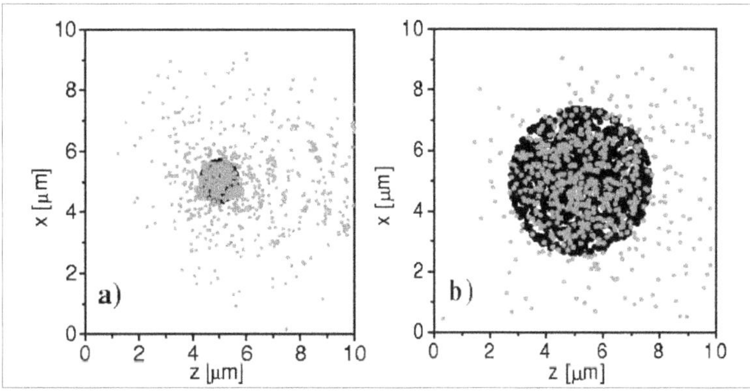


**FIGURE 8.6**
Coulomb Explosion of an atomic cluster: particle-in-cell simulations using an intensity of $I_0 = 10^{20}\,Wcm^{-2}$, at two instants $t = 50fs$ and $t = 100fs$. The ions are represented as back dots and the electrons as grey dots [Adapted from [153]].

wakefields, there is a strong defocusing effect on positrons, expelling them easily from the central region and interrupts the longitudinal acceleration process (see for a review [154]). For this reason, it is necessary to find a physical configuration where this defocusing effect is compensated by a stronger focusing force. This is provided by the donut wakefield configuration created by a LG laser pulse, if the positron beam is launched at the axis. A strong focusing field for positively charged particles exists, and is provided by the accumulation of electrons at the center of the donut, which creates an attractive force for positrons, as demonstrated in [121].

To understand this effect, we go back to the donut configuration, discussed previously, and consider the field distribution in the central region, near the axis. The transverse field structure is able to focus a positron beam, while the longitudinal field in the same region provides acceleration. If we start from the Lorentz equation for particle trajectory in this field, we can identify the electric field component $E_z$ as the accelerating field and the transverse field $W_r = E_r - cB_\theta$ as the focusing field. As previously shown, for an intense laser beam described by the electromagnetic potential $a_L(r, \eta)$, where $\eta = z - ct$, the electric field associated with the bubble is determined by

$$E_z(\eta) = -\frac{1}{4}\int_{-\infty}^{\eta}\left(\frac{\partial a_L^2}{\partial \eta'}\right)\sin(\eta - \eta')d\eta' \,. \qquad (8.109)$$

On the other hand, the transverse field $W_r$ can be obtained using the so-called *Panofsky-Wenzel theorem* discussed above. Applying this theorem to a positron beam injected into the center of a donut wake and co-moving with a laser wake, we obtain

$$\frac{\partial E_z}{\partial r} = \frac{\partial W_r}{\partial \eta}, \quad W_r = E_r - cB_\theta . \tag{8.110}$$

This can be illustrated for a simple flat-top laser pulse, described by

$$a_L(r,\eta) = a_0 f_r(r) \left[H(\eta) - H(\eta - \Delta\eta)\right] , \tag{8.111}$$

where $f_r(r)$ is the transverse laser form function and $\Delta\eta$ is the pulse length. Optimized conditions for wakefield generation correspond to a resonant pulse length of $\Delta\eta \sim \lambda_p/4$, where $\lambda_p = 2c/\omega_p$. In this case, eqs. (8.109) and (8.110) lead to the following simple results

$$E_z(r,\eta) = -\frac{1}{2} a(r)^2 \sin\eta\,, \quad W_r(r,\eta) = \frac{1}{r} a(r)^2 \left(1 - \frac{2r^2}{w_0}\right) \cos\eta\,, \tag{8.112}$$

where $a(r) \equiv a_0 f_r(r)$ and $w_0$ is the laser beam waist at the focus. These results show that the longitudinal accelerating field $E_z$ is maximum at the laser maximum. It means that it forms a donut shape for $r = w_0/2$, as expected. This is the typical region for cylindrical hollow electron beam acceleration. Although the field is much lower near the axis $r \simeq 0$, the nonlinear wakefield structure it is still able to produce a significant axial field $E_z$, with the right sign for positron acceleration in this region, as demonstrated by simulations. On the other hand, eq. (8.112) allow us to determine the focusing effect. If we assume a relativistic bubble in the blowout regime, as described above, we can estimate the maximum focusing force, as

$$W_r(r) \simeq \frac{r}{2}\left(1 - \frac{r_b^2}{2r^2}\right), \quad (\delta < r < r_b)\,, \tag{8.113}$$

where $r_b$ is the bubble radius. Near the axis, this expression is replaced by

$$W_r(r) \simeq \frac{r}{2}\left(1 - \frac{r_b^2}{2\delta^2}\right), \quad (r < \delta)\,, \tag{8.114}$$

where $\delta$ is the width of the electron layer, formed outside the bubble. The first term in these expressions is due to the ion cloud existing inside the bubble, and the second term is provided by the thin electron layers located at the bubble walls. In contrast with simple bubbles created by Gaussian laser pulses, in the case of donut bubbles a thin electron layer also exists around $r = 0$ (see Figure 8.7 for an illustration). In the usual case of $\delta \ll r_b$, the second term dominates in eq. (8.114), and provides a strong focusing force acting on a positron bunch. These features are confirmed by numerical simulations. A similar focusing force and acceleration could also eventually apply to protons, although much less efficient because of the mass difference.

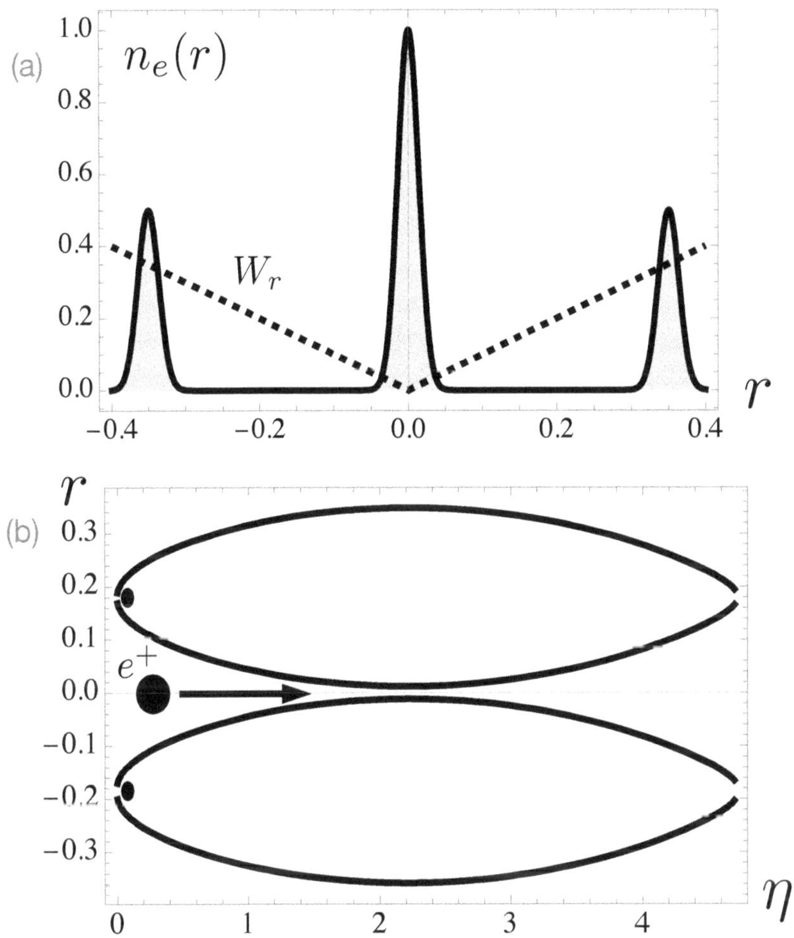


**FIGURE 8.7**
Positron acceleration in a donut wakefield: (a) the radial focusing force $W_r$ is shown (dotted line), as well as the radial dependence of the electron density $n_e(r)$ and (b) the donut wakefield, where the accelerated positron bunch $e^+$ is represented at the center.

Other alternative configurations have been proposed, for possible positron acceleration. Of particular interest is the use of a hollow electron beam in vacuum, propagating ahead of the positron bunch, in replacement of the wakefield structure. Notice however that the donut wakefield configuration is able to accelerate simultaneously a cylindrical hollow electron beam with radius $r = w_0/2$, and a positron bunch at the center $r \simeq 0$, which is quite remarkable.

# 9

# *Radiation Processes*

The emission of radiation in a plasmas contains two different kinds of effects. The first kind is single particle emission, which is eventually modified by plasma dispersion. It is mainly due to particle acceleration. Typical examples are Bremsstrahlung and cyclotron emission. But emission can also occur for non-accelerated particles, such as Cherenkov emission and transition radiation. These four basic radiation processes will be discussed in the present chapter. We also include a discussion of radiation reaction, which is a recurrent and still unsolved problem, still requiring additional theoretical consistency. Most of this discussion can be found in well-known books on electromagnetic theory [15, 16] and plasma radiation processes [155].

The second kind is due to collective processes that involve the simultaneous action of the plasma particle population. A typical example is the emission of high harmonic cascades, already discussed in Chapter 5. Other kinds of collective processes are those associated with stimulated scattering, to be discussed in the next chapter.

## 9.1 Potentials and Fields

Before discussing the different radiation processes, we review the general formulas related to single particle emission, starting with the retarded potentials and the Lienard-Wiechert formula.

### 9.1.1 Retarded Potentials

Let us go back to Maxwell's equations and to the formulation of Chapter 2. Using eq. (2.33), that relates the electromagnetic field tensor $F^{\mu\nu}$ with the 4-current $J^\mu \equiv (c\rho, \mathbf{J})$, we can establish an equation for the 4-potentials $A^\mu = (V/c, \mathbf{A})$, in the general form

$$\Box A^\nu - \partial^\nu(\partial_\mu A^\mu) = \mu_0 J^\nu \, , \tag{9.1}$$

DOI: 10.1201/9781003514701-9

where $\Box \equiv \partial^2/\partial_\mu \partial^\mu$. It is useful to assume that the potentials satisfy the *Lorentz condition*

$$\partial_\mu A^\mu = 0\,, \quad or \quad \nabla \cdot \mathbf{A} = -\frac{1}{c}\frac{\partial V}{\partial t}\,. \tag{9.2}$$

This allows us to simplify this potential equation, reducing it to

$$\Box A^\nu = \mu_0 J^\nu\,, \tag{9.3}$$

Writing more explicitly in terms of the scalar and vector potentials, $V$ and $\mathbf{A}$, we have

$$\nabla^2 \mathbf{A} - \frac{1}{c^2}\frac{\partial^2 \mathbf{A}}{\partial t^2} = -\mu_0 \mathbf{J}\,, \qquad \nabla^2 V - \frac{1}{c^2}\frac{\partial^2 V}{\partial t^2} = -\frac{\rho}{\epsilon_0}\,. \tag{9.4}$$

Let us first consider the simple case of a point charge $q$ at rest, located at position $\mathbf{r}'$, such that $\rho(\mathbf{r}) = q\delta(\mathbf{r} - \mathbf{r}')$. In this case, the potential created by the particle will only depend on the distance $R \equiv |\mathbf{R}|$, between the observation point and the particle, with $\mathbf{R} = \mathbf{r} - \mathbf{r}'$. We can use spherical coordinates, $\mathbf{R} = (R, \theta, \varphi)$, and write the potential equation as

$$\frac{1}{R^2}\frac{\partial}{\partial R}\left(R^2 \frac{\partial V}{\partial R}\right) - \frac{1}{c^2}\frac{\partial^2 V}{\partial t^2} = 0\,, \tag{9.5}$$

valid for any $R \neq 0$. We can use $V = U/R$, and rewrite this equation as

$$\left(\frac{\partial^2}{\partial R^2} - \frac{1}{c^2}\frac{\partial^2}{\partial t^2}\right) U = 0\,, \tag{9.6}$$

We now introduce the *d'Alembert coordinates*, defined as

$$\xi = t - R/c\,, \quad \eta = t + R/c\,, \tag{9.7}$$

and reduce the potential equation to

$$\frac{\partial^2 U}{\partial \xi\, \partial \eta} = 0\,, \tag{9.8}$$

which is solved as

$$U(\xi, \eta) = f(\xi) + g(\eta)\,. \tag{9.9}$$

This solution defines the well-known retarded and advanced potentials, corresponding to the first and second terms. Physically, due to causality, only the retarded solutions make sense, and this allows us to write

$$V(\mathbf{r}, t) = \frac{f(t - R/c)}{|\mathbf{r} - \mathbf{r}'|} = \int \frac{f(t')}{|\mathbf{r} - \mathbf{r}'|}\, \delta(t' - t + R/c)\, dt'\,. \tag{9.10}$$

Obviously, the time $t'$ represents the time where the particle is at a position $\mathbf{r}'$, only meaningful for a moving particle (see below). In the following, this will be called the *retarded time*. Comparing this with eq. (9.4), we realize that the explicit form of this solution for a single charge such that $\rho(\mathbf{r}) = q\delta(\mathbf{r}-\mathbf{r}')$ should be given by

$$V(\mathbf{r},t) = \frac{1}{4\pi\epsilon_0}\frac{q}{|\mathbf{r}-\mathbf{r}'|}\,. \tag{9.11}$$

In the general case of an arbitrary charge distribution $\rho(\mathbf{r},t)$ this can be generalized to

$$V(\mathbf{r},t) = \frac{1}{4\pi\epsilon_0}\int \frac{\rho(\mathbf{r}',t-R/c)}{R}d\mathbf{r}'\,. \tag{9.12}$$

Let us now assume a moving charge, with velocity $\mathbf{v}(t)$, such that

$$\rho(\mathbf{r}',t') = q\,\delta\left(\mathbf{r}'-\mathbf{v}t'\right)\,. \tag{9.13}$$

Replacing this in eq. (9.10) and integrating in $t'$, we get

$$V(\mathbf{r},t) = \frac{1}{4\pi\epsilon_0}\frac{q}{R(1-\mathbf{v}\cdot\mathbf{n}/c)}\,, \tag{9.14}$$

where $\mathbf{n} = \mathbf{R}/R$ is the unit vector pointing to the observation point. A straightforward generalization of this result for the 4-potential $A^\mu \equiv (V/c,\mathbf{A})$ gives

$$A^\mu = \frac{\mu_0}{4\pi}\frac{qv^\mu}{(R_\nu v^\nu)}\,, \tag{9.15}$$

where $R^\nu \equiv (c(t-t'),\mathbf{r}-\mathbf{r}') \equiv (R/c,\mathbf{R})$, and $v^\nu = (c,\mathbf{v})$. More explicitly, this can be written as

$$V = \frac{1}{4\pi\epsilon_0}\left.\frac{q}{R(1-\mathbf{v}\cdot\mathbf{n}/c)}\right|_{t'}\,, \qquad \mathbf{A} = \frac{\mu_0}{4\pi}\left.\frac{q\mathbf{v}}{R(1-\mathbf{v}\cdot\mathbf{n}/c)}\right|_{t'}\,, \tag{9.16}$$

These are the celebrated *Lienard-Wiechert potentials*, determining the classical electromagnetic field of a moving charge.

A more formal derivation of these potentials can be obtained using the *Green's function method*. To understand this method, we start from eq. (9.3), describing the evolution of the 4-potential $A^\mu \equiv A^\mu(x)$, and define the corresponding Green's function, $G(x,x')$, as solution of the equation

$$\Box_x G(x,x') = \delta(x-x')\,, \tag{9.17}$$

where $x$ and $x'$ represent the 4-positions $x^\mu$ and $x'^\mu$, and

$$\delta(x-x') \equiv \frac{1}{c}\delta(t-t')\,\delta(\mathbf{r}-\mathbf{r}') \tag{9.18}$$

is a four-dimensional delta-function. Given the uniformity of space-time, the Green's function only depends on the difference $z^\mu = x^\mu - x'^\mu$, allowing us to re-write eq. (9.17) as

$$\Box_z G(z) = \delta(z)\,. \tag{9.19}$$

This can be solved using Fourier analysis. For that purpose, we introduce the Fourier integral

$$G(z) = \int G(k) \exp(-ik \cdot z) \frac{dk}{(2\pi)^4}, \tag{9.20}$$

with $(k \cdot z) \equiv k_\nu z^\nu = \omega t - \mathbf{k} \cdot \mathbf{r}$, where the time component of the 4-wavevector is $k_0 \equiv \omega/c$. We also should note that the delta-function in eq. (9.19) can be replaced by the integral

$$\delta(z) = \int \exp(-ik \cdot z) \frac{dk}{(2\pi)^4}. \tag{9.21}$$

This means that the Fourier components of the Green's function are determined by

$$G(k) = -\frac{1}{(k \cdot k)} \equiv -\frac{1}{k_\nu k^\nu}. \tag{9.22}$$

Replacing this in eq. (9.20), we get

$$G(z) = -\int \frac{\exp(-ik \cdot z)}{(k \cdot k)} \frac{dk}{(2\pi)^4}. \tag{9.23}$$

Given the singularity in the denominator, it is useful to write this expression more explicitly, as

$$G(z) = -\int \exp(-i\mathbf{k} \cdot \mathbf{r}) \frac{d\mathbf{k}}{(2\pi)^3} \int \frac{\exp(-ik_0 x^0)}{\omega^2 - \kappa^2} \frac{dk_0}{2\pi}, \tag{9.24}$$

where we have now defined $\kappa = |\mathbf{k}|c$ in the denominator. This integral has two poles, defined by $\omega = \pm\kappa$, and can be solved by making the replacement

$$G_\pm(z) = -\int \exp(-i\mathbf{k} \cdot \mathbf{r}) \frac{d\mathbf{k}}{(2\pi)^3} \int \frac{\exp(-ik_0 x^0)}{(\omega \pm i\epsilon)^2 - \kappa^2} \frac{dk_0}{2\pi}, \tag{9.25}$$

where $\epsilon$ represents an infinitesimal displacement above and below the real axis, when we integrate over $k_0$. We then use

$$\frac{1}{(\omega \pm i\epsilon)^2 - \kappa^2} = \frac{1}{2\kappa} \left\{ \frac{1}{(\omega - \kappa \pm i\epsilon)} - \frac{1}{(\omega + \kappa \pm i\epsilon)} \right\}. \tag{9.26}$$

Integration then leads to

$$G_\pm(z) = \frac{H(\pm z_0)}{4\pi R} \int_{-\infty}^{\infty} \left\{ e^{i(z_0 - R)\kappa} - e^{i(z_0 + R)\kappa} \right\} \frac{d\kappa}{2\pi}, \tag{9.27}$$

where $H(z)$ is the step function and $R = (\mathbf{r} - \mathbf{r}')$ is the spatial distance, as defined before. We should note that, for $z_0 > 0$ and $R > 0$, the second integral is always equal to zero. For $z_0 < 0$, the opposite situation occurs. This allows us to write

$$G_\pm(z) = \frac{H(\pm z_0)}{4\pi R} \delta(z_0 \mp R). \tag{9.28}$$

Or, in a more explicit form

$$G_{\pm}(x-x') = \frac{1}{2\pi} H[\pm(x_0 - x_0')]\delta[(x-x')^2] \,. \tag{9.29}$$

The Green's functions can be used to establish the solutions of the covariant wave equation (9.3), as

$$A^\mu(x) = A^\mu_\pm(x) + \mu_0 \int G_\pm(x-x') J^\mu(x') dx' \,, \tag{9.30}$$

where $A^\mu_\pm(x)$ are the solutions of the homogeneous wave equation. The *radiated field* is determined by the difference between the outgoing $A^\mu_-(x)$ and the incoming $A^\mu_+(x)$ potentials, as

$$A^\mu_{rad}(x) \equiv A^\mu_-(x) - A^\mu_+(x) = \mu_0 \int \left[G_+(z) - G_-(z)\right] J^\mu(x') dx' \,, \tag{9.31}$$

where $z = x - x'$. In the absence of any incoming radiation, we have $G_-(z) = 0$. In the case of a single charge moving with velocity $\mathbf{v}$ along a given trajectory $x(\tau)$ in four-dimensional spacetime, where $\tau$ is the particle proper time, the 4-current takes the form

$$J^\mu(x') = q \int v^\mu(\tau)\delta[x' - x'(\tau)] d\tau \,, \tag{9.32}$$

with $v^\mu \equiv (c, \mathbf{v})$. Replacing this in eq. (9.31), we get

$$A^\mu(x) = \mu_0 \int G_+(z) J^\mu(x') dx' \,. \tag{9.33}$$

Or,

$$A^\mu(x) = \frac{q\mu_0}{2\pi} \int v^\mu(\tau) H[\pm(x_0 - x_0(\tau)]\delta[(x - x(\tau))^2] d\tau \,, \tag{9.34}$$

where $x_0(\tau)$ defines the initial position, associated with the light-cone condition $[x - x(\tau)]^2 = 0$. Noting that

$$\delta[f(x)] = \sum_j \frac{\delta(x - x_j)}{|df/dx|_{x_j}} \,, \tag{9.35}$$

where $x_j$ are the zeros of $f(x) = 0$, we need to calculate the derivative

$$\frac{d}{d\tau}[x - x(\tau)]^2 = -2[x - x(\tau)_\nu] v^\nu(\tau) \,. \tag{9.36}$$

The potential (9.34) becomes

$$A^\mu(x) = \frac{q\mu_0}{4\pi} \left[ \frac{v^\mu(\tau)}{v_\nu [x - x(\tau)]^\nu} \right]_{\tau = \tau_0} \,. \tag{9.37}$$

These are the Lienard-Wiechert potentials (9.15), written in terms of the particle trajectory in spacetime.

### 9.1.2 Radiation Fields

The retarded potentials discussed above can be used to calculate the fields produced by a moving charge, in order to determine the radiated energy, and also its spectrum. For this purpose we use the formulae

$$\mathbf{E} = -\nabla V - \frac{\partial \mathbf{A}}{\partial t}, \quad \mathbf{B} = \nabla \times \mathbf{A}, \tag{9.38}$$

where the potentials are determined by eqs. (9.15), and depend on a retarded time $t'$. We should therefore have in mind that

$$\frac{\partial R}{\partial t} = \frac{\partial R}{\partial t'}\frac{\partial t'}{\partial t} = -\frac{\mathbf{R}\cdot\mathbf{v}}{R}\frac{\partial t'}{\partial t}. \tag{9.39}$$

On the other hand, we know that

$$R = c(t-t'), \quad \frac{\partial R}{\partial t} = c\left(1 - \frac{\partial t'}{\partial t}\right). \tag{9.40}$$

From these two expressions, we then get

$$\frac{\partial t'}{\partial t} = \frac{1}{(1-\mathbf{v}\cdot\mathbf{n}/c)}, \tag{9.41}$$

with $\mathbf{n} = \mathbf{R}/R$. Similarly, for the space derivatives, we could write

$$\nabla t' = -\frac{\mathbf{n}}{c - \mathbf{v}\cdot\mathbf{n}}. \tag{9.42}$$

Using this in eqs. (9.38), we obtain for the electric field an expression of the form

$$\mathbf{E} = \frac{q}{4\pi\epsilon_0}\left[\frac{\mathbf{n}-\mathbf{v}/c}{\gamma^2(1-\mathbf{v}\cdot\mathbf{n}/c)^3R^2} - \frac{\mathbf{n}\times(\mathbf{n}-\mathbf{v}/c)\times\dot{\mathbf{v}}/c}{(1-\mathbf{v}\cdot\mathbf{n}/c)^3R}\right]_{t'} \tag{9.43}$$

where, in the second term, we have used the notation $\dot{\mathbf{v}} = \partial\mathbf{v}/\partial t'$. For the magnetic field, we simply have

$$\mathbf{B} = \frac{1}{c}\left[\mathbf{n}\times\mathbf{E}\right]_{t'}. \tag{9.44}$$

This result shows that the first term of the field depends on the particle velocity $\mathbf{v}$ and decays rapidly in space, as $1/R^2$. This is a local field that can be called the *static field*. In contrast, the second term depends on the particle acceleration $\dot{\mathbf{v}}$ and decays slowly with distance, as $1/R$. This is the *radiation field*. Let us focus on the second term. The radiation emitted by the accelerated particle is determined by the energy flux, $\mathbf{S}$, the so-called *Poynting's vector*, defined by

$$\mathbf{S} = (\mathbf{E}\times\mathbf{H}). \tag{9.45}$$

In vacuum, or in a non-magnetic medium, we simply have $\mathbf{H} = \mathbf{B}/\mu_0$. The electromagnetic energy flux detected at an observation point $\mathbf{r}$ is the flux component pointing in the $\mathbf{R}$ direction, or $(\mathbf{S} \cdot \mathbf{n})$. We can then state that the energy radiated per unit of angle solid $d\Omega$, as observed at this point, is determined by

$$\frac{dW}{d\Omega} = \int^t R^2(\mathbf{S} \cdot \mathbf{n}) dt = \int^t R^2(\mathbf{S} \cdot \mathbf{n}) \frac{dt}{dt'} dt' . \tag{9.46}$$

This means that the total power $P(t') = \partial W/\partial t'$, emitted per solid angle, is determined by

$$\frac{dP}{d\Omega} = \frac{q^2}{16\pi^2 \epsilon_0 c} \frac{\left[\mathbf{n} \times (\mathbf{n} - \mathbf{v}/c) \times \dot{\mathbf{v}}/c\right]^2}{(1 - \mathbf{v} \cdot \mathbf{n}/c)^5} . \tag{9.47}$$

Let us illustrate this result with two useful examples. The first is *linear motion*, such that $\mathbf{v} \parallel \dot{\mathbf{v}}$. Defining $\theta$ as the angle between the direction of observation $\mathbf{n}$ and the direction of motion, we get

$$\frac{dP}{d\Omega} = \frac{q^2 \dot{v}^2}{16\pi^2 \epsilon_0 c^3} \frac{\sin^2 \theta}{(1 - \beta \cos \theta)^5} , \tag{9.48}$$

with $\beta = v/c$. It can easily be seen that the maximum intensity of the emitted radiation occurs at an angle $\theta_{max}$, determined by

$$\cos \theta_{max} = \frac{1}{3\beta} \left( \sqrt{1 + 15\beta^2} - 1 \right) . \tag{9.49}$$

For ultra-relativistic motion, such that $\beta \to 1$, this angle tends to zero, which means that radiation is emitted mainly in the direction of the particle motion. In contrast, for slow particles, $\beta \ll 1$, the angle of maximum radiation is nearly perpendicular to the direction of acceleration, $\theta_{\text{max}} \sim \pi/2$. This is illustrated in Figure 9.1. If we integrate this equation over the solid angle $d\Omega = \sin \theta d\theta d\phi$, and note that the particle momentum is $\mathbf{p} = m\gamma\mathbf{v}$, we can write the total radiated power as

$$P(t') = \frac{1}{6\pi\epsilon_0} \frac{q^2}{m^2 c^3} \left( \frac{dp}{dt'} \right)^2 , \tag{9.50}$$

where $d\mathbf{p}/dt = m\gamma\mathbf{a}$.

The other important example is that of *circular motion* at a constant speed. In this case $\mathbf{v} \perp \dot{\mathbf{v}}$, and eq. (9.47) becomes

$$\frac{dP}{d\Omega} = \frac{q^2 \dot{v}^2}{4\pi\epsilon_0 c^3} \frac{1}{(1 - \beta \cos \theta)^3} \left[ 1 - \frac{\sin^2 \theta \cos^2 \phi}{\gamma^2 (1 - \beta \cos \theta)^2} \right] , \tag{9.51}$$

where $\theta$ is again the angle between $\mathbf{n}$ and $\mathbf{v}$, and $\phi$ is the second polar angle.

Integration over the solid angle $d\Omega = \sin \theta d\theta d\phi$, now leads to the following radiated power

$$P(t') = \frac{1}{6\pi\epsilon_0} \frac{q^2}{m^2 c^3} \gamma^2 \left( \frac{dp}{dt'} \right)^2 , \tag{9.52}$$

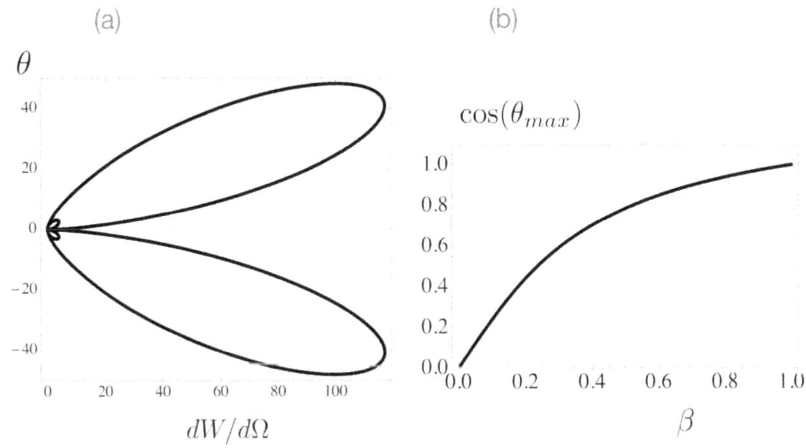


**FIGURE 9.1**
Linear acceleration, with $\mathbf{v} \parallel \dot{\mathbf{v}}$: (a) Angular intensity radiated per solid angle $dW/d\Omega$ (in arbitrary units), for $\beta = 0,5$ (internal curve) and $\beta = 0,8$ (external courve) and (b) Angle of maximum intensity, $\cos(\theta_{max})$, as a function of $\beta = v/c$.

to be compared with eq. (9.50). This shows that radiation due to circular motion is $\gamma^2$ times larger than that of linear acceleration.

If we now go back to the general expression eq. (9.48), valid for arbitrary motion, and integrate over the solid angle, we obtain

$$P(t') = \frac{q^2}{6\pi\epsilon_0 c^3}\gamma^6 \left[a^2 - (\mathbf{v} \times \mathbf{a})^2/c^2)\right] . \tag{9.53}$$

This general expression for the radiated power by a particle with arbitrary acceleration $\mathbf{a}$ is known as the *relativistic Larmor formula*.

### 9.1.3 Radiation Spectrum

In order to describe the spectral properties of the emitted radiation, we need to integrate eq. (9.47) over time, leading to the expression

$$W(\Omega) \equiv \int^{t'} \frac{dP}{d\Omega} dt = \frac{q^2}{16\pi^2\epsilon_0 c} \int^{t'} \left| \mathbf{n} \times (\mathbf{n} - \vec{\beta}) \times \dot{\vec{\beta}} \right|^2 \frac{1}{g^4} dt' . \tag{9.54}$$

where we have used $\vec{\beta} = \mathbf{v}/c$, and $g = (1 - \vec{\beta} \cdot \mathbf{n})$. Let us consider the *Parceval theorem* for a generic spectral intensity $W(\omega)$, as given by

$$\int_0^\infty W(\omega) d\omega \equiv \frac{1}{\pi} \int_0^\infty V(\omega) V^*(\omega) d\omega = \int_{-\infty}^\infty |V(t)|^2 dt . \tag{9.55}$$

Comparing this expression with eq. (9.54), we can identify the quantity $V(t)$, and write the spectral components of $W(\Omega)$ as

$$W(\omega,\Omega) = \frac{q^2}{16\pi^2\epsilon_0 c}\left|\int_{-\infty}^{\infty}\left[\mathbf{n}\times(\mathbf{n}-\vec{\beta})\times\dot{\vec{\beta}}\right]e^{-i\omega(t-\mathbf{n}\cdot\mathbf{r}(t)/c)}\frac{1}{g^2}dt\right|^2 . \quad (9.56)$$

where we have used $t = t' + R(t')/c$, with $\mathbf{R}(t') = \mathbf{r}_0 - \mathbf{n}\cdot\mathbf{r}(t')$. Notice that, at large distances from the emitting particle, the unit vector is essentially independent of time. In the non-relativistic limit, where $g \to 1$ and $dt' \equiv dt$, this will be reduced to

$$W(\omega,\Omega) = \frac{q^2}{16\pi^2\epsilon_0 c^3}\left|\int_{-\infty}^{\infty}\mathbf{n}\times(\mathbf{n}\times\mathbf{a})e^{-i\omega t}dt\right|^2 . \quad (9.57)$$

We can also write the relativistic expression (9.56) in a more adequate version, by noting that the integrand is a perfect differential

$$\frac{1}{g^2}\left[\mathbf{n}\times(\mathbf{n}-\vec{\beta})\times\dot{\vec{\beta}}\right] = \frac{d}{dt}\left[\frac{\mathbf{n}\times(\mathbf{n}\times\vec{\beta})}{g}\right] . \quad (9.58)$$

We can then integrate by parts and obtain

$$W(\omega,\Omega) = \frac{q^2\omega^2}{16\pi^2\epsilon_0 c}\left|\int_{-\infty}^{\infty}\mathbf{n}\times(\mathbf{n}\times\vec{\beta})\,e^{-i\omega(t-\mathbf{n}\cdot\mathbf{r}(t)/c)}dt\right|^2 . \quad (9.59)$$

Or, equivalently

$$W(\omega,\Omega) = \frac{q^2\omega^2}{16\pi^2\epsilon_0 c}\left|\mathbf{n}\times[(\mathbf{n}\times\vec{\beta}(\omega))]\right|^2 , \quad (9.60)$$

with

$$\vec{\beta}(\omega) = \int_{-\infty}^{\infty}\vec{\beta}(t)\,e^{-i\omega(t-\mathbf{n}\cdot\mathbf{r}(t)/c)}dt . \quad (9.61)$$

An alternative, and more general derivation of the radiated power can be obtained, if we start directly from Maxwell's equations with a source current that describes an ensemble of point charges moving in the medium. This approach is particularly useful in the context of scattering problems, as discussed in the next chapter.

## 9.2 Bremsstrahlung

This is the German name for breaking radiation, and refers to radiation emitted by acceleration or deceleration of a charged particle. For a plasma in equilibrium at a given temperature $T$, Bremsstrahlung is mainly due to electron-ion collisions. This leads to emission of broadband radiation that eventually contributes to considerable plasma cooling, as it occurs in magnetic fusion devices. In laser-plasma interactions, accelerated particles can coexist with a cold plasma and therefore radiation from fast particles can be dominant. We will examine these two cases separately. Furthermore, in laser plasma interactions, a significant fraction of laser energy is absorbed by *inverse Bremsstrahlung*, which accounts for dissipation of the electron oscillations induced by the incident laser field, due to electron-ion collisions. In contrast with the (direct) Bremsstrahlung, where the energy flows from the plasma particles to the radiated field, in the inverse process, the energy flows from the field to the particles. A discussion of the inverse Bremsstrahlung with be given in a later chapter.

### 9.2.1 Coulomb Collisions

Let us go back to eq. (9.57). Integrating over the solid angle, we get

$$W(\omega) = \int W(\omega, \Omega) d\Omega = \frac{q^2}{6\pi^2 \epsilon_0 c^3} \left| \int_{-\infty}^{\infty} \mathbf{a}(t) e^{-i\omega t} dt \right|^2 . \tag{9.62}$$

This expression is particularly useful when we want to determine the radiation emitted during an electron-ion collision. Calculation of the electron acceleration $\mathbf{a}(t)$ using the non-relativistic trajectories associated with a Coulomb collision, in terms of their distance $r$ and energy $\epsilon$, is determined by the relations

$$r = b_0 \frac{\epsilon^2 - 1}{1 - \epsilon \cos\theta} , \quad \epsilon^2 = 1 + \left( \frac{b}{b_0} \right)^2 , \tag{9.63}$$

where $b$ is the *impact parameter*, and $b_0$ is defined as

$$b_0 = \frac{Ze^2}{4\pi\epsilon_0 m v^2} , \tag{9.64}$$

with the particle charges, $q_e = -e$ and $q_i = Ze$. The result is

$$W(\omega) = \frac{2e^2\omega^2}{3\pi^2\epsilon_0 c^3} b_0^2 \left\{ \left[K_\nu'(\nu\epsilon)\right]^2 + \frac{\epsilon^2 - 1}{\epsilon^2} \left[K_\nu(\nu\epsilon)\right]^2 \right\} e^{\pi\nu} , \tag{9.65}$$

where $\nu = \omega b_0 / v$, and $K_\nu$ are the modified Hankel functions of non-integer order $\nu$. Integration of this equation over the impart parameter $b$, from 0 to $\infty$, leads to what is called the *differential emissivity* of the medium, defined as

$$\eta(\omega) \equiv N_i \int_0^\infty W(\omega)\, b db = N_i \frac{e^2 b_0^2}{6\pi^2 \epsilon_0 c^3 v} \left\{ -\nu K_\nu'(\nu) K_\nu(\nu) \right\} e^{\pi\nu} , \tag{9.66}$$

where $N_i$ is the ion density. It should be noticed that, for distant collisions, such that the incident electrons only suffer very small deviations, and their trajectories are very close to a straight line, we can use the approximation

$$\nu \ll 1 , \quad \nu\epsilon \simeq \frac{\omega b}{v} . \tag{9.67}$$

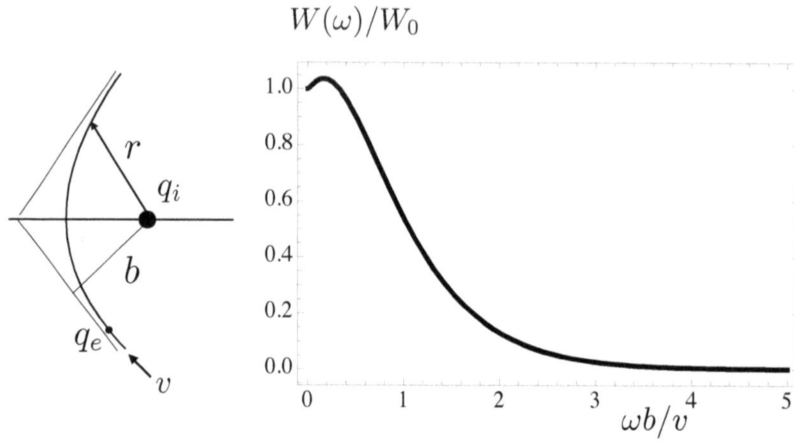


**FIGURE 9.2**
Spectrum of Bremsstrahlung, $W(\omega)/W_0$, in the limit of small deviations, as a function of the normalized frequency $\omega b/v$, as determined by eq. (9.68).

In this case, eq. (9.65) reduces to

$$W(\omega) = W_0 \left(\frac{\omega b}{v}\right)^2 \left\{ \left[K_0'(\omega b/v)\right]^2 + \left[K_0(\omega b/v)\right]^2 \right\}, \quad W_0 = \frac{2e^2 v}{3\pi^2 \epsilon_0 c^3} \frac{b_0^2}{b^2}. \tag{9.68}$$

In this approximation, the emission of radiation is small at large frequencies, such that $\omega > v/b$, as illustrated in Figure 9.2. The emissivity (9.66) is now reduced to

$$\eta(\omega) = N_i \frac{e^2 b_0^2}{6\pi^2 \epsilon_0 c^3 v} \ln\left(\frac{2}{1.78} \frac{v}{\omega b_0}\right), \tag{9.69}$$

This logarithmic dependence on the ratio $(v/\omega b_0)$ is a characteristic signature of the Coulomb collisions, as already noticed in a previous chapter.

### 9.2.2 Accelerated Beam

Let us now consider the case of emission by an accelerated particle beam. The simplest possible situation of an acceleration process is that due to a constant electric field. In this case, the relativistic gamma factor of a single particle is determined by

$$\gamma(t) = \frac{1}{\sqrt{1 - v^2(t)/c^2}} = 1 + \frac{at}{c}, \tag{9.70}$$

where $a$ is the constant acceleration produced by the applied field. Using the relativistic Larmor formula, we then get for the radiated power,

$$P(t') = \frac{e^2 a^2}{6\pi \epsilon_0 c^3} \left(1 + \frac{at}{c}\right)^6. \tag{9.71}$$

Noting that

$$P(t') = \frac{dW}{dt'} = \frac{dW}{dt}\frac{dt}{dt'} = P(t')\bar{g}\,, \tag{9.72}$$

where $P(t)$ is the power measured in laboratory time and $\bar{g} = (1 - \vec{\beta} \cdot \mathbf{n})$, we obtain

$$P(t) = \frac{e^2 a^2}{6\pi\epsilon_0 c^3}\left(1 + \frac{at}{c}\right)^4 . \tag{9.73}$$

Now using the expression for the energy radiated per solid angle, as stated in eq. (9.51), for $q = -e$, we arrive at

$$\frac{dW}{d\Omega} = \frac{e^2 a^2}{16\pi^2\epsilon_0 c^3}\sin^2\theta\left\{1 - \beta\frac{\cos\theta}{[1+\beta^2]^{1/2}}\right\}^{-5} , \tag{9.74}$$

with $\beta = at/c$. From here, we conclude that the angle for maximum emission of radiation is determined by

$$\cos\theta_{max} = \frac{1}{3\beta(t)}[1+\beta^2]^{1/2}\left[\sqrt{1 + 15\frac{\beta^2}{(1+\beta^2)}} - 1\right] . \tag{9.75}$$

It would be a useful exercise to determine the corresponding frequency spectrum, and to extend the results for a beam defined by a charged particle bunch $\rho(t) = n_b f(\mathbf{r}, t)$, where $n_b$ is the total particle number and $f(\mathbf{r}, t)$ is a form function, representing the extension of the beam in space and time.

## 9.3 Cyclotron Radiation

### 9.3.1 Particle Radiation

Let us now consider the emission of a charged particle due to its circular motion in a static magnetic field. To be specific, we consider an electron, with charge $q = -e$. We have seen, in Chapter 2, that cyclotron motion in a static magnetic field $\mathbf{B}_0$ is characterized by the Larmor radius, $r_L$, and the cyclotron frequency $\omega_c$, defined by

$$\omega_c = \frac{eB_0}{\gamma m}\,, \qquad r_L = \frac{v_\perp}{\omega_c} . \tag{9.76}$$

In the absence of a parallel velocity, $v_\parallel = 0$, we have $v \equiv v_\perp$, and $\gamma = 1/\sqrt{1 - v_\perp^2/c^2}$. The corresponding acceleration is

$$\mathbf{a} = \frac{e}{\gamma m}(\mathbf{v} \times \mathbf{B}_0) . \tag{9.77}$$

Replacing this in eq. (9.51), and integrating over a cyclotron orbit with $\phi = \omega_c t$, we get

$$\frac{dP}{d\Omega} = \frac{e^2 \omega_c^2}{16\pi^2 \epsilon_0 c^3} \int_0^{2\pi} d\phi \left[ \frac{\sin^2\theta/\gamma^2 + (\beta - \cos\theta\cos\phi)^2}{(1 - \beta\cos\theta\cos\phi)^5} \right] , \tag{9.78}$$

with $\beta = v/c$. This leads to

$$\frac{dP}{d\Omega} = \frac{e^2 \omega_c^2}{8\pi\epsilon_0 c^3} f(\beta, \theta) , \tag{9.79}$$

where

$$f(\beta, \theta) = \frac{2 + \beta^2 \cos^2\theta}{(1 - \beta^2\cos^2\theta)^{5/2}} - \frac{(4 + \beta^2\cos^2\theta)\cos^2\theta}{4\gamma^2(1 - \beta^2\cos^2\theta)^{7/2}} . \tag{9.80}$$

We can see that radiation is predominantly emitted in the plane of the cyclotron orbit, defined by $\theta = 0$. The ratio between radiation emitted in this plane and in the perpendicular plane $\theta = \pi/2$ is given by a simple expression

$$\frac{(dP/d\Omega)_0}{(dP/d\Omega)_{\pi/2}} = \frac{4 + 3\beta^2}{3(1 - \beta^2)^{5/2}} , \tag{9.81}$$

On the other hand, the cyclotron orbit introduces an oscillating phase in the radiated field, as described by the phase factor $\exp(i\mathbf{k}\cdot\mathbf{r} - i\omega t)$, where the electron position is defined by $\mathbf{r} = \mathbf{r}_0 + \mathbf{r}_L(t)$, with

$$\mathbf{r}_L(t) = \frac{v_\perp}{\omega_c} \left[\sin(\omega_c t)\mathbf{e}_x - \cos(\omega_c t)\mathbf{e}_y)\right] + v_\parallel t \, \mathbf{e}_z . \tag{9.82}$$

This periodic phase modulation creates an infinite spectrum of harmonics, as described by

$$\exp\left[-i\omega\left(t - \mathbf{k}\cdot\mathbf{r}_L(t)/c\right)\right] = \sum_{n=-\infty}^{\infty} J_n(\kappa) e^{-i(\omega - n\omega_c)t} e^{i\omega\beta_\parallel \cos\theta t} , \tag{9.83}$$

where $J_n(\kappa)$ are Bessel functions, with $\kappa = (\omega/\omega_c)\beta_\perp \sin\theta$. From here we get $dP = \sum_n dP_n$, such that, for each harmonic $\omega = n\omega_c$, we have the following radiation intensity

$$\frac{dP_n}{d\Omega} = n^2 \frac{e^2}{8\pi\epsilon_0 c} \frac{a_c^2}{\gamma m} \left[\tan^2\theta J_n^2(\kappa) + \beta_\perp^2 J_n'^2(\kappa)\right] . \tag{9.84}$$

Here, we have used the cyclotron acceleration $a_c = evB_0/\gamma m$, and $J_n'(\kappa) = dJ_n/d\kappa$. Integration in solid angle, and retaining the contribution of the parallel motion, leads to

$$P_n = P_0 \frac{(1 - \beta_0^2)}{\beta_0} \left[ n\beta_0^2 J_{2n}'(2n\beta_0) - n^2(1 - \beta_0^2) \int_0^{\beta_0} J_{2n}(2n\xi) d\xi \right] , \tag{9.85}$$

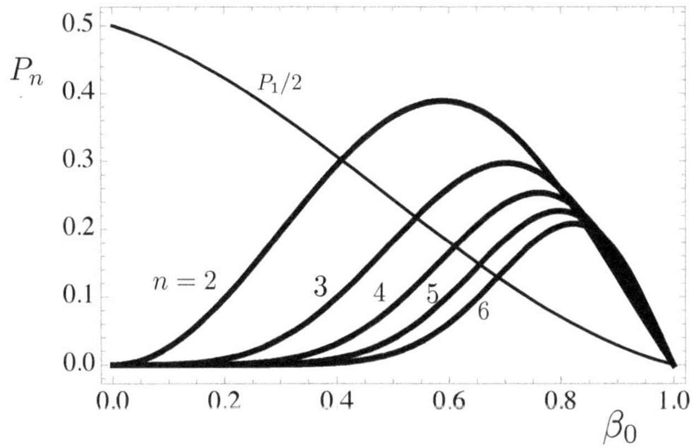


**FIGURE 9.3**
Cyclotron Radiation: Harmonic Power spectrum, $P_n$, as a function of $\beta_0 = \beta_\perp$, as determined by eq. (9.85), for different harmonic orders, $n = 2, 3, 4, 5, 6$. Power radiated at the fundamental frequency, $P_1/2$, as also shown for comparison (thin line). All values are normalized by the same factor, $P_0$.

with $\beta_0 = \beta_\perp \sqrt{1-\beta_\parallel^2}$. Adding all the harmonic contributions, we arrive at the total intensity radiated by the particle, as

$$P_{tot} = \frac{P_0}{3}\left[(1-\beta_\parallel^2)\gamma^2 - 1\right], \quad P_0 = \frac{e^2\omega_c^2}{2\pi\epsilon_0 c}. \tag{9.86}$$

We can see that, for an ultra-relativistic particle, the radiated intensity increases with $\gamma^2$, as $P_{tot} \propto P_0\gamma^2$. The harmonic power spectrum is illustrated in Figure 9.3.

### 9.3.2 Cyclotron Instability

In order to stress the qualitative differences between single particle emission processes and collective emission of radiation in a plasma, we briefly discuss the *cyclotron instability*, which results from non-equilibrium particle distribution functions, and play an important role in space and astrophysics [156, 157]. For that purpose, let us consider a plasma where the electrons are characterized by distributions known as *ring-shaped distributions*, $f_0(\mathbf{p})$. A typical example of these distributions is given by

$$f(p_\perp, p_z) = F_0 e^{-(p_\perp - p_0)^2/2mT_\perp} e^{-p_z^2/2mT_z}, \tag{9.87}$$

where $F_0$ is an normalization factor, the momentum $p_0$ characterizes the ring distribution, $T_\perp$ is the ring temperature, $T_z$ is the parallel temperature and $p_z$ is the parallel momentum with respect to the static magnetic field $\mathbf{B}_0 = B_0\mathbf{e}_z$.

The name results from the fact that they define a ring in the perpendicular momentum plane.

Due to relativistic effects, the cyclotron frequency of an electron with momentum $\mathbf{p}$ can be written as

$$\omega_c = \omega_{c0} \left( 1 + \frac{p^2}{m^2 c^2} \right)^{-1/2} , \quad \omega_{c0} = \frac{eB_0}{m} . \tag{9.88}$$

It is known (see Chapter 4) that the refractive index $N$ of an extraordinary mode that propagates in a direction perpendicular to the static magnetic field $\mathbf{B}_0$, and is polarized perpendicularly to this field, is determined by

$$N^2 = \frac{1}{\kappa_1} \left( \kappa_1^2 - \kappa_2^2 \right) , \tag{9.89}$$

where the dielectric tensor elements $\kappa_1$ and $\kappa_2$ were defined in Section 4.3. We can write these quantities in a simple form if we assume the weakly relativistic limit, valid for $p^2/m^2c^2 \ll 1$. In this limit, the dielectric tensor elements become

$$\kappa_1 = 1 - \frac{1}{2} \frac{\omega_p^2}{\omega(\omega + \omega_{c0})} + A , \quad \kappa_2 = \frac{1}{2} \frac{\omega_p^2}{\omega(\omega + \omega_{c0})} + A , \tag{9.90}$$

where $A$ represents the relativistic kinetic correction, defined by

$$A = \frac{\pi}{2} \frac{\omega_p^2}{\omega^2} \int_0^\infty p_\perp dp_\perp \int_{-\infty}^\infty \frac{p_\perp \partial f_0 / \partial p_\perp}{(\omega - \omega_c)} dp_z . \tag{9.91}$$

We should notice that, in the denominator of the parallel momentum integral, we can use the expansion

$$\omega_c \simeq \omega_{c0} \left( 1 - \frac{1}{2} \frac{p^2}{m^2 c^2} \right) . \tag{9.92}$$

From here, it is then possible to derive an approximate expression for the kinetic correction, of the form

$$A = -\frac{1}{2} \frac{\omega_p^2}{\omega(\omega - \omega_{c0})} - i\alpha , \tag{9.93}$$

where the quantity $\alpha$ is determined by the integral

$$\alpha = -\frac{\pi^2}{2} \frac{\omega_p^2}{\omega_{c0}} \int_0^\infty p^3 dp \int_{-1}^1 \delta(\omega - \omega_c) \left( \frac{\partial f_0}{\partial p} - \frac{\mu}{p} \frac{\partial f_0}{\partial \mu} \right) d\mu , \tag{9.94}$$

and $\omega_c$ is given by eq. (9.92). Here, we have introduced polar coordinates in momentum space, such that $\mathbf{p} \equiv (p, \mu, \phi)$, with $\mu = p_\perp / p_z = \cos\theta$. After integration, this reduces to

$$\alpha = -\frac{\pi^2}{2} \frac{\omega_p^2}{\omega_{c0}^2} m^2 c^2 \int_{-1}^1 (1 - \mu^2) \left( \frac{\partial f_0}{\partial p} - \frac{\mu}{p} \frac{\partial f_0}{\partial \mu} \right)_{p_r} d\mu , \tag{9.95}$$

where the integrand is calculated at the resonant momentum

$$p_r = mc\sqrt{2(\omega_{c0} - \omega)\omega_{c0}} \tag{9.96}$$

This resonant momentum is dictated by the delta function in eq. (9.94). When we neglect the kinetic correction $\alpha$, and take $\alpha = 0$ in eq. (9.93), this reduces the expected dispersion of an extraordinary mode in a cold plasma, namely

$$N^2 = \frac{(\omega^2 - \omega_p^2)^2 - \omega^2\omega_{c0}^2}{\omega^2(\omega^2 - \omega_{uh}^2)}, \quad \omega_{uh}^2 = (\omega_p^2 + \omega_{c0}^2), \tag{9.97}$$

where $\omega_{uh}$ is the upper-hybrid frequency. Let us then focus on the perturbation of the refractive index, $\delta N$, due to the existence of the kinetic corrections, when $\alpha \neq 0$. This perturbation is determined by the expression

$$2N\delta N = -i\alpha\frac{(\omega - \omega_{c0})^2}{(\omega^2 - \omega_{uh}^2)^2}\left(\omega + \omega_{c0} - \frac{\omega_p^2}{\omega}\right)^2. \tag{9.98}$$

For $\omega \simeq \omega_{c0}^2$, this can be approximated by

$$2N\delta N = -i\alpha\frac{(\omega - \omega_{c0})^2}{\omega_p^4\omega_{c0}^2}\left(2\omega_{c0} - \omega_p^2\right)^2. \tag{9.99}$$

From this expression, we can see that, for $\alpha < 0$ the perturbation $\delta N$ is imaginary positive, which means that the extraordinary wave is stable. This is true for Maxwellian distributions, where the derivative $\partial f_0/\partial p$ is always positive, which implies that $\alpha < 0$, as seen from eq. (9.95). In contrast, $\alpha$ can become positive, for certain distributions, of the type

$$f_0(p, \mu) = F(p)g(\mu), \tag{9.100}$$

when $F(p)$ peaks at a value $p \neq 0$. This is the case of the ring distributions (see Figure 9.4). In this case, we get

$$\alpha = \frac{\pi^2}{2}\frac{\omega_p^2}{\omega_{c0}^2}m^2c^2p^2\left(P\frac{\partial F}{\partial p} + Q\frac{F}{p}\right)_{p=p_r}, \tag{9.101}$$

where

$$P = \int_{-1}^{1}\left(1 - \mu^2\right)g(\mu)d\mu, \quad Q = \int_{-1}^{1}\left(1 - 3\mu^2\right)g(\mu)d\mu. \tag{9.102}$$

We can see that the quantity $P$ is always positive, which leads to $\alpha > 0$. In this case, we have a radiative instability of the extraordinary mode, at a frequency very close to the electron cyclotron frequency, $\omega \simeq \omega_{c0}$. This instability is usually called the *cyclotron maser*. We should, however, point out that the quantity $Q$ can be negative, which means that a scenario of stable to unstable transition can occur, depending on the exact balance of the Maxwellian and ring-shaped components of the electron population that eventually co-exist in the medium. The existence of a cyclotron maser can be relevant to the interpretation of space and astrophysical observations.

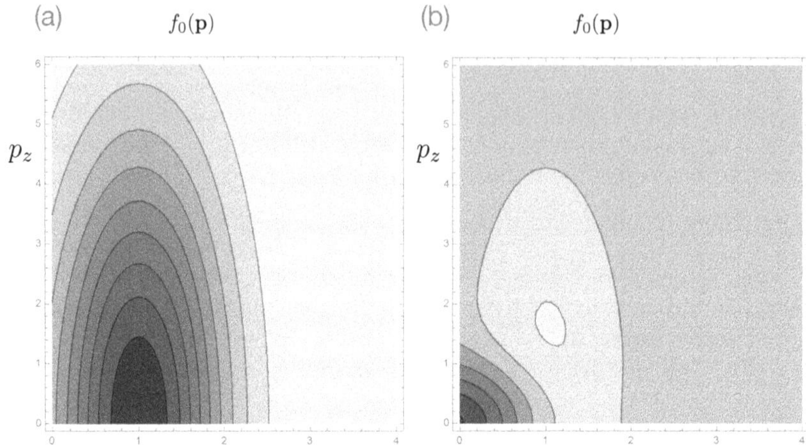


**FIGURE 9.4**
Ring-shaped distribution, leading to cyclotron maser effect: (a) simple ring-shaped distribution and (b) Composite Maxwell plus ring-shaped distribution.

## 9.4 Transition Radiation

Transition radiation is emitted by charged particles moving at constant velocity, $\dot{v} = 0$, when crossing the boundary between two dielectric media. This type of radiation was first described theoretically by Ginzburg and Franck in 1945 [158], and later developed by Garibian [159]. First observations date from 1959 [160]. The intensity of transition radiation produced at a single boundary is very low, but it can significantly be enhanced using a stratified or periodic medium [161]. It has been successfully used for high-energy particle detection [162].

Transition radiation emitted by relativistic electron populations was also observed in laser-plasma interactions [163]. In this case, the medium is not periodic, but radiation can be enhanced by short electron bunches crossing a solid target. Instead of resonant emission by a single particle, we have here the coherent emission by a modulated beam. The theory of this coherent process was developed in [164, 165].

### i) Uniform Medium

In order to calculate the radiation field, we define the total electric charge and current as the sum of two different terms, due to the plasma and to the particle beam. We can therefore write

$$\rho = \rho_{pl} + \rho_b \, , \quad \mathbf{J} = \mathbf{J}_{pl} + \mathbf{J}_b \, , \tag{9.103}$$

where the plasma terms are defined in the usual way (see below), and the beam terms are associated with a number $n_b$ of fast electrons, and can be written as

$$\rho_b = -e \sum_{j=1}^{n_b} \delta(\mathbf{r} - \mathbf{r}_j) \, , \quad \mathbf{J}_b = -e \sum_{j=1}^{n_b} \mathbf{v}_j \delta(\mathbf{r} - \mathbf{r}_j) \, . \tag{9.104}$$

Here, $\mathbf{r}_j = \mathbf{r}_{j0} + \mathbf{v}_j t$ is the trajectory of each particle $j$, and $\mathbf{v}_j$ the respective velocity. Using Poisson's equation, we get

$$\nabla \cdot \mathbf{E} = \frac{1}{\epsilon_0} \left( \rho_b - e\tilde{n} \right) \, , \tag{9.105}$$

where $\tilde{n} - n - n_0$ are the density perturbations induced by the beam, and $n_0$ the equilibrium plasma density. For a cold plasma with immobile ions, we simply have

$$\frac{\partial \tilde{n}}{\partial t} = -\nabla \cdot (n_0 \mathbf{v}) \, , \quad \frac{\partial \mathbf{v}}{\partial t} = -\frac{e}{m} \mathbf{E} \, . \tag{9.106}$$

Using $\tilde{\rho} = -e\tilde{n}$, this leads to

$$\left( \frac{\partial^2}{\partial t^2} + \omega_p^2 \right) \tilde{\rho} = -\omega_p^2 \rho_b \, , \tag{9.107}$$

where we have used the electron plasma frequency, $\omega_p = (e^2 n_0 / \epsilon_0 m)^{1/2}$. The total electric field $\mathbf{E}$ can be determined by the wave equation

$$\left( \nabla^2 - \frac{1}{c^2} \frac{\partial^2}{\partial t^2} \right) \mathbf{E} - \nabla \left( \nabla \cdot \mathbf{E} \right) = \mu_0 \frac{\partial \mathbf{J}}{\partial t} \, . \tag{9.108}$$

We also have, for the two components of the electron current, the obvious continuity relations

$$\frac{\partial \mathbf{J}_{pl}}{\partial t} = -e n_0 \frac{\partial \mathbf{v}}{\partial t} \, , \quad \nabla \cdot \mathbf{J}_b = -\frac{\partial \rho_b}{\partial t} \, . \tag{9.109}$$

Assuming perturbed quantities that evolve in time as $\exp(-i\omega t)$, we obtain from these equations

$$\mathbf{J}_{pl} = -\frac{\epsilon_0}{i\omega} \omega_p^2 \mathbf{E} \, , \quad \rho_b = \frac{1}{i\omega} \left( \nabla \cdot \mathbf{J}_b \right) \, . \tag{9.110}$$

From eq. (9.107), we also get

$$\tilde{\rho} = \frac{\omega_p^2}{\left( \omega^2 - \omega_p^2 \right)} \rho_b \, . \tag{9.111}$$

Using these expressions in the time Fourier transformed wave equation (9.108), we obtain

$$\left[ \nabla^2 + \frac{\omega^2}{c^2} \epsilon(\omega) \right] \mathbf{E} = \frac{1}{i\omega \epsilon_0 \epsilon(\omega)} \left[ \nabla \left( \nabla \cdot \mathbf{J}_b \right) + \frac{\omega^2}{c^2} \mathbf{J}_b \right] \, . \tag{9.112}$$

where $\epsilon(\omega) = (1 - \omega_p^2/\omega^2)$ is the plasma dielectric function. This equation is actually valid for any background medium. In vacuum, we simply have $\epsilon(\omega) = 1$. Its solution can be written as the sum of two terms: $\mathbf{E} = \mathbf{E}_b + \mathbf{E}_h$. The first term is the particular solution associated with the beam current ($\mathbf{J}_b \neq 0$), and the second term is the general solution of the homogeneous equation (with $\mathbf{J}_b = 0$). Performing a space Fourier transform, such that $\mathbf{E} \propto \exp(i\mathbf{k} \cdot \mathbf{r})$, we can then see that the homogeneous solution is transverse, $\mathbf{k} \cdot \mathbf{E}_h = 0$, and that the field

$$\mathbf{E}_h = \mathbf{E}_h(\omega, \mathbf{k}) \exp(i\mathbf{k} \cdot \mathbf{r} - i\omega t)\,, \tag{9.113}$$

satisfies the dispersion relation, $k^2c^2 = \omega^2 \epsilon(\omega)$, valid for isotropic media.

In order to derive the particular solution associated with the beam, $\mathbf{E}_b$, we now need to use the Fourier transform of the beam current (9.104). This is given by

$$\mathbf{J}_b(\omega, \mathbf{k}) = -2\pi e \sum_j \mathbf{v}_j \delta(\omega - \mathbf{k} \cdot \mathbf{v}_j) \exp(i\mathbf{k} \cdot \mathbf{r}_{j0})\,. \tag{9.114}$$

Replacing this in eq. (9.112), we obtain the beam field

$$\mathbf{E}_b(\omega, \mathbf{k}) = \frac{2\pi}{\omega \epsilon_0} \frac{ie}{\epsilon(\omega)} \sum_j \left[ (\mathbf{k} \cdot \mathbf{v}_j) - \frac{\omega^2}{c^2} \epsilon(\omega) \right] \frac{\delta(\omega - \mathbf{k} \cdot \mathbf{v}_j)}{[k^2c^2 - \omega^2 \epsilon(\omega)]} e^{i\mathbf{k} \cdot \mathbf{r}_{j0}}\,. \tag{9.115}$$

This result will allow us to determine the field continuity relations at the boundary, as shown next.

### ii) Sharp Boundary

We assume a sharp boundary between the plasma and vacuum, located at $z = 0$, as illustrated in the Figure 9.5. The direction of observation of the emitted radiation is defined by the wave vector $\mathbf{k}$, and the position of the $j$-th particle is given by $\mathbf{r}_j = \mathbf{r}_\perp + z\mathbf{e}_z$. In general, the particle velocity would have a perpendicular component, as $\mathbf{v}_j = \mathbf{v}_{j\perp} + v_{jz}\mathbf{e}_z$. However, in order to simplify the discussion, we assume that all the particles move perpendicularly to the boundary. This means that the angle $\theta_j$ indicated in the Figure, is equal to zero.

From Maxwell's equations we know that the boundary conditions at $z = 0$ imply the continuity of the electric field component parallel to the boundary, and the continuity of the displacement vector perpendicular to the boundary. This last condition can be replaced by the continuity of the magnetic field parallel to the boundary. Of particular interest is the case of a very dense plasma, which allows us to use the approximation $|\epsilon(\omega)| \ll 1$, to describe the spectrum radiated in the vacuum region $z \geq 0$. The boundary conditions are then reduced to

$$(\mathbf{e}_z \times \mathbf{E})_{z=0} = 0\,, \tag{9.116}$$

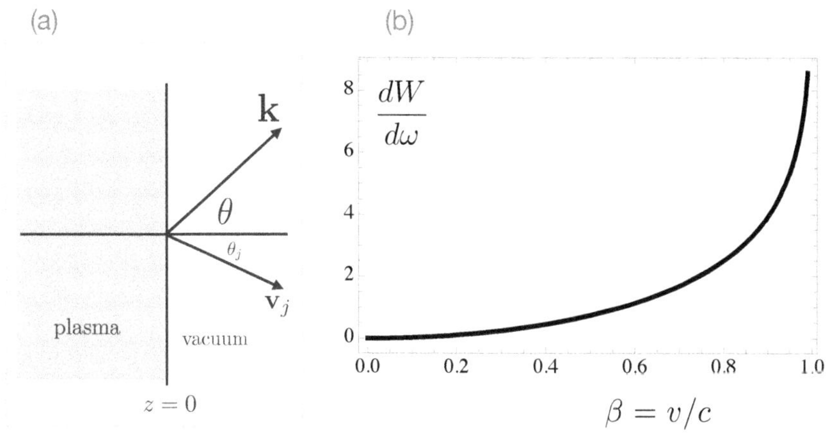


**FIGURE 9.5**
Transition Radiation: (a) A sharp boundary between the plasma and vacuum is located at $z = 0$. The fast particles cross the plasma boundary, and move into the vacuum region and (b) Radiation intensity for a single particle, in arbitrary units, as a function of the particle velocity, as given by eq. (9.124).

where, as we have seen, $\mathbf{E} = \mathbf{E}_b + \mathbf{E}_h$. Using eqs. (9.113) and (9.115), we then get

$$\mathbf{E}_h(\omega, \mathbf{k}_\perp, z \geq 0) = \frac{ie}{\omega\epsilon_0}\mathbf{e}_z \sum_j \mathcal{E}_j(\mathbf{k}_\perp)\exp(i\varphi_j)\,, \tag{9.117}$$

where the new quantities are defined as

$$\mathcal{E}_j(\mathbf{k}_\perp) = \frac{\beta_j \sin\theta}{\gamma_j(1 - \beta_j^2\cos^2\theta)}\,, \quad \varphi_j = \mathbf{k}_\perp \cdot \mathbf{r}_j - \omega(z_j/v_j)\,, \tag{9.118}$$

with $\beta_j = v_j/c$, and $\gamma_j = (1 - \beta_j^2)^{-1/2}$. The field radiated in the vacuum region $z \geq 0$ is given by the inverse Fourier transform

$$\mathbf{E}_h(\mathbf{r}, t) = \int \frac{d\omega}{2\pi} \int \frac{d\mathbf{k}_\perp}{(2\pi)^2} \mathbf{E}_h(\omega, \mathbf{k}_\perp, z) e^{i\mathbf{k}_\perp \cdot \mathbf{r}_\perp - i\omega t}\,, \tag{9.119}$$

and the energy radiated by the transition into the vacuum region $z \geq 0$, is determined by

$$\frac{dW}{d\omega d\Omega} = \frac{\omega^2\cos^2\theta}{(2\pi)^4 c}\left|\mathbf{E}_h(\omega, \mathbf{k}_\perp, z)\right|^2\,, \tag{9.120}$$

Summation over the particle beam, gives terms proportional to the total number $N_b$ as well as terms proportional to its square $N_b^2$. They correspond to the incoherent and coherent parts of the emitted radiation, where the incoherent part results from the simple addition of individual contributions, and the second and more interesting part results from the constructive interference of individual fields. They are described by

$$\frac{dW}{d\omega d\Omega} = \frac{e^2}{\pi^2 c^2} \left\{ \sum_i^{N_b} |\mathcal{E}_j|^2 + \sum_{i,j}^{N_b} \mathcal{E}_i \mathcal{E}_j^* \exp(i\varphi_i - \varphi_j) \right\} . \tag{9.121}$$

This can also be written in the form

$$\frac{dW}{d\omega d\Omega} = \frac{e^2}{\pi^2 c^2} \left\{ N_b \left\langle \mathcal{E}^2 \right\rangle + N_b(N_b - 1) | \left\langle \mathcal{E} F \right\rangle |^2 \right\} , \tag{9.122}$$

where the brackets indicate an average over all the particles in the beam, and $F \equiv F(\mathbf{v}_j)$. The averaged quantities can easily be obtained by comparison with the previous formula. It is particularly useful to consider the case of a single particle, $N_b = 1$, as in the original theory. In this case, we are reduced to

$$\frac{dW}{d\omega d\Omega} = \frac{e^2}{\pi^2} \frac{\beta^2 \sin^2\theta}{(1 - \beta^2 \cos^2\theta)^2} , \tag{9.123}$$

where $\beta = v/c$ determines the velocity of the particle. After integration over the solid angle $d\Omega = 2\pi \sin\theta d\theta$, we obtain

$$\frac{dW}{d\omega} = \frac{e^2}{2\pi\beta} \left[ (1 + \beta^2) \ln\left( \frac{1+\beta}{1-\beta} \right) - 2\beta \right] , \tag{9.124}$$

which is the classical result for a single particle. Equation (9.124) is illustrated in Figure 9.5.

## 9.5 Cherenkov Radiation

Another basic mechanism of emission of radiation by particles moving with constant velocity is the celebrated *Cherenkov effect*, first observed in 1934, and explained theoretically by Tamm and Franck soon after. In contrast with transition radiation, which occurs for any value of the particle velocity, here we need to consider particles that move with a velocity larger than the phase velocity of the emitted radiation.

In order to describe this effect, we can start from the wave equation for the potentials, eqs. (9.5) using

$$\rho = \rho_m + \rho_q , \quad \mathbf{J} = \mathbf{J}_m + \mathbf{J}_q , \tag{9.125}$$

where $\rho_m$ and $\mathbf{J}_m$ describe the medium, and

$$\rho_q = q\,\delta(\mathbf{r} - \mathbf{v}t) , \quad \mathbf{J}_q = \mathbf{v}\rho_q , \tag{9.126}$$

describe a single particle with charge $q$, moving with constant velocity $\mathbf{v}$ across the medium. Using the space and time Fourier transformation of all quantities, generically represented by $f(\mathbf{r}, t)$, as

$$f(\mathbf{r}, t) = \int \frac{d\omega}{2\pi} \int \frac{d\mathbf{k}}{(2\pi)^3} f(\omega, \mathbf{k}) \exp(i\mathbf{k} \cdot \mathbf{r} - i\omega t) , \tag{9.127}$$

we transform the potential equations into

$$\left[k^2 - \frac{\omega^2}{c^2}\epsilon(\omega)\right] V(\mathbf{k}, \omega) = \frac{1}{\epsilon_0 \epsilon(\omega)} \rho_q(\omega, \mathbf{k}) , \tag{9.128}$$

and

$$\left[k^2 - \frac{\omega^2}{c^2}\epsilon(\omega)\right] \mathbf{A}(\mathbf{k}, \omega) = \mu_0 \mathbf{J}_q((\omega, \mathbf{k}) , \tag{9.129}$$

where $\epsilon(\omega) \equiv \epsilon(\omega, \mathbf{k})$ is the dielectric function of a generic dispersive medium, and

$$\rho_q(\omega, \mathbf{k}) = 2\pi q \, \delta(\omega - \mathbf{k} \cdot \mathbf{v}) , \quad \mathbf{J}_q(\omega, \mathbf{k}) = \mathbf{v} \, \rho_q(\omega, \mathbf{k}) . \tag{9.130}$$

This leads to

$$V(\omega, \mathbf{k}) = \frac{2\pi}{\epsilon_0} \frac{q}{\epsilon(\omega)} \frac{\delta(\omega - \mathbf{k} \cdot \mathbf{v})}{[k^2 - (\omega^2/c^2)\epsilon(\omega)]} , \tag{9.131}$$

and

$$\mathbf{A}(\omega, \mathbf{k}) = \frac{\mathbf{v}}{c} \epsilon(\omega) V(\omega, \mathbf{k}) . \tag{9.132}$$

From here, we can easily derive the Fourier transform of the electromagnetic field as

$$\mathbf{E}(\omega, \mathbf{k}) = i \left[\frac{\omega}{c^2} \epsilon(\omega) \mathbf{v} - \mathbf{k}\right] V(\omega, \mathbf{k}), \tag{9.133}$$

and

$$\mathbf{B}(\omega, \mathbf{k}) = \frac{i}{c^2} \epsilon(\omega) \, (\mathbf{k} \times \mathbf{v}) \, V(\omega, \mathbf{k}) \tag{9.134}$$

It is now useful to specify the direction of motion and the direction of observation, as in Figure 9.6. Here, $b$ is a kind of impact parameter, defining the point of observation $P_0$ with respect to the direction of motion, and $\theta$ is the emission angle. We calculate the field amplitude at a position $P_0(\mathbf{r} = \mathbf{r}_0)$, performing an inverse space Fourier transformation of the field created by the particle. This gives

$$\mathbf{E}(\omega, \mathbf{r}_0) = \int \frac{d\mathbf{k}}{(2\pi)^3} \mathbf{E}(\omega, \mathbf{k}) \exp(i\mathbf{k} \cdot \mathbf{r}_0) , \tag{9.135}$$

For an appropriate choice of the origin of coordinates, we can simply write $(\mathbf{k} \cdot \mathbf{r}_0) \equiv (k_y b)$. Let us first consider the field component parallel to the particle velocity $\mathbf{v} = v\mathbf{e}_x$. Using the above equation, we can easily get

$$E_x(\omega, \mathbf{r}_0) = \frac{2\pi i}{\epsilon_0} \frac{q}{\epsilon(\omega)} e^{ik_y b} \int \frac{d\mathbf{k}}{(2\pi)^3} \left[\frac{\omega}{c^2} \epsilon(\omega) v - k_x\right] \frac{\delta(\omega - k_x v)}{k^2 - (\omega^2/c^2)\epsilon(\omega)} , \tag{9.136}$$

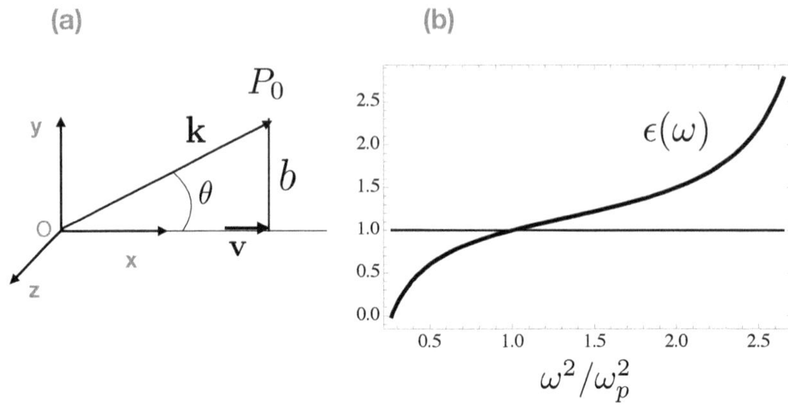


**FIGURE 9.6**
Cherenkov Radiation: (a) Geometry of the problem. (b) Dispersion function $\epsilon(\omega)$ of the extraordinary mode, as a function of $\omega^2/\omega_p^2$, for $\omega_c = \sqrt{2}\omega_p$, showing the region of Cherenkov emission between $\omega_p$ and the upper-hybrid resonance $\omega_{uh}$.

Now, introducing the auxiliary variable $\eta = k_x v$, and using $k^2 = k_\perp^2 + k_x^2$, with $k_\perp^2 = k_y^2 + k_z^2$, we perform the integration in $dk_x = d\eta/v$, as

$$\frac{1}{v}\int \frac{d\eta}{2\pi}\left[\frac{\omega}{c^2}\epsilon(\omega)v - k_x\right]\frac{\delta(\omega-\eta)}{k_\perp^2 + k_x^2 - (\omega^2/c^2)\epsilon(\omega)} \tag{9.137}$$
$$= \frac{1}{2\pi}\left[\beta^2\epsilon(\omega) - 1\right]\frac{1}{k_\perp^2 + (\omega^2/v^2)[1-\beta^2\epsilon(\omega)]}\,.$$

where $\beta = v/c$. Equation (9.136) can then be written as

$$E_x(\omega, \mathbf{r}_0) = -\frac{iq\omega}{\epsilon_0 v^2}\left[\frac{1}{\epsilon(\omega)} - \beta^2\right]\int \frac{dk_y}{2\pi}e^{ik_y b}\int \frac{dk_z}{2\pi}\frac{1}{k_y^2 + k_z^2 + \lambda^2}\,, \tag{9.138}$$

where, to simplify the expression, we have used

$$\lambda^2 = \frac{\omega^2}{v^2}\left[1 - \beta^2\epsilon(\omega)\right]\,. \tag{9.139}$$

Noting that

$$\int_{-\infty}^{\infty}\frac{dk_z}{k_y^2 + k_z^2 + \lambda^2} = \frac{\pi}{\sqrt{k_y^2 + \lambda^2}}\,, \tag{9.140}$$

we can also write this electric field component as

$$E_x(\omega, \mathbf{r}_0) = -\frac{iq\omega}{4\pi\epsilon_0 v^2}\left[\frac{1}{\epsilon(\omega)} - \beta^2\right]\int_{-\infty}^{\infty}\frac{e^{ik_y b}}{\sqrt{k_y^2 + \lambda^2}}dk_y\,, \tag{9.141}$$

At this point, we should compare this last integral with the modified Bessel function of zero-order, which can be defined as [87]

$$K_0(\kappa) = \frac{1}{2} \int_{-\infty}^{\infty} \frac{\exp(i\kappa z)}{z^2 + 1} dz \,. \tag{9.142}$$

Using $\kappa = b\lambda$, and $z = k_y/\lambda$, we finally obtain

$$E_x(\omega, \mathbf{r}_0) = -\frac{iq\omega}{2\pi\epsilon_0 v^2} \left[ \frac{1}{\epsilon(\omega)} - \beta^2 \right] K_0(b\lambda) \,. \tag{9.143}$$

A similar calculation can be made for the perpendicular component of the electric field, $E_y$, leading to the following result

$$E_y(\omega, \mathbf{r}_0) = \frac{q\omega}{2\pi\epsilon_0 v} \frac{\lambda}{\epsilon(\omega)} K_1(b\lambda) \,, \tag{9.144}$$

which is now determined by the first-order modified Bessel function $K_1(b\lambda)$. Finally, for the magnetic field, we obtain ($B_x = 0, B_z = 0$), and

$$B_y(\omega, \mathbf{r}_0) = \frac{\beta}{c} \epsilon(\omega) E_y(\omega, \mathbf{r}_0) \,. \tag{9.145}$$

Let us now discuss the physical relevance of the results obtained so far. For this purpose, we consider very large impact parameters $b$, because we are interested in radiation fields that can be observed far away from the moving particle. To be more specific, we assume that $|b\lambda| \gg 1$, and use the asymptotic expressions for the Bessel functions of order $\nu$, as [87]

$$K_\nu(\kappa) \simeq \sqrt{\frac{\pi}{2\kappa}} e^{-\kappa} \left\{ 2 + \frac{4\nu^2 - 1}{8\kappa} + \dots \right\} . \tag{9.146}$$

The electric field components (9.143) and (9.144) can then be approximately written as

$$E_x(\omega, b) \simeq \frac{iq\omega}{2\sqrt{2\pi}\epsilon_0 c^2} \left[ 1 - \frac{1}{\beta^2 \epsilon(\omega)} \right] \frac{e^{-\lambda b}}{\sqrt{\lambda b}} \,, \tag{9.147}$$

and

$$E_y(\omega, b) \simeq \frac{q}{2\sqrt{2\pi}\epsilon_0} \frac{1}{v\epsilon(\omega)} \sqrt{\frac{\lambda}{b}} \,. \tag{9.148}$$

This allows us to calculate the energy radiated by the particle, $dW$ when it travels over an infinitesimal distance $dx$ at constant velocity. Noting that the power $P = dW/dt$, radiated by the particle through the surface of an infinite cylinder of arbitrary radius $a$, located around the trajectory of the particle, is determined by

$$\frac{dW}{dx} = \frac{1}{v} \frac{dW}{dt} = \frac{c}{4\pi} \int_{-\infty}^{\infty} (\mathbf{E} \times \mathbf{H}) \, dx \,, \tag{9.149}$$

we can write

$$\left(\frac{dW}{dx}\right)_{b>a} = -\frac{ca}{\mu_0}\int_{-\infty}^{\infty}[E_x(t)B_z(t)]\,dt = -\frac{ca}{\mu_0}\Re\int_0^{\infty}[E_x(\omega)B_z^*(\omega)]\,\frac{d\omega}{2\pi}\,, \tag{9.150}$$

where the symbol $\Re$ means the real part. Now, using eqs. (9.147) and (9.148), we obtain

$$-\frac{ca}{\mu_0}E_x(\omega)B_z^*(\omega) = \frac{q^2\omega}{c^2}\left(-i\sqrt{\frac{\lambda^*}{\lambda}}\right)\left[1-\frac{1}{\beta^2\epsilon(\omega)}\right]\exp[-(\lambda+\lambda^*)a]\,. \tag{9.151}$$

If $\lambda$ is real and positive, the exponential factor in this expression decays for large distances $b > a$, and no radiation is emitted. However, when $\lambda$ is imaginary, the exponential factor is equal to one, and the result is independent of $a$. This means that the energy associated with the particle will propagate indefinitely in space. This is Cherenkov radiation. As can be seen from eq. (9.139), in order to have an imaginary value for $\lambda$, the condition $\beta^2\epsilon(\omega) > 1$ has to be satisfied. This necessary condition can be written as

$$v > \frac{c}{\sqrt{\epsilon(\omega)}} \equiv \frac{c}{n(\omega)}\,, \tag{9.152}$$

where $n(\omega)$ is the refractive index of the medium, assumed real. In this case, we have $\sqrt{\lambda^*/\lambda} = i$, and the energy radiated by the particle will be determined (in radians) by

$$\left(\frac{dW}{dx}\right)_{rad} = \frac{q^2}{c^2}\int\left[1-\frac{1}{\beta^2\epsilon(\omega)}\right]\omega d\omega\,. \tag{9.153}$$

where integration is extended to the region where $\epsilon(\omega) > 1/\beta^2$. This condition also allows us to define a characteristic angle, $\theta_c$, such that

$$\cos\theta_c = \frac{1}{\beta\sqrt{\epsilon(\omega)}}\,. \tag{9.154}$$

Radiation is only emitted inside the cone with an angle $\theta_c$, the famous *Cherenkov cone*, such that

$$\mathbf{k}\cdot\mathbf{v} = kv\cos\theta > \frac{kc}{\sqrt{\epsilon(\omega)}}\,. \tag{9.155}$$

satisfying the inequality $\theta < \theta_c$. Let us apply this general discussion to the case where a relativistic particle moves across a cold plasma. In this case, for isotropic plasmas, we have

$$\epsilon(\omega) = 1 - \frac{\omega_p^2}{\omega^2} \leq 1\,. \tag{9.156}$$

This prevents Cherenkov radiation, as seen from eq. (9.152). However, for magnetized plasmas, Cherenkov radiation can indeed occur. As shown in a previous chapter, the extraordinary mode propagating perpendicularly to a static magnetic field in a plasma, satisfies the dispersion relation $(kc/\omega) = \epsilon(\omega)$, where the dielectric function is

$$\epsilon(\omega) = 1 - \frac{\omega_p^2}{\omega^2} \frac{\omega^2 - \omega_p^2}{\omega^2 - \omega_{uh}^2} , \tag{9.157}$$

and the upper-hybrid frequency $\omega_{uh}$ is defined as twice the geometric mean between the electron plasma frequency and the electron cyclotron frequency, $\omega_{uh} = (\omega_p^2 + \omega_c^2)^{1/2}$. An equivalent, and more suggestive expression of the dielectric function is

$$\epsilon(\omega) = \frac{(\omega^2 - \omega_-^2)(\omega^2 - \omega_+^2)}{\omega^2(\omega^2 - \omega_{uh}^2)} , \tag{9.158}$$

This shows that the extraordinary mode has a resonance frequency, defined by $\omega = \omega_{uh}$, and two cut-off frequencies at $\omega_\pm = ((\omega_c^2 + 4\omega_p^2)^{1/2} \pm \omega_c)/2$. This example shows that, in a magnetized plasma, the Cherenkov condition (9.152) is satisfied in the frequency range: $(\omega_p < \omega < \omega_{uh})$, as illustrated in Figure 9.6).

## 9.6 Radiation Reaction

Until now, all radiative processes have been described assuming that the particle motion is not affected by radiation. This is valid when the radiated energy is negligible with respect to the particle energy. But this is not necessarily true for relativistic particles, because the radiated energy increases with the relativistic gamma-factor $\gamma$. In order to refine our description, we therefore need to include a radiation damping force in the particle equations of motion. However, this is not an easy task, and several different versions of this force have been proposed along the years [166, 167].

Here, we review the radiation force that needs to be added to the Lorentz equation, in order to account for the particle energy losses by radiation. We start with a non-relativistic model, first considered by Lorentz [168]. We have seen that, in this limit, the power radiated by a particle with charge $q$ moving with velocity $\mathbf{v}$ is determined by

$$P \equiv \frac{dW}{dt} = \frac{q^2}{6\pi\epsilon_0 c^3} \dot{v}^2 , \tag{9.159}$$

where $\dot{\mathbf{v}}$ is the acceleration. The total energy of the system "charge+radiation" must be conserved. This means that $P$, plus the work done by the field on the particle, must be equal to zero. This can be written as $P + (\mathbf{F} \cdot \mathbf{v}) = 0$, where

$F$ is the unknown force to be determined. Integration of this equality over a time interval $\Delta t = t_1 - t_0$ is then given by

$$\int_{t_0}^{t_1} (\mathbf{F} \cdot \mathbf{v})\, dt = - \int_{t_0}^{t_1} \frac{q^2}{6\pi\epsilon_0 c^3} \dot{v}^2 dt\,. \tag{9.160}$$

Integration by parts gives

$$\int_{t_0}^{t_1} \dot{v}^2 dt \equiv \int_{t_0}^{t_1} (\mathbf{a} \cdot \mathbf{a})\, dt = [\mathbf{v} \cdot \mathbf{a}]_{t_0}^{t_1} - \int_{t_0}^{t_1} (\dot{\mathbf{a}} \cdot \mathbf{v})\, dt\,. \tag{9.161}$$

In the case of periodic motion, we can choose the instants $t_0$ and $t_1$ such that their velocities and accelerations are equal, $\mathbf{v}(t_0) = \mathbf{v}(t_1) = 0$, and $\mathbf{a}(t_0) = \mathbf{a}(t_1) = 0$. The first term is then zero, and we are left with

$$\int_{t_0}^{t_1} \left( \mathbf{F} - \frac{q^2}{6\pi\epsilon_0 c^3} \dot{\mathbf{a}} \right) \cdot \mathbf{v}\, dt = 0\,, \tag{9.162}$$

from where we derive an expression of the radiation force, as

$$\mathbf{F} = \frac{q^2}{6\pi\epsilon_0 c^3} \dot{\mathbf{a}}\,, \tag{9.163}$$

Adding this force to the Lorentz equation, we obtain

$$\frac{d\mathbf{v}}{dt} = \frac{q}{m} (\mathbf{E} + \mathbf{v} \times \mathbf{B}) + \kappa \frac{d^2\mathbf{v}}{dt^2}\,, \tag{9.164}$$

where the constant $\kappa$ is determined by

$$\kappa = \frac{q^2}{6\pi\epsilon_0 m c^3}\,. \tag{9.165}$$

For an electron, $(q = -e)$, this is simply equal to $2r_e/3c$, where $r_e$ is the classical electron radius. This equation can be accurately applied to any non-relativistic periodic motion, such as that of an electron under the action of an oscillating electric field $\mathbf{E} = E_0 \mathbf{e}_x \exp(-i\omega_0 t)$. Neglecting the radiation force, the particle will perform a one-dimensional motion along the $x$-axis with angular frequency $\omega_0$, as described by $x(t) = A \exp(-i\omega_0 t)$, with $A = eE_0/m\omega_0^2$. According to eq. (9.159), the radiated power will be equal to

$$P = \frac{e^2}{6\pi\epsilon_0 c^3} \omega_0^4 A^2\,, \tag{9.166}$$

If we now include the radiation force, as in eq. (9.164), the equation of motion will become

$$\frac{d^2 x}{dt} + \omega_0^2 x = \kappa \frac{d^3 x}{dt^3}\,, \tag{9.167}$$

The last term can also be approximately written as

$$\kappa \frac{d^3 x}{dt^3} \simeq -\Gamma \frac{dx}{dt}\,, \quad \Gamma = \kappa \omega_0^2 = \frac{2\omega_0^2}{3c} r_e\,. \tag{9.168}$$

This reduces the particle equation of motion to that of a damped harmonic oscillator

$$\frac{d^2 x}{dt} + \omega_0^2 x + \Gamma \frac{dx}{dt} = 0\,, \tag{9.169}$$

with an solution $x(t) = A \exp(-i\omega_0 t - \Gamma t/2)$, valid for small $\Gamma$. The spectral distribution of radiation $W(\omega)$, emitted by the particle will then be given by a Lorentzian function $L(\omega)$ as

$$W(\omega) \propto |x(\omega)|^2 = A^2 L(\omega)\,, \quad L(\omega) = \frac{1}{\pi} \frac{(\Gamma/2)}{(\omega - \omega_0)^2 + (\Gamma/2)^2}\,. \tag{9.170}$$

This provides a classical description of the natural lifetime $1/\Gamma$ of a spectral transition, showing that the above solution for the radiation force is physically plausible.

However, there is a fundamental difficulty associated with eq. (9.164), which is related to the existence of unphysical solutions, called *runaway solutions*. This can easily be seen when we consider the motion of a particle in the absence of fields, ($\mathbf{E} = 0, \mathbf{B} = 0$). The equation of motion then reduces to

$$\frac{d\mathbf{v}}{dt} = \kappa \frac{d^2 \mathbf{v}}{dt^2} \;\rightarrow\; \frac{d\mathbf{v}}{dt} = \mathbf{a}_0 \exp(t/\kappa)\,. \tag{9.171}$$

After integration, this leads to a solution of the form

$$\mathbf{v}(t) = \kappa \mathbf{a}_0 \exp(t/\kappa) + \mathbf{b}_0\,, \tag{9.172}$$

where $\mathbf{a}_0$ and $\mathbf{b}_0$ are constants. This unphysical result represents a free particle that is accelerated by the radiation force, in the absence of any applied field. According to this solution, the particle would attain unlimited velocities after extremely short times, of the order of $\kappa \simeq r_2/c \sim 10^{-23} seconds$.

Covariant versions of the above derivation were formulated by Abraham and Dirac [169, 170]. This led to the so-called *Lorentz-Abraham-Dirac equation*, or *LAD equation* in short, which can be written as

$$\frac{du^\mu}{d\tau} = \frac{q}{m} F^{\mu\nu} u_\nu + \kappa \left( \frac{d^2 u^\mu}{d\tau^2} + u^\mu \frac{du_\nu}{d\tau} \frac{du^\nu}{d\tau} \right)\,, \tag{9.173}$$

where $u^\mu \equiv \gamma(c, \mathbf{v})$ is the 4-velocity, $F^{\mu\nu}$ is the electromagnetic field tensor and $\tau$ is the particle proper time. The term with curved brackets, on the left, is the covariant version of the radiation force discussed above. Unfortunately, the second derivative appearing in this covariant radiation force still leads to runaway solutions. This inconvenience is avoided if, in the radiation force term, we replace $(du_\mu/d\tau)$ by the Lorentz force $(q/m)F^{\mu\nu}u_\nu$. This leads to the *Landau-Lifshitz equation*, or simply the *LL equation*, as

$$\frac{du^\mu}{d\tau} = \frac{q}{m}F^{\mu\nu}u_\nu + g^\mu \,, \tag{9.174}$$

where the radiation force term takes the form

$$g^\mu = \kappa\frac{q^2}{m^2}\left[\frac{m}{q}(\partial_\alpha F^{\mu\nu})u_\nu u_\alpha + F^{\mu\nu}F_{\nu\alpha}u^\alpha + (F^{\mu\alpha}u_\alpha)^2 u^\mu\right] . \tag{9.175}$$

It should be stressed that this new quantity was chosen to satisfy the relation $q^\mu u_\mu = 0$ [16]. It is useful to write the LL equation in a more explicit form. For the three space components of the 4-velocity $\mathbf{u} = \gamma\mathbf{v}$, and for electrons $(q = -e)$, we get [171]

$$\begin{aligned}\frac{du}{dt} = & -\frac{e}{m}(\mathbf{E} + \mathbf{v}\times\mathbf{B}) - \kappa\frac{e}{m}\gamma\left(\frac{d\mathbf{E}}{dt} + \mathbf{v}\times\frac{d\mathbf{B}}{dt}\right) \\ & + \frac{2}{3}\frac{e}{cE_{cr}}\left[\mathbf{E}(\mathbf{v}\cdot\mathbf{E}) - c\mathbf{B}\times + (\mathbf{E} + \mathbf{v}\times\mathbf{B})\right] \\ & - \frac{2}{3}\frac{q}{cE_{cr}}\gamma^2\left[(\mathbf{E} + \mathbf{v}\times\mathbf{B})^2 - (\mathbf{v}\cdot\mathbf{E})^2\right]\mathbf{v} \,,\end{aligned} \tag{9.176}$$

where $E_{cr} = 4\pi\epsilon_0 m^2c^4/e^3 \equiv mc^2/(er_e)$ is the critical field. From here, we can immediately see that radiation reaction is only relevant for fields of the order to the critical field, which is two orders of magnitude higher than the Schwinger field: $E_{cr} = E_S/\alpha$, where $\alpha$ is the fine-structure constant. This means that, even for extremely high intensity laser fields, of the order of $E_S \simeq 1.32\times 10^{18}V/m$, we still expect very small (if not negligible) radiation reaction corrections, of the order of a few percent. However, at such intensities, quantum effects associated with the emission of single photons would have become relevant, and this classical force loses its sense. But, even in this limit, it would still be possible to use a classical description, refined by the inclusion of stochastic forces that simulate the quantum transitions [171].

# 10

# *Scattering*

In this chapter, we consider scattering processes, which are secondary radiation processes that result from the existence of incident electromagnetic radiation, acting on particles in the medium. Specific examples of scattering effects are included. We first study Thompson scattering, in the linear and nonlinear regimes. This is basically a single particle process, but plays an important role in plasma physics. We then consider stimulated Raman and Brillouin scattering, which are essentially collective processes in a plasma. We also pay attention to scattering by generic plasma density fluctuations. This chapter ends with the discussion of other specific processes, which include scattering from relativistic electron bunches, plasma bubbles and a classical version of the Unruh radiation.

## 10.1 Thomson Scattering

We start with scattering of electromagnetic radiation by a free electron, first observed by Thomson at the beginning of the last century [172]. It can be seen as the low-energy limit of the Compton effect [173], to be discussed in Chapter 12, and plays an important role in plasma physics. We first give the basic linear description, leading to the famous Thomson cross-section. We then consider the nonlinear effects due to the relativistic electron response to intense electromagnetic radiation.

### 10.1.1 Linear Model

Let us assume an incident electromagnetic wave, propagating in vacuum or in a uniform plasma with wavevector $\mathbf{k}$, as described by the electric field

$$\mathbf{E}(\mathbf{r}, t) = \mathbf{E}_0 \exp(i\mathbf{k} \cdot \mathbf{r} - i\omega t) . \tag{10.1}$$

where the wave frequency is given by $\omega = kc/\sqrt{\epsilon(\omega)}$. But we usually consider vacuum, where we simply have $\omega = kc$. The case of particles in a plasma is more complicated, because it involves the study of density fluctuations, as discussed later.

DOI: 10.1201/9781003514701-10

A free electron in free space will oscillate under the action of this field, and these oscillations will eventually create a scattered field. For low intensities, the electron response is linear, and is simply given by

$$\mathbf{a} \equiv \frac{d\mathbf{v}}{dt} = -\frac{e}{m}\mathbf{E}. \tag{10.2}$$

Using the expression for the radiated power per unit of angle solid, given by eq. (9.47), for an electron in the non-relativistic regime, or

$$\frac{dP}{d\Omega} = \frac{e^2}{16\pi^2\epsilon_0 c^3}\left|\mathbf{n}\times\mathbf{n}\times\mathbf{a}\right|^2 , \tag{10.3}$$

where $\mathbf{n} = \mathbf{R}/R$ is the unit vector defining the scattered direction, we get

$$\frac{dP}{d\Omega} = \frac{e^4}{16\pi^2\epsilon_0 m^2 c^3}\left|\mathbf{n}\times\mathbf{E}\right|^2 , \tag{10.4}$$

We can see that

$$\left|\mathbf{n}\times\mathbf{E}\right|^2 = |E_0|^2\left[1-(\mathbf{n}\cdot\mathbf{e}_0)^2\right] , \tag{10.5}$$

where $\mathbf{e}_0$ is the unit polarization vector of the incident wave. Defining $\theta$ as the scattering angle, we have $|\mathbf{n}\cdot\mathbf{e}_0| = \sin\theta$, which is equal to zero when $\mathbf{e}_0$ is in the plane perpendicular to the scattering plane. It is sometimes useful to assume incident unpolarized radiation, for which we have

$$\left[1-(\mathbf{n}\cdot\mathbf{e}_0)^2\right] = \frac{1}{2}(1+\cos^2\theta). \tag{10.6}$$

In this case, eq. (10.4) can be reduced to

$$\frac{dP}{d\Omega} = \frac{c}{2}\epsilon_0|E_0|^2 r_e^2(1+\cos^2\theta), \tag{10.7}$$

where we have used *classical electron radius* $r_e$, defined as

$$r_e = \frac{1}{4\pi\epsilon_0}\frac{e^2}{mc^2} \simeq 2.818\times 10^{-15}\,(m). \tag{10.8}$$

As we have seen in Chapter 1, this quantity is the distance at which the electrostatic self-energy of the electron equals its rest-mass energy. We now introduce the *scattering cross-section* per unit solid-angle, $d\sigma/d\Omega$, as the ratio between the scattered power and the incident energy flux $\mathbf{S}$, or

$$\frac{d\sigma}{d\Omega} = \frac{1}{|\mathbf{S}|}\frac{dP}{d\Omega}, \tag{10.9}$$

where

$$\mathbf{S} = |\mathbf{E}\times\mathbf{H}| = \frac{1}{\mu_0 c}|E_0|^2. \tag{10.10}$$

This leads to the simple expression

$$\frac{d\sigma}{d\Omega} = \frac{1}{2} r_e^2 (1 + \cos^2\theta) , \tag{10.11}$$

Finally, integration of this differential cross-section over the entire solid angle, using $d\Omega = 2\pi \sin\theta d\theta$, leads to the famous formula for the *Thomson cross-section*

$$\sigma_T = \frac{8\pi}{3} r_e^2 \equiv \frac{8\pi}{3} \left( \frac{e^2}{4\pi\epsilon_0 mc^2} \right)^2 , \tag{10.12}$$

valid for incident unpolarized light. It is also useful to write this quantity in terms of the fine structure constant $\alpha$, and the *Compton wavenumber*, $k_C$, as

$$\sigma_T = \frac{8\pi}{3} \left( \frac{\alpha}{k_C} \right)^2 \simeq 6.652 \times 10^{-29} \, (m^2) . \tag{10.13}$$

Although very small, this cross-section is a very important quantity. As shown in the next chapter, the electrons in a plasma can be coupled to the electromagnetic radiation, leading to collective processes that include the possible occurrence of Bose-Einstein condensation.

It should also be noticed that the above results are only valid for low incident frequencies, such that the wavenumber of the incident radiation is small when compared with the Compton wavenumber, $k \equiv (\omega/c) \ll k_C$. When this inequality is not satisfied, we need to take into account the loss of photon momentum associated with the scattering process. As a result of this loss, the scattered wavenumber is different from the incident one, $k' \neq k$. In this case, the differential cross-section (10.11) is replaced by

$$\frac{d\sigma}{d\Omega} = \frac{1}{2} r_e^2 \left( \frac{k'}{k} \right)^2 \left[ \frac{k'}{k} + \frac{k}{k'} - \sin^2\theta \right] , \tag{10.14}$$

where the ratio $(k'/k)$ is determined by the quantum formula

$$\left( \frac{k'}{k} \right) = \left[ 1 + \frac{\hbar\omega}{mc^2} (1 - \cos\theta) \right]^{-1} . \tag{10.15}$$

This is known as the *Klein-Nishina formula* [174]. A more detailed discussion of Compton scattering, which includes the influence of plasma dispersion, is given in Chapter 12.

### 10.1.2 Nonlinear Description

At relativistic intensities, the scattered radiation no longer occurs at a single frequency, which would be equal to the incident wave frequency, $\omega' = \omega$ for electrons initially at rest. Instead, the scattered spectrum includes harmonics of the incident wave frequency, eventually coalescing into a broad spectrum.

This is very similar to what occurs in cyclotron radiation, but here the static magnetic field disappears and is replaced by the incident electromagnetic wave.

The first experimental observation of the nonlinear Thomson scattering was reported in 1998 [175], and many authors contributed to the theory of this process (see [176, 177, 178]). In order to describe the nonlinear scattering regime, we start with the Lorentz equation determining the relativistic electron response as

$$\frac{dp}{dt} = -e\left(\mathbf{E} + \mathbf{v} \times \mathbf{B}\right) , \tag{10.16}$$

and assume an intense incident wave with electric field $\mathbf{E} = \mathbf{E}_0 \cos(\mathbf{k}_0 \cdot \mathbf{r} - \omega_0 t)$ propagating in vacuum. The wave magnetic field is $\mathbf{B} = (\mathbf{k}_0 \times \mathbf{E})/\omega_0$, and its amplitude simply $B_0 = E_0/c$. We have seen in Chapter 2 that the electron orbit in this field has a characteristic 8-shape, as illustrated in Figure 2.2. For the present purposes, we assume that the wave propagates along the $z$-axis, and rewrite the equation of motion as

$$\frac{d}{d\tau}(\gamma\vec{\beta}) = \left(\mathbf{e}_x + \vec{\beta} \times \mathbf{e}_y\right) a_0 \cos\eta \, , \tag{10.17}$$

where $\vec{\beta} = \mathbf{v}/c$ is the normalized particle velocity, $a_0 = (eE_0/m\omega_0 c)$ is the normalized field amplitude, time is the dimensionless variable $\tau = \omega_0 t$ and $\eta = \tau - (\omega_0/c)z$. We consider an electron satisfying the initial conditions

$$\mathbf{r}(\tau = 0) \equiv (0, 0, z_0) \, , \quad \vec{\beta}(\tau = 0) \equiv (0, 0, \beta_0 \equiv v_0/c) \, , \tag{10.18}$$

The solution of eq. (10.17) can be written in parametric form as

$$\mathbf{r} = \mathbf{r}(\eta) \, , \quad \vec{\beta} = \vec{\beta}(\eta) \, , \quad \tau = \tau(\eta) \, , \tag{10.19}$$

where $\eta$ is the wave phase. Noting that, in the present configuration, the particle velocity in the $z$-direction is always zero, $\beta_z = 0$, we get a solution of the form

$$\gamma\beta_x = \frac{a_0}{\gamma_0(1-\beta_0)} (\sin\eta - \sin\eta_0) \, , \quad \gamma\beta_y = \gamma - \gamma_0(1-\beta_0) \, , \tag{10.20}$$

where $\eta_0 = -(\omega_0/c)z_0$ and $\gamma_0 = (1-\beta_0^2)^{-1/2}$. From here, we can also retrieve the relativistic factor, as

$$\gamma = \gamma_0 + \frac{a_0^2}{2\gamma_0(1-\beta_0)} (\sin\eta - \sin\eta_0)^2 \, , \tag{10.21}$$

In the particular case of a particle initially at rest, such that $z_0 = 0$ and $\beta_0 = 0$, this would simplify to

$$\beta_x = \frac{a_0}{\gamma} \sin\eta \, , \quad \beta_y = 1 - \frac{1}{\gamma} \, , \quad \gamma = 1 + \frac{a_0^2}{2} \sin^2\eta \, . \tag{10.22}$$

The particle trajectories corresponding to eqs. (10.20)–(10.21) can be written in parametric form, as

$$x(\eta) = -\frac{a_0}{\gamma_0(1-\beta_0)}\left[(\cos\eta - \cos\eta_0) + (\eta-\eta_0)\sin\eta_0\right]\frac{c}{\omega_0}, \tag{10.23}$$

and

$$\tau(\eta) = -\frac{(\eta-\eta_0)}{(1-\beta_0)}\left[1+\frac{a_0^2}{4}(1+\beta_0)\left(1+2\sin^2\eta_0\right)\right] + \frac{a_0^2}{2}\frac{(1+\beta_0)}{(1-\beta_0)}\left[2\cos\eta\sin\eta_0 - \frac{1}{4}\sin 2\eta - \frac{3}{4}\sin 2\eta_0\right]. \tag{10.24}$$

Again, for the particular case of $z_0 = 0$ and $\beta_0 = 0$, these trajectories would reduce to

$$x(\eta) = \frac{ca_0}{\omega_0}(1-\cos\eta), \quad \tau(\eta) = 1 + \frac{a_0^2}{4}\left[1 + 4\cos\eta - \frac{1}{2}\sin 2\eta\right]. \tag{10.25}$$

Using the above solutions, we can define the period $T$ of the 8-shaped orbital motion, as a time interval where $\tau$ increases by $2\pi$, or

$$T = \frac{2\pi}{(1-\beta_0)}\left[1+\frac{a_0^2}{4}(1+\beta_0)\left(1+2\sin^2\eta_0\right)\right]. \tag{10.26}$$

As we can see, the period of the electron motion depends not only on the normalized wave amplitude $a_0$, but also on the initial conditions. It should be noticed that, after one period, the particle position moves from its initial value $\mathbf{r}_0 \equiv (0, 0, z_0)$ to the new position $\mathbf{r}_T \equiv (x_T, 0, z_T)$, where $x_T$ is simply determined by an increase of $\eta$ by $2\pi$ in eq. (10.23). This gives

$$x_T \equiv x(\eta_0 + 2\pi) = -\frac{2\pi c}{\omega_0}\frac{a_0}{\gamma_0(1-\beta_0)}\sin\eta_0. \tag{10.27}$$

Similarly, the value of $z_T$ is determined from eq. (10.25), as

$$z_T = (T - 2\pi)\frac{c}{\omega_0}. \tag{10.28}$$

Given the periodicity of the electron orbit, we expect to observe a scattered signal with an infinity of harmonics, $\omega_\nu = \nu\omega_1$, where $\omega_1 = 2\pi/T$ is the basic frequency of the electron motion, and $\nu$ is an integer. In order to understand the shape and intensity of the emitted spectrum, we go back to eq. (9.60), which states that

$$\frac{dW}{d\omega d\Omega} = \frac{e^2\omega^2}{16\pi^2\epsilon_0 c}\left|\mathbf{n}\times\left[(\mathbf{n}\times\vec{\beta}(\omega)\right]\right|^2, \tag{10.29}$$

with

$$\vec{\beta}(\omega) = \int_{-\infty}^{\infty}\vec{\beta}(t)\, e^{-i\omega[t-\mathbf{n}\cdot\mathbf{r}(t)/c)]}dt. \tag{10.30}$$

Noting that $\mathbf{r}(\tau)$ is periodic, with period $T = 2\pi/\omega_1$, we can use a series expansion of the form

$$\vec{\beta}(\omega) = \sum_{\nu=-\infty}^{\infty} \vec{\beta}_\nu \delta(\omega - \nu\omega_1) , \tag{10.31}$$

with amplitudes

$$\vec{\beta}_\nu = \frac{1}{T} \int_0^T \vec{\beta}(t)\, e^{-i\omega_1[t - \mathbf{n}\cdot\mathbf{r}(t)/c]} dt . \tag{10.32}$$

According to eq. (10.26), we can use

$$\omega_1 = \frac{\omega_0}{1 + a_0^2/2} \left(1 - \beta_0^2\right) \gamma_0^2 . \tag{10.33}$$

The power emitted by each harmonic will then be given by

$$P_\nu = \frac{e^2 \nu^2 \omega_1^2}{16\pi^3 \epsilon_0 c} \left| \mathbf{n} \times \vec{\beta}_\nu \right|^2 , \tag{10.34}$$

After integration of eq. (10.32), we obtain $P_\nu = 0$, for even harmonics $\nu = (0, 2n)$, where $n$ is integer, and

$$P_\nu = \frac{e^2 \nu^2 \omega_0^2}{4\pi \epsilon_0 c} \left( \frac{\omega_1}{\omega_0} \right)^4 a_0^2 \left\{ J_{(\nu-1)/2}(\nu\xi) - J_{(\nu+1)/2}(\nu\xi) \right\} , \tag{10.35}$$

for odd values $\nu = 2n + 1$, where we have used Bessel functions of the first kind, with arguments that depend on the wave intensity, as specified by

$$\xi = \frac{a_0^2}{4(1 + a_0^2/2)} . \tag{10.36}$$

It is useful to consider some asymptotic expressions for $P_\nu$, valid for weak and large incident fields. For weak fields, such that $a_0^2 \ll 1$, the scattered power is negligible for all harmonics, $P_{\nu\neq 1} \simeq 0$, and the scattered signal reduces to a single frequency $\omega_1$, such that

$$(\omega_1/\omega_0) \simeq \gamma_0^2 (1 - \beta_0)^2 \simeq 1 , \quad P_{tot} \simeq P_1 \propto a_0^2 \omega_1^3 . \tag{10.37}$$

In contrast, for large fields, $a_0^2 > 1$, we observe a large number of harmonics up to a limit $\nu_{lim}$, and the total scattered intensity decays with the incident field according to

$$\nu_{lim} = 0.32 a_0^3 , \quad P_{tot} \propto \frac{1}{a_0} . \tag{10.38}$$

We can see from these estimates that, although the total scattered power increases with the incident wave intensity for $a_0 \ll 1$, as predicted from the linear model of Thomson scattering, it nevertheless starts to decrease as $1/a_0$, for large intensities above $a_0 = 1$. This is due to an increase of the electron effective mass. This behavior is illustrated in Figure 10.1.

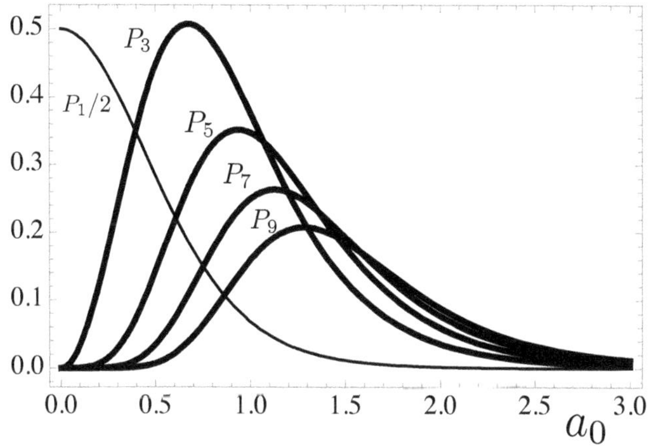


**FIGURE 10.1**
Nonlinear Thomson Scattering: Scattered Power $P_\nu$, emitted at as a function of the incident field amplitude $a_0$, for successive harmonic numbers, $n \to 3, 5, 7, 9$. For comparison, the fundamental frequency power $P_1$, divided by 2, is also represented (thin line). The curves are represented in arbitrary units (same normalization factor for all harmonics)

## 10.2 Stimulated Raman Scattering

We have seen that incident electromagnetic waves can be scattered by individual particles. However, in homogeneous plasmas, this single particle effect is canceled, due to phase mixing of the fields radiated by particles located at different parts of the medium. For every electron, we can always find another electron at a distance $\pi c/\omega$ from the first one, such that it produces a scattered field with the opposite phase, thus canceling the field of the first one. This means that only density fluctuations that deviate from uniformity, can contribute to a finite value of Thomson scattering. These density fluctuations can simply be of thermal origin. The effect of scattering by an arbitrary spectrum of fluctuations will be considered in a later section.

More interestingly, collective scattering can occur due to plasma waves and oscillations. Particularly important are the cases of Raman scattering, due to coupling of the incident radiation with electron plasma waves, and Brillouin scattering due to coupling with ion-acoustic waves. If the incident radiation is sufficiently intense, an interesting situation occurs where these electrostatic waves can be excited by the incident field, and the scattered wave amplitude grows exponentially. This process of self-excitation is called *stimulated scattering*.

Here, we examine the case of *Raman instability*, or Raman stimulated scattering. Originally, Raman scattering was discovered in atomic physics [179], when it was shown that incident light in an atomic gas could lead to the appearance of sidebands of the scattered frequency, distant from the incident frequency by an amount that is equal to the transition frequency between two energy states of the atom, $\omega_a$. This is an inelastic process, where photons of the incident light lose energy to the atoms (Stokes sideband), or gain energy from the atom (anti-Stokes sideband). The concept of stimulated Raman scattering appeared also in atomic physics, in 1962 [180], related with the excitation of internal energy levels of the atoms (see the review [181]).

In plasmas, the same name is given to scattered light with sidebands where, instead of the atomic frequency $\omega_a$, we have the electron plasma frequency $\omega_p$. Raman scattering in plasmas is therefore similar, but not identical, to the atomic Raman scattering. For the present purpose, we assume an incident electromagnetic wave propagating on a homogeneous and isotropic plasma, where coupling between density perturbations and the incident field is included. This is described by the wave equation

$$\left( \frac{\partial^2}{\partial t^2} - c^2 \nabla^2 + \frac{\omega_p^2}{\gamma} \right) \mathbf{A} = - \frac{\omega_p^2}{\gamma} \left( \frac{\tilde{n}}{n_0} \right) \mathbf{A} \, , \tag{10.39}$$

where $\mathbf{A} \equiv \mathbf{A}(\mathbf{r}, t)$ represents the vector potential of the transverse electromagnetic waves, and $\tilde{n} = n - n_0$ are the density perturbations eventually excited by the wave in the medium. We also include the relativistic factor, $\gamma$. For a cold plasma, we can write $\gamma \simeq \gamma_a \equiv \sqrt{1 + a^2}$, with $\mathbf{a} = e\mathbf{A}/mc$. Let us assume a wave solution of the form

$$\mathbf{A} = \mathbf{A}_0 \exp[i(\mathbf{k}_0 z - \omega_0 t)] + c.c. \tag{10.40}$$

representing wave propagation along $Oz$, with a slowly varying amplitude $\mathbf{A}_0$. Assuming an underdense plasma, such that $\omega_0^2 \gg \omega_p^2/\gamma_a$, and using the nonlinear dispersion relation $\omega_0^2 = k_0^2 c^2 + \omega_p^2/\gamma_a$, we can derive from eq. (10.39) an envelope equation for the normalized amplitude, of the form

$$2i\omega_0 \frac{\partial}{\partial t} \mathbf{a} = \frac{\omega_p^2}{\gamma_a} \left( \frac{\tilde{n}}{n_0} \right) \mathbf{a} \, . \tag{10.41}$$

The density perturbation $\tilde{n}$ is coupled to the high-frequency electromagnetic wave by the ponderomotive force. Using the cold electron fluid equations, where the ponderomotive force is included, we obtain an evolution equation for $\tilde{n}$, given by

$$\left( \frac{\partial^2}{\partial t^2} + \frac{\omega_p^2}{\gamma} \right) \tilde{n} = n_0 \frac{c^2}{2\gamma_a} \nabla^2 |\mathbf{a}|^2 \, . \tag{10.42}$$

We now assume that the incident electromagnetic wave is modulated by the electron plasma oscillations and develops two sidebands, as described by envelope solutions of the form

$$\mathbf{a}(\mathbf{r},t) = a_0 + \sum_{\pm} \mathbf{a}_{\pm} \exp\left[\pm i\left(\mathbf{k}' \cdot \mathbf{r} - \omega' t\right)\right] , \tag{10.43}$$

where $(\mathbf{k}', \omega')$ pertain to low-frequency oscillations of the envelope. From this, we get

$$\frac{1}{2}|\mathbf{a}|^2 = a_0^2 + \left\{(a_+ + a_-^*)a_0 e^{i(\mathbf{k}'\cdot\mathbf{r}-\omega' t)} + c.c.\right\} , \tag{10.44}$$

We can also use, for the density fluctuations

$$\tilde{n} = n' e^{i(\mathbf{k}'\cdot\mathbf{r}-\omega' t)} + c.c. , \tag{10.45}$$

where $n'$ is an amplitude. Replacing this in the above eqs. (10.41) and (10.42), it is then possible to derive a dispersion equation for the sidebands $\omega_\pm = \omega_0 \pm \omega'$, given by

$$\omega'^2 = \frac{\omega_p^2}{\gamma_0}\left\{1 + a_0^2 k'^2 c^2 \frac{D_+ + D_-}{D_+ D_-}\right\} , \tag{10.46}$$

where we use $\gamma_0 = \sqrt{1 + a_0^2}$ , and where $D_\pm$ are the dispersion relations of the scattered waves (sidebands), valid in the limit of infinitesimal density perturbations, $n' \simeq 0$, and defined as

$$D_\pm = \omega_\pm^2 - \left\{k_\pm^2 c^2 + \frac{\omega_p^2}{\gamma_0}\right\} . \tag{10.47}$$

At this point, it should be noticed that $\mathbf{k}' = \mathbf{k}_\pm - \mathbf{k}_0$. This allows us to approximately write

$$k'^2 c^2 = \frac{\omega_p^2}{\gamma_0} + 2\left[\omega_0 \frac{\omega_p}{\sqrt{\gamma_0}} - (\mathbf{k}_0 \cdot \mathbf{k}')c^2\right] . \tag{10.48}$$

Inserting this in eqs. (10.47), we obtain

$$D_- \simeq -2\left(\omega_0 - \frac{\omega_p}{\sqrt{\gamma_0}}\right)\left(\omega' - \frac{\omega_p}{\sqrt{\gamma_0}}\right) , \tag{10.49}$$

and

$$D_+ \simeq 2\left(\omega_0 + \frac{\omega_p}{\sqrt{\gamma_0}}\right)\left(\omega' - \frac{\omega_p}{\sqrt{\gamma_0}}\right) + 2\left(\frac{\omega_p^2}{\gamma_0} - k'^2 c^2\right) , \tag{10.50}$$

We now consider the direction of propagation of the scattered waves, noting that $\mathbf{k}' = \mathbf{k}_\pm - \mathbf{k}_0$, and defining $\theta$ as the angle between $\mathbf{k}_\pm$ and $\mathbf{k}_0$. This allows us to rewrite eq. (10.48) in the form

$$k'^2 c^2 \simeq \frac{\omega_p^2}{\gamma_0}\left(1 + \frac{\omega_p^2}{\omega_0^2 \gamma_0}\right) + 4\omega_0^2 \sin^2(\theta/2) \,. \tag{10.51}$$

Similarly, we get

$$D_+ \simeq 2\left(\omega_0 + \frac{\omega_p}{\sqrt{\gamma_0}}\right)\left(\omega' - \frac{\omega_p}{\sqrt{\gamma_0}}\right) - 4\omega_0^2 \sin^2(\theta/2) - 2\frac{\omega_p^4}{\gamma_0^2 \omega_0^2} \,. \tag{10.52}$$

Let us first consider the case of forward scattering, corresponding to $\theta = 0$. Noting that $|D_+| \gg |D_-|$, we can approximately write eq. (10.46 ) as

$$\left(\omega'^2 - \frac{\omega_p^2}{\gamma_0}\right) = -\frac{\omega_p^4}{\gamma_0^2} \frac{a_0^2}{2\omega_0(\omega' - \omega_p/\sqrt{\gamma_0})} \,. \tag{10.53}$$

This leads to

$$\left(\omega' - \frac{\omega_p}{\sqrt{\gamma_0}}\right)^2 \left(\omega' + \frac{\omega_p}{\sqrt{\gamma_0}}\right) \simeq -\frac{\omega_p^4}{\gamma_0^2} \frac{a_0^2}{2\omega_0} \tag{10.54}$$

We now use $\omega' = (\omega_p/\gamma_0) + i\Gamma$, where $\Gamma$ is the growth rate. From here, we easily get

$$\Gamma \simeq \frac{1}{2}\left(\frac{\omega_p}{\omega_0 \sqrt{\gamma_0}}\right)^{1/2} \frac{\omega_p}{\sqrt{\gamma_0}} a_0 \,. \tag{10.55}$$

Apart from the factor $\gamma_0 \geq 1$, this corresponds to the standard growth rate for forward Raman instability that can be found in the literature. This result is illustrated in Figure 10.2.

To complete our discussion, let us consider the case where the angular term becomes dominant in eq. (10.51). Replacing this in eq. (10.46), and still ignoring $D_+$, we obtain

$$\left(\omega'^2 - \frac{\omega_p^2}{\gamma_0}\right) \simeq -\frac{\omega_p^2}{\gamma_0} \frac{2\omega_0 \sin^2(\theta/2)}{\omega_0(\omega' - \omega_p/\sqrt{\gamma_0})} a_0^2 \,. \tag{10.56}$$

Using $\omega' = (\omega_p/\gamma_0) + i\Gamma$, this then leads to the growth rate for Raman side-scattering, as a function of the scattering angle, given by

$$\Gamma(\theta) \simeq \frac{1}{2}\left(\frac{\omega_p}{\omega_0 \sqrt{\gamma_0}}\right)^{1/2} \omega_p a_0 \sin(\theta/2) \,. \tag{10.57}$$

Comparing this with eq. (10.55), we can see that for $\sin(\theta/2) > (\omega_p/2\omega_0\sqrt{\gamma_0})$, side-scattering will dominate over forward scattering. A more detailed analysis would necessarily involve the neglected upper-sideband $D_+$.

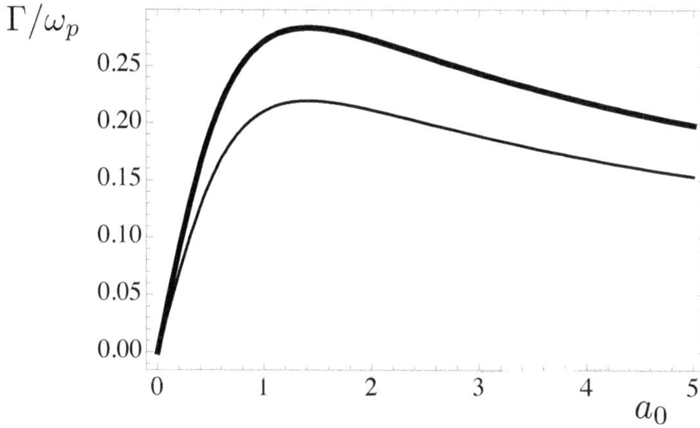


**FIGURE 10.2**
Stimulated Raman Scattering: Instability growth rate for forward scattering, normalized to the plasma frequency $\Gamma/\omega_p$, as a function of the incident field amplitude $a_0$, for $\omega_0/\omega_p = 1.2$ (bold curve) and $\omega_0/\omega_p = 2$ (thin curve).

## 10.3 Stimulated Brillouin Scattering

We now consider scattering of electromagnetic waves by ion-acoustic oscillations. Here, again, an instability can occur, which is now called *Brillouin instability*. The ion response needs to be added, and we write the electron plasma frequency as $\omega_{pe}$, instead of simply $\omega_p$, in order to introduce the ion plasma frequency $\omega_{pi}$, ignored in the previous section. We also include wave losses, in order to determine the instability thresholds.

The electromagnetic field is now described by the wave equation

$$\left( \frac{\partial^2}{\partial t^2} - c^2 \nabla^2 + \omega_{pe}^2 + 2\gamma_{coll} \frac{\partial}{\partial t} \right) \mathbf{A} = -\omega_{pe}^2 \left( \frac{\tilde{n}_e}{n_0} \right) \mathbf{A} \,, \tag{10.58}$$

where, in contrast with eq. (10.39), we ignore the electron relativistic factor (which can easily be added), and introduce an electromagnetic damping rate, $\gamma_{coll}$ associated with particle collisions. This quantity is defined as

$$\gamma_{coll} = \nu_{ei} \frac{\omega_{pe}^2}{2\omega_0^2} \,, \tag{10.59}$$

where $\nu_{ei}$ is the electron-ion collision frequency, and $\omega_0$ the incident wave frequency. The high-frequency electromagnetic field couples directly to the electrons, as before, because the ion response is much weaker due to the mass difference. But the electrons couple themselves to the ions through the collective plasma motions. We therefore need to consider both electron and ion oscillations, which are described by the following coupled equations

$$\left( \frac{\partial^2}{\partial t^2} - S_e^2 \nabla^2 + 2\gamma_e \frac{\partial}{\partial t} \right) \tilde{n}_e + \omega_{pe}^2 \left( \tilde{n}_e - \tilde{n}_i \right) = \frac{e^2 n_0}{2 m_e^2 c^2} \nabla^2 |\mathbf{A}|^2 , \tag{10.60}$$

and

$$\left( \frac{\partial^2}{\partial t^2} - S_i^2 \nabla^2 \right) \tilde{n}_i + \omega_{pi}^2 \left( \tilde{n}_i - \tilde{n}_e \right) = 0 . \tag{10.61}$$

Here, $S_j^2$ are the electron and ion thermal velocities ($j = e, i$), and $\gamma_e$ is the electron damping rate (eventually including Landau damping). We assume an incident electromagnetic wave with frequency $\omega_0$ and wavevector $\mathbf{k}_0 = k_0 \mathbf{e}_x$, and consider a backscattered signal that propagates with frequency $\omega_s = \omega_0 - \omega$, and wavevector $\mathbf{k}_s = (\mathbf{k}_0 - \mathbf{k}) = (k_0 - k)\mathbf{e}_x$. The total electromagnetic field will be described as

$$\mathbf{A} = \mathbf{A}_0 \exp(i k_0 x - i \omega_0 t) + \mathbf{A}_s \exp[i(k - k_0)x - i(\omega_0 - \omega)] . \tag{10.62}$$

The density perturbations have a similar space-time behavior. Use of eqs. (10.58)–(10.61) then leads to the following dispersion relation

$$\epsilon(\omega, \mathbf{k}) \equiv 1 + \chi_e + \chi_i + \chi_{NL} = 0 , \tag{10.63}$$

where the electron and ion susceptibilities can be written as

$$\chi_e = - \frac{\omega_{pe}^2}{(\omega - i\gamma_e)^2 - k^2 S_e^2} , \quad \chi_i = - \frac{\omega_{pi}^2}{(\omega^2 - k^2 S_i^2)} , \tag{10.64}$$

and the nonlinear susceptibility

$$\chi_{NL} = \frac{1}{2} \frac{\chi_e (1 + \chi_i) k^2 c^2}{\left[ (\omega_s - i\gamma_e)^2 - \omega_{pe}^2 - k^2 c^2 \right]} a_0^2 , \tag{10.65}$$

with $a_0 = eA_0/m_e c$. In order to discuss the stability of the oscillations satisfying this dispersion relation, we take $\omega \ll \omega_0$, and use $\chi_e \simeq 1/k^2 \lambda_{De}^2$. This then leads to

$$(\omega - \delta) \left( \omega^2 - k^2 v_{ac}^2 - 2i\gamma_e \omega \right) = - \frac{k^2 c^2}{4\omega_0} \left( \omega_{pi}^2 - \omega^2 \right) a_0^2 , \tag{10.66}$$

where $v_{ac} = \sqrt{T_e/m_i}$ is the ion acoustic velocity. The new quantity $\delta$ is defined by

$$\delta = i\gamma_s + \frac{c^2}{2\omega_0} \left( k_0^2 - k^2 \right) , \tag{10.67}$$

where $\gamma_s$ is the collisional damping rate (10.59) for the scattered wave. It should be noticed that backscattering implies that $k \simeq 2k_0$, and that

$\omega \sim \omega_{ac} \equiv k v_{ac}$. The nonlinear term will become larger than the loss terms associated with $\gamma_e$ and $\gamma_s$ when the incident field amplitude is sufficiently large, such that

$$a_0^2 \geq a_{thr}^2 = 8 \frac{\gamma_e}{\omega_{ac}} \frac{\gamma_s}{\omega_0} \frac{\omega_0^2}{\omega_{pe}^2} \frac{S_e^2}{c^2} , \tag{10.68}$$

where $\gamma_s \propto \nu_{ei}$, and Landau damping contributions to $\gamma_e$ could eventually be dominant. This condition defines the intensity threshold $a_{thr}^2$ for stimulated backscattering. Above threshold, the growth rate will then be given by $\Gamma$, as the imaginary part of the frequency $\omega$, such that

$$\Gamma \simeq -\frac{1}{(\gamma_e + \gamma_s)} \left\{ 2\gamma_e\gamma_s - \frac{k_0^2 c^2}{\omega_{ac}\omega_0} \omega_{pi}^2 a_0^2 \right\} . \tag{10.69}$$

The instability will therefore grow as $\Gamma \propto a_0^2$. In contrast, for intensities well above threshold, such that $a_0^2 \gg a_{thr}^2$, wave damping will be negligible and the instability growth rate becomes

$$\Gamma \simeq \frac{k c \omega_{pi}}{2\sqrt{2\omega_{ac}\omega_0}} a_0 . \tag{10.70}$$

In this high-intensity regime, the growth rate is then proportional to $a_0$. There is also a third regime, where $|\omega| \gg \omega_{ac}$, where the ion-acoustic oscillations are no longer proper modes satisfying the dispersion relation of the medium, but forced oscillations sometimes called *quasi-modes*. In this case, eq. (10.66) will reduce to

$$\omega^3 = -\frac{k_0^2 c^2}{\omega_0} \omega_{pi}^2 a_0^2 , \tag{10.71}$$

with the solution

$$\omega = \frac{1}{2}(1 - i\sqrt{3}) \left( \frac{k_0^2 c^2}{\omega_0} \omega_{pi}^2 a_0^2 \right)^{1/3} . \tag{10.72}$$

In this case, we have $\Gamma \propto a_0^{2/3}$. These are the three different growth rate regimes for the decay instabilities, first identified by Nishikawa in 1968 [182]. Apart from wave damping, another source of limitation for the Raman and Brillouin instabilities is the finite size and duration of the incident radiation (the so-called finite pump effects). This will have to be included in a more complete analysis.

## 10.4 Scattering by Density Fluctuations

We have seen previously that Thomson scattering by a single free electron is given, in the linear regime by eq. (10.3). This can be rewritten as

$$\frac{dP}{d\Omega} = r_e^2 \left| \mathbf{n} \times \mathbf{n} \times \mathbf{e}_0 \right|^2 |S_0| , \tag{10.73}$$

where $\mathbf{n}$ is the unit vector in the scattering direction, $\mathbf{e}_0 \equiv \mathbf{E}_0/|E_0|$ is the unit polarization vector of the incident field, $r_e$ is the classical electron radius and $S_0$ the incident energy flux, as defined by

$$r_e = \frac{e^2}{4\pi\epsilon_0 mc^2}\,, \quad S_0 = |\mathbf{E}_0 \times \mathbf{H}_0| = \epsilon_0 c|E_0|^2\,. \tag{10.74}$$

We now generalize the single-particle analysis of Section 10.1, and consider scattering by a large number of electrons per unit volume, $n_0 \gg 1$, located at positions $\mathbf{r} = \mathbf{r}_j(t)$, described by the microscopic density

$$N(\mathbf{r}, t) = \sum_j^{n_0} \delta\left(\mathbf{r} - \mathbf{r}_j(t)\right)\,. \tag{10.75}$$

In this case, the total scattered electric field, seen at a distance $R$ from the emitting region is

$$\mathbf{E}_s(\mathbf{r}, t) = \frac{r_e}{R} E_0 \sum_j \left[\mathbf{n} \times \mathbf{n} \times \mathbf{e}_0\right] e^{i(\mathbf{k}_0\cdot\mathbf{r}_j - \omega_0 t')}\,, \tag{10.76}$$

where $t'$ is the retarded time. Or, using (10.75)

$$\left(\frac{\mathbf{E}_s}{E_0}\right) = \frac{r_e}{R}\int N(\mathbf{r}, t)\left[\mathbf{n} \times \mathbf{n} \times \mathbf{e}_0\right] e^{i(\mathbf{k}_0\cdot\mathbf{r} - \omega_0 t')} d\mathbf{r}\,. \tag{10.77}$$

We can now define the spectrum of density fluctuations as the double Fourier integral of the microscopic density

$$N(\omega', \mathbf{k}') = \int d\mathbf{r} \int dt\, N(\mathbf{r}, t) e^{-i\mathbf{k}'\cdot\mathbf{r} + i\omega' t}\,. \tag{10.78}$$

The spectral component of the scattered intensity can also be written as

$$\frac{dP(\omega)}{dtd\Omega} = |S_0| r_e^2 \lim_{T\to\infty} \frac{1}{2\pi T}\left|\int_V d\mathbf{r} N(\mathbf{r}, t) e^{i(\mathbf{k}'\cdot\mathbf{r}_j - \omega' t)}\left[\mathbf{n} \times \mathbf{n} \times \mathbf{e}_0\right]\right|^2\,, \tag{10.79}$$

where $(\omega = \omega_0 - \omega')$ is the frequency of the scattered radiation. The corresponding time-averaged quantity is

$$\left\langle \frac{dP(\omega)}{dtd\Omega} \right\rangle = |S_0| r_e^2 \left|\mathbf{n} \times \mathbf{n} \times \mathbf{e}_0\right|^2 n_0 S(\omega', \mathbf{k}')\,, \tag{10.80}$$

where we have introduced the *spectral power density* of the particle density fluctuations, defined as

$$S(\omega', \mathbf{k}') = \lim_{T\to\infty} \frac{1}{T n_0} \left|N(\omega', \mathbf{k}')\right|^2\,, \tag{10.81}$$

and $N(\omega', \mathbf{k}')$ is defined by (10.78). This result is quite general, and can be applied to any spectrum. In particular, for thermal fluctuation, this quantity can be calculated from the microscopic plasma theory of Chapter 3 (see [183], for a discussion of early work on this subject). The relativistic kinetic theory of thermal fluctuations can be found in [184].

It is useful to show that the scattered power by density fluctuations can be derived from a more general approach, where propagation of the scattered radiation in the plasma (and not in vacuum as above) is taken into account. For this purpose, we consider that the scattered field $E_s$ is described by a wave propagating in the medium, where plasma dispersion and a nonlinear source term are included. For waves scattered at a frequency $\omega$ and wavevector $\mathbf{k}$, we have

$$\left[\mathbf{k} \times (\mathbf{k} \times \mathbf{e}_s) + \frac{\omega^2}{c^2}\bar{\bar{\epsilon}} \cdot \mathbf{e}_s\right] E_s = -i\omega\mu_0 \mathbf{J}_{NL}(\omega, \mathbf{k}), \tag{10.82}$$

where $\mathbf{e}_s = \mathbf{E}_s/|E_s|$ is the unit polarization vector and $\bar{\bar{\epsilon}}$ is the plasma dielectric tensor. Multiplying this equation by $\mathbf{e}_s^*$, and noting that $(\mathbf{e}_s \cdot \mathbf{e}_s^*) = 1$, we get

$$E_s(\omega, \mathbf{k}) = -\frac{i}{\epsilon_0 \omega} \frac{\mathbf{e}_s^* \cdot \mathbf{J}_{NL}(\omega, \mathbf{k})}{D(\omega, \mathbf{k})}, \tag{10.83}$$

where

$$D(\omega, \mathbf{k}) = \left(k^2 - |\mathbf{k} \cdot \mathbf{e}_s|^2\right) \frac{c^2}{\omega^2} + \mathbf{e}_s^* \cdot \bar{\bar{\epsilon}} \cdot \mathbf{e}_s, \tag{10.84}$$

is the plasma dielectric function. In order to determine the scattered power, we need to calculate the current $\mathbf{J}_{NL}$. For arbitrary density fluctuations, $n(\mathbf{r}, t)$, the nonlinear current associated with the interaction between incident field and fluctuations is $\mathbf{J}_{NL}(\mathbf{r}, t) = -eN(\mathbf{r}, t)\mathbf{v}_0(\mathbf{r}, t)$. For an incident field of the form $\mathbf{E}_0(\mathbf{r}, t) = E_0 \mathbf{e}_0 \exp(i\mathbf{k}_0 \cdot \mathbf{r} - i\omega_0 t)$, and using the Fourier spectrum of the density fluctuations, $N(\omega', \mathbf{k}')$, we can then write

$$\mathbf{J}_{NL}(\mathbf{r}, t) = -eE_0 \int \frac{d\omega'}{2\pi} \int \frac{d\mathbf{k}'}{(2\pi)^3} N(\omega', \mathbf{k}')[\mathbf{e}_0 \cdot \bar{\bar{\mu}}(\omega_0)] e^{i(\mathbf{k}_0 + \mathbf{k} \cdot \mathbf{r} - i(\omega_0 + \omega')t}, \tag{10.85}$$

where $\bar{\bar{\mu}}(\omega_0)$ is the mobility tensor, defined generically by the relation $\mathbf{v} = \bar{\bar{\mu}} \cdot \mathbf{E}$. From the definition of the Fourier transform

$$\mathbf{J}_{NL}(\omega, \mathbf{k}) = \int dt \int d\mathbf{r} \mathbf{J}_{NL}(\mathbf{r}, t) e^{i\mathbf{k} \cdot \mathbf{r} - i\omega t}, \tag{10.86}$$

we then get

$$\mathbf{J}_{NL}(\omega, \mathbf{k}) = -eE_0 \int d\omega' \int d\mathbf{k}' N(\omega', \mathbf{k}')[\mathbf{e}_0 \cdot \bar{\bar{\mu}}(\omega)] \delta(\mathbf{k} - \mathbf{k}_0 - \mathbf{k})\, \delta(\omega - \omega_0 - \omega'). \tag{10.87}$$

This expression can now be used to calculate the power scattered by density fluctuations in the $\mathbf{k}$ direction. We define the scattered power as the work performed by the scattered field on the plasma charges, namely

$$P(\omega, \mathbf{k}) = -\mathbf{J}_{NL}(\omega, \mathbf{k}) \cdot \mathbf{E}_s^*(\omega, \mathbf{k}). \tag{10.88}$$

Using eq. (10.83), we then arrive at the expression

$$P(\omega, \mathbf{k}) = \frac{1}{\epsilon_0 \omega} \frac{|\mathbf{e}_s \cdot \mathbf{J}_{NL}(\omega, \mathbf{k})|^2}{D^*(\omega, \mathbf{k})|} . \tag{10.89}$$

And, using (10.87), we get

$$P(\omega, \mathbf{k}) = \frac{e|E_0|^2}{\epsilon_0 \omega} \frac{|N(\omega', \mathbf{k}')|^2}{D^*(\omega, \mathbf{k})} \left| \mathbf{e}_0 \cdot \bar{\bar{\mu}}(\omega_0) \cdot \mathbf{e}_s \right|^2 , \tag{10.90}$$

where $\omega = \omega_0 + \omega'$ and $\mathbf{k} = \mathbf{k}_0 - \mathbf{k}'$. A final step has to be taken, because we need to integrate this expression over all the possible frequencies $\omega$ associated with a given scattering wavevector. It can easily be recognized that only the frequencies satisfying the linear dispersion relation $D(\omega, \mathbf{k}) = 0$ will survive. This can be stated by the following relation

$$\Re \left[ \frac{i}{D^*(\omega, \mathbf{k})} \right] = -\pi \frac{\delta(\omega - \omega_k)}{(\partial D / \partial \omega_{\mathbf{k}})} . \tag{10.91}$$

Such a relation is demonstrated in the following way. We first define a complex frequency $\omega = \omega_{\mathbf{k}} + (\delta + i\gamma)$, where $\omega_{\mathbf{k}}$ is a solution of the linear dispersion relation, $\delta \ll \omega_{\mathbf{k}}$ is a small deviation with respect to this solution, and $\gamma$ is a small dissipation factor. We then develop $D(\omega, \mathbf{k})$ around the linear solution, according to

$$\omega D^*(\omega, \mathbf{k}) = \omega_{\mathbf{k}} D^*(\omega, \mathbf{k}) + (\delta + i\gamma) \left( \frac{\partial \omega D^*}{\partial \omega} \right)_{\omega_k} . \tag{10.92}$$

This allows us to write

$$\Im \left[ \frac{1}{\omega D^*(\omega, \mathbf{k})} \right] = \frac{\gamma}{(\gamma^2 + \delta^2)} \frac{1}{(\partial \omega D / \partial \omega)_{\omega_k}} . \tag{10.93}$$

If we integrate this expression in frequency, we get

$$\int \frac{\gamma}{\delta^2 + \gamma^2} d\omega \equiv \gamma \int \frac{dx}{x^2 + \gamma^2} = \gamma 2\pi i \left( \frac{1}{2i\gamma} \right) = \pi . \tag{10.94}$$

If we now take $\gamma \to 0$, we obtain

$$\Im \left[ \frac{1}{\omega D^*(\omega, \mathbf{k})} \right] \to \frac{\pi \delta(\omega - \omega_{\mathbf{k}})}{(\partial \omega D / \partial \omega)_{\omega_k}} . \tag{10.95}$$

This is valid for every solution $\omega = \omega_{\mathbf{k}}$ of the plasma dispersion relation, and can be written in the form of eq. (10.91). Going back to eq. (10.90), we can now write

$$P(\omega, \mathbf{k}) = \pi \frac{e^2 |E_0|^2}{\epsilon_0 \omega} \frac{|N(\omega', \mathbf{k}')|^2}{(\partial D / \partial \omega)_{\omega_k}} \left| \mathbf{e}_0 \cdot \bar{\bar{\mu}}(\omega_0) \cdot \mathbf{e}_s \right|^2 \delta(\omega - \omega_k) . \tag{10.96}$$

It is then possible to show that this general expression reduces to eq. (10.80) in the appropriate limit. If scattering occurs in vacuum, the plasma mobility tensor and the dispersion function reduce to

$$\bar{\bar{\mu}}(\omega_0) = \frac{ie}{m\omega_0}\bar{\bar{1}}, \quad D(\omega, \mathbf{k}) = \frac{k^2c^2}{\omega^2} - 1, \tag{10.97}$$

such that $\partial D/\partial \omega = -2/\omega$.

## 10.5 Classical Unruh Scattering

We now consider scattering by uniformly accelerated charged particles. This can be seen as extension of Thomson scattering, and at the same time, can be seen as the classical equivalent of the *Unruh radiation*.

The original Unruh effect is a purely quantum process, which corresponds to scattering of quantum vacuum fluctuations by an accelerated particle [185, 186]. In the proper reference frame of the particle, vacuum fluctuations are perceived as a thermal medium. Therefore, scattering by quantum fluctuations carry the signature of an effective temperature, called the *Unruh temperature*, which is proportional to the particle acceleration. Proposals to observe Unruh radiation using intense laser fields have made by several authors [187, 188].

In the classical problem considered here, the zero-point fluctuations of the original process are replaced by the monochromatic field of an incident wave. Because of the uniform acceleration, this monochromatic field is perceived in the particle frame, as a thermal spectrum, as shown below. Our physical geometry therefore includes two fields. First, a static electric field along some direction $z$-axis that accelerates the particle. Second, the oscillating field of an electromagnetic wave that simulates the effect of quantum fluctuations. For simplicity, we assume propagation in the same direction. See the geometry of the model in Figure 10.3.

### i) Scattering Model

We start from the Lorentz equation for a particle with charge $q$ and mass $m$ in electric and magnetic fields $\mathbf{E}$ and $\mathbf{B}$, as

$$\frac{d\mathbf{u}}{d\tau} = \frac{q}{m}\left(\gamma\mathbf{E} + \mathbf{u} \times \mathbf{B}\right), \tag{10.98}$$

where $\tau = t/\gamma$ is the particle proper time, $\mathbf{u} = \gamma\mathbf{v}$ is the covariant velocity, and $\gamma = \sqrt{1 + u^2/c^2} = (1 - v^2/c^2)^{-1/2}$. The fields can be described by the scalar and vector potentials $V(\mathbf{r}, t)$ and $\mathbf{A}(\mathbf{r}, t)$, as $\mathbf{E} = -\nabla V - \partial A/\partial t$, where the first term describes the static field $\mathbf{E}_s = E_s\mathbf{e}_z$, and the second corresponds

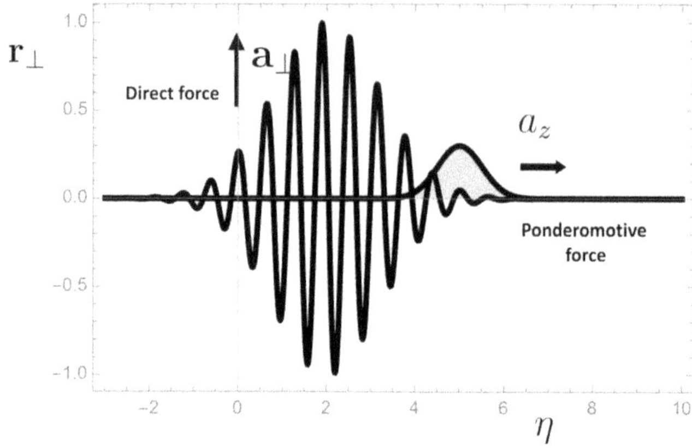


**FIGURE 10.3**
Geometry of a classical Unruh configuration: constant field + transverse field oscillations. Snowplow laser acceleration could provide the two fields: the ponderomotive force responsible for longitudinal acceleration $a_z$ and the laser field for transverse acceleration $a_\perp$.

to the wave field. The covariant acceleration resulting from the static field is determined by

$$\frac{du_z}{d\tau} = \frac{q}{m}\gamma E_s(z)\,, \quad E_s(z) = -\frac{\partial V}{\partial z}\,. \tag{10.99}$$

At the same time, we have perpendicular acceleration, determined by the vector potential, **A**, as

$$\frac{d\mathbf{u}_\perp}{d\tau} = \frac{q}{m}\gamma \mathbf{E}_\perp(\mathbf{r},t)\,, \quad \mathbf{E}_\perp(\mathbf{r},t) = -\frac{\partial \mathbf{A}}{\partial t}\,. \tag{10.100}$$

For an arbitrary transverse field spectrum, we have

$$\mathbf{u}_\perp = -\frac{q}{m}\mathbf{A}(\mathbf{r},t) = -\frac{q}{m}\int \mathbf{A}(\omega)\exp(i\mathbf{k}\cdot\mathbf{r} - i\omega t)\frac{d\omega}{2\pi}\,. \tag{10.101}$$

Assuming circular polarization, to simplify, and for a particle that initially is not moving in the perpendicular direction, the relativistic gamma factor is

$$\gamma = \sqrt{1 + \frac{u_z^2}{c^2} + \left|\frac{qA_\perp}{m}\right|^2}\,. \tag{10.102}$$

On the other hand, if we assume a particle initially at rest, the particle trajectory in the parallel direction is determined by

$$v_z(t) = \frac{q}{m\gamma}\int E_s(z)dt\,, \quad z(t) = z_0 + \int_0^t v_z(t')dt'\,. \tag{10.103}$$

Replacing this in eq. (10.100), the perpendicular acceleration becomes

$$\mathbf{a}_\perp(t) = \int \mathbf{a}_\omega \exp\left[i\varphi(t)\right] \frac{d\omega}{2\pi}\,, \tag{10.104}$$

with

$$\mathbf{a}_\omega = -i\frac{\omega e}{m}\mathbf{A}(\omega)\,, \quad \varphi(t) = \varphi_0 + k\int^t v_z(t')dt' - \omega t\,. \tag{10.105}$$

Let us take the simple case of a constant static field, $E_s(z) = E_0 = const$. We get

$$v_z(t) = \frac{at}{\sqrt{1+(at/c)^2}}\,, \quad z(t) = \frac{c^2}{a}\sqrt{1+(at/c)^2}\,, \tag{10.106}$$

with $a = qE_0/m$. We notice that, for short times $t \ll c/a$, we have constant acceleration, $v_z(t) \simeq at$. In contrast, for longer times, $t \gg c/a$, the particle acceleration decreases and ultimately will tend to zero.

### ii) Thermal Spectrum

Let us first focus on the constant acceleration regime. The velocity of the accelerated particle in the rest frame can be written in terms of particle proper time $\tau$ as

$$v_z(\tau) = c\,tanh\left(\frac{a\tau}{c}\right)\,, \quad a = \frac{qE_0}{m}\,. \tag{10.107}$$

This expression suggests the use of the so-called *Rindler coordinates* [190, 191] $(\xi, \eta)$, which are natural coordinates in the accelerated frame. They are defined by

$$z = \xi\cosh\eta\,, \quad ct = \xi\sinh\eta\,. \tag{10.108}$$

From here, we can easily get

$$\xi = \sqrt{z^2 - c^2t^2}\,, \quad \eta = \frac{1}{2}\ln\frac{z+ct}{z-ct}\,. \tag{10.109}$$

In order to give appropriate dimensions to these new coordinates, we use $(\xi = c^2/a,\ \eta = a\tau/c)$. The phase function can be written in these coordinates as

$$\varphi(t) = (kz - \omega t) = \frac{\omega\xi}{c}\exp(-\eta) = \frac{\omega c}{a}\exp\left(-\frac{a\tau}{c}\right)\,. \tag{10.110}$$

Replacing this in the expression for the perpendicular acceleration, eq. (10.104), we get

$$\mathbf{a}_\perp(\tau) = \int \mathbf{a}_\omega \exp\left[i\frac{\omega c}{a}e^{-a\tau/c}\right]\frac{d\omega}{2\pi}\,, \tag{10.111}$$

We can now Fourier transform this quantity with respect to the particle proper time $\tau$, using the expression

$$\mathbf{a}_\perp(\Omega) = \int \mathbf{a}_\perp(\tau)\exp(i\Omega\tau)d\tau\,, \tag{10.112}$$

where $\Omega$ is the frequency of the scattered spectrum, as seen in the accelerated frame. Using eq. (10.111), this can also be written as

$$\mathbf{a}_\perp(\Omega) = \int \mathbf{a}_\omega F(\omega, \Omega) \frac{d\omega}{2\pi}, \tag{10.113}$$

where we have introduced the auxiliary function

$$F(\omega, \Omega) = \int_{-\infty}^{\infty} e^{i\Omega\tau - i\omega\bar{\tau}} d\tau, \quad \bar{\tau} = -\frac{c}{a} e^{-a\tau/c}. \tag{10.114}$$

On the other hand, we also have

$$\mathbf{a}_\perp(\Omega)^* \equiv \mathbf{a}_\perp(-\Omega) = \int \mathbf{a}_\omega F(\omega, -\Omega) \frac{d\omega}{2\pi}, \tag{10.115}$$

where $F(\omega, -\Omega) = F^*(-\omega, \Omega)$. For a single frequency $\omega$ of the incident wave spectrum, we should have

$$\mathbf{a}_\perp(\Omega) = \mathbf{a}_\omega F(\omega, \Omega) + \mathbf{a}_{-\omega} F(-\omega, \Omega). \tag{10.116}$$

We also have

$$\mathbf{a}_\perp(-\Omega) = \mathbf{a}_\omega F(\omega, -\Omega) + \mathbf{a}_{-\omega} F(\omega, \Omega), \tag{10.117}$$

where we have used $F(\omega, \Omega) = F^*(-\omega, -\Omega)$. This then leads to

$$|\mathbf{a}_\perp(\Omega)|^2 = |\mathbf{a}_\omega|^2 \, |F(\omega, -\Omega)|^2. \tag{10.118}$$

This quantity is proportional to the number of photons scattered by the particle and due to the perpendicular oscillations. In order to find an explicit expression, we notice that the function $F(\omega, \Omega)$ can be written in terms of the Gamma function, $\Gamma(z)$, which is defined by the integral [192]

$$\Gamma(z) = \int_0^{\infty} e^{-t} t^{z-1} dz, \quad \Re(z) > 0. \tag{10.119}$$

This can be related to eq. (10.114), using an auxiliary variable $y = \exp(-a\tau/c)$, such that

$$F(\omega, \Omega) = -\frac{1}{a} \int y^{(i\Omega c/a - 1)} \exp\left(i\frac{\omega c}{a}\right) dy, \tag{10.120}$$

A further transformation of variables, defined by $t = -i(\omega c/a)y$, and eq. (10.119), then leads to

$$F(\omega, \Omega) = -\frac{1}{a} \left(-i\frac{\omega c}{a}\right)^{-z} \Gamma(z), \tag{10.121}$$

with $z = i\Omega c/a$. This result should be taken with some caution, because of the existence of a pole near the origin, $z = 0$. Regularization of the Gamma function in the complex plane is then necessary, and implies the replacement of $\Gamma(z)$ by $\Gamma(z)\exp(-i\pi z/2)$. After this regularization, we finally get

$$F(\omega,\Omega) = -\frac{1}{a}\left(-i\frac{\omega c}{a}\right)^{-i\Omega x/a} \Gamma(i\Omega c/a)\exp\left(\pi\frac{\Omega c}{2a}\right). \qquad (10.122)$$

At this point, it should be noticed that the Gamma function $\Gamma(z)$ also satisfies the identity

$$|\Gamma(iy)|^2 = \frac{\pi}{y\sinh(\pi y)}. \qquad (10.123)$$

Using this identity in eqs. (10.122) and (10.118), we are finally able to state that

$$|\mathbf{a}_\perp(\Omega)|^2 = |\mathbf{a}_\omega|^2 \frac{2\pi c}{\Omega a}\frac{1}{\exp(\hbar\Omega/T_a)-1}, \quad T_a = \frac{\hbar a}{2\pi c}, \qquad (10.124)$$

where the temperature is written in energy units. This explicit result is somewhat surprising, because it shows that the radiation observed in the accelerated frame of the particle is given by the Planck spectrum at a temperature, $T_a$, that is proportional to the acceleration $a$, and coincides with the definition of *Unruh temperature*. This is also analogous to the *Hawking temperature*, $T_h$, characterising the horizon of a black hole, and defined as

$$T_h = \frac{1}{8\pi M}\frac{\hbar c^3}{G}, \qquad (10.125)$$

where $G$ is the gravitational constant. In the present model however, no gravitational field and no quantum fluctuations are present. At this point, we should notice that the energy density of the scattered radiation is proportional to

$$\Omega^3|\mathbf{a}_\perp(\Omega)|^2 \propto \left(\frac{\hbar\Omega}{T_a}\right)^2 \frac{1}{\exp(\hbar\Omega/T_a)-1}. \qquad (10.126)$$

This model shows that, a single frequency component $\omega$, of the incident wave spectrum will be perceived by an accelerated particle as a Planck spectrum. Furthermore, the temperature of this spectrum is proportional to the acceleration $a$. This thermal spectrum is able to produce scattered secondary radiation, due to the existence of perpendicular acceleration, with components $a_\perp(\Omega)$. Notice that, in a quantum description of the Unruh radiation, the incident field would be replaced by zero-point fluctuations of the vacuum field. What is remarkable is that a similar broad spectrum is produced by incident waves with a single frequency $\omega$, and that this single frequency field is able to replace in this classical model the infinitely broad spectrum of quantum fluctuations.

### iii) Spectrum in the Lab Frame

It is also useful to study the scattered spectrum as observed in the laboratory frame. For this purpose, we go back to eq. (10.113) and assume a monochromatic wave with amplitude $\mathbf{a}_0$ and frequency $\omega_0$. In this case, we are reduced to

$$\mathbf{a}_\perp(t) = \mathbf{a}_0 \exp[i\varphi_0(t)] + c.c. , \quad \varphi_0(t) = \varphi_0 + k_0[z(t) - ct] , \tag{10.127}$$

where $z(t)$ is the particle trajectory. For a particle with constant acceleration, the Fourier spectrum of the transverse acceleration (and therefore scattered radiation), is given by

$$\mathbf{a}_\perp(\omega) = \mathbf{a}_0 \int_{-\infty}^{\infty} \exp\left[i\omega_0\left(at^2 - t\right) + i\omega t\right] dt , \tag{10.128}$$

After integration, we get

$$\mathbf{a}_\perp(\omega) = \mathbf{a}_0 \left(\frac{\pi}{-ia\omega_0}\right)^{1/2} \exp\left[-i\frac{(\omega_0 - \omega)^2}{4a}\right] . \tag{10.129}$$

This shows a Gaussian spectrum with amplitude $\mathbf{a}_0$, such that

$$|\mathbf{a}_\perp(\omega)|^2 = |\mathbf{a}_0|^2 \frac{\pi}{a\omega_0} . \tag{10.130}$$

But this solution is not completely correct, because the particle motion starts at $t = 0$. We therefore need to insert a Heaviside function $H(t)$ in the integrand of eq. (10.128). Analytical integration is still possible and gives

$$\mathbf{a}_\perp(\omega) = \mathbf{a}_0 \frac{\pi}{2\sqrt{-ia\omega_0}} \exp\left[-i\frac{(\omega_0 - \omega)^2}{4a}\right] \left(1 - iErf\left[\frac{\omega_0 - \omega}{2\sqrt{-ia\omega_0}}\right]\right) . \tag{10.131}$$

The spectral components are not significantly modified by the Heaviside function, and tend to a continuum for $\omega \gg \omega_0$. Notice that this continuum is generated by a single incident frequency $\omega_0$, and results from the temporal variation of the instantaneous frequency $\omega(t)$, perceived by the accelerated particle. Taking the time derivative of the phase, we get

$$\omega(t) \equiv -\frac{\partial}{\partial t}\varphi(t) = \omega_0 \left[1 - \frac{1}{c} v_z(t)\right] . \tag{10.132}$$

For acceleration by a constant electric field $E_s$, we get

$$\omega(t) = \omega_0 \left[1 - \frac{(qE_0/mc)t}{\sqrt{1 + (qE_0 t/mc)^2}}\right] . \tag{10.133}$$

For short times, in the initial stages of the acceleration, this frequency increases with time, as $\omega(t) \sim \omega_0[(qE_0/mc)t - 1]$. In contrast, due to the increase of the relativistic gamma factor, this frequency saturates for long times, $\omega(t \to \infty)$, and eventually tends to zero, as illustrated in Figure 10.4. This can be seen as a consequence of the growing particle effective mass.

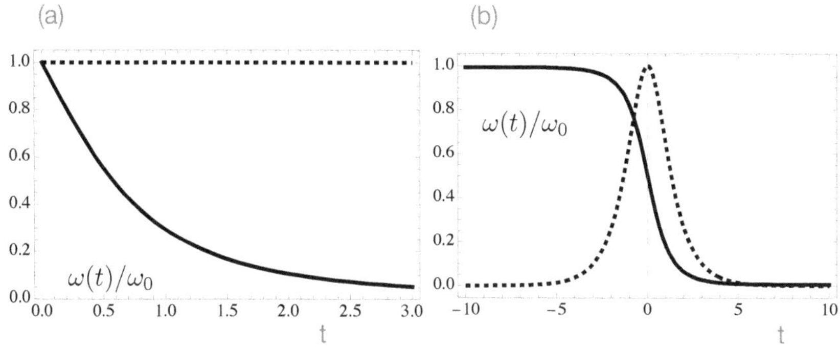


**FIGURE 10.4**
Frequency spectrum $\omega(t)/\omega_0$ created by: (a) A constant electric field, $\mathbf{E}_0 = E_0\mathbf{e}_z$; (b) A local electric field, $\mathbf{E}_0 = E_0 sech(z)\mathbf{e}_z$. The field amplitudes are indicated by dashed curves, and the frequency in bold. In the accelerated frame, this can be seen as a thermal spectrum. Reversing the direction of the field, we get frequency up-shifts, symmetric to the represented downshifts.

Another interesting configuration is the acceleration by a local static field, as described by the function

$$E_s(z) = E_0 sech(t/\tau) \,. \tag{10.134}$$

The parallel covariant velocity is now determined by

$$u_z(t) = \frac{2qE_0}{m}\tau \, tan^{-1}\left[\tanh(t/2\tau)\right] \,. \tag{10.135}$$

Using the corresponding value for the $\gamma$-factor, we then get the instantaneous frequency

$$\omega(t) = \omega_0 \left[1 - \frac{v_\tau}{c} f_\tau(t)\right] \,, \quad f_\tau(t) = \frac{tan^{-1}\left[\tanh(t/2\tau)\right] + cte.}{\sqrt{1 + (v_\tau^2/c^2) tan^{-1}\left[\tanh(t/2\tau)\right]^2}} \,, \tag{10.136}$$

where $v_\tau = (2qE_0/m)\tau$. Comparison between the constant field and the local field configurations is also shown in Figure 10.4.

## 10.6 Gamma-Ray Bursts

We now consider the emission of *gamma-ray bursts*, by collective scattering of near-infrared laser pulses on energetic electron bunches. From the point of view of the incident radiation, these electron bunches can be seen as *imperfect relativistic mirrors*, leading to a double-Doppler shift. We call them *imperfect*, because total reflection is not achieved. But, in addition to scattering, the nonlinear electron response will produce harmonics of the incident field frequency, which are double-Doppler shifted by the imperfect mirror reflection. The whole process can be described as collective backscattering. Conditions for an efficient gamma-ray emission can be established [193].

We start with an electron beam propagating in vacuum with relativistic velocity along the $z$-direction, $\mathbf{v}_b = -v_b \mathbf{e}_z$, with $v_b \sim c$. The beam density can be defined by $n(\mathbf{r}, t) = n_b f(z+v_b t)$, where $n_b$ is the beam density, and $f(z+v_b t)$ is the parallel form function defining the shape of the particle beam. For cold beams, their energy spread is small and we can consider that this function remains constant. Let us also assume that the laser pulse propagates in the opposite direction, with wavevector $\mathbf{k}_0 = k_0 \mathbf{e}_z$, and frequency $\omega_0 = k_0 c$. The interaction between the laser field and the electron beam leads to a scattered field $\mathbf{E}_s$, described by the wave equation

$$\left( \nabla^2 - \frac{1}{c^2} \frac{\partial^2}{\partial t^2} \right) \mathbf{A}_s = -\mu_0 \mathbf{J}_s \, , \tag{10.137}$$

where $\mathbf{A}_s$ is the vector potential, and $\mathbf{E}_s = -\partial \mathbf{A}_s / \partial t$. We also have the current $\mathbf{J}_s = -en(\mathbf{r}, t) \, \delta \mathbf{v}_b$,where $\delta \mathbf{v}_b$ is the perturbation of the beam velocity due to the incident field. In order to calculate this current, we use the electron fluid equations to determine the electron momentum $\mathbf{p} = m\gamma \mathbf{v}$, with $\gamma = (1 - \beta^2)^{1/2}$, and $\beta = v/c$. The momentum equation can be written in terms of $\beta$, as

$$\left( \gamma \frac{\partial \beta}{\partial t} + \beta \frac{\partial \gamma}{\partial t} \right) = -\frac{e\mathbf{E_0}}{mc} + \frac{c}{2} \nabla \gamma |\beta|^2 \, . \tag{10.138}$$

The last term in this equation represents the ponderomotive force, which is important for very short pulses. We focus on the one-dimensional problem, where the transverse particle and laser beam can be ignored. For an incident field polarized in the $x$-direction, $\mathbf{E}_0 = E_0 \mathbf{e}_x$, we get $\mathbf{p} = \mathbf{p}_b + \tilde{p} \, \mathbf{e}_x$, and $\mathbf{v} = \mathbf{v}_b + \tilde{v} \, \mathbf{e}_x$. The unperturbed quantities are determined by the initial electron beam properties: $\beta_b = v_b/c$, and $\mathbf{p}_b = m\gamma \mathbf{v}_b$. In contrast, the perturbed quantities, $\tilde{p}$ and $\tilde{\beta} = \tilde{v}/c$, occur in the perpendicular direction, and are determined by the incident laser pulse. This allows us to write

$$\frac{\partial \gamma}{\partial t} = \gamma^3 \tilde{\beta} \frac{\partial \tilde{\beta}}{\partial t} \, , \quad \gamma = (1 - \beta_b^2 - \tilde{\beta}^2)^{-1/2} \, . \tag{10.139}$$

Equation (10.138) can then be written as

$$\left( 1 + \gamma^2 \tilde{\beta}^2 \right) \frac{\partial \beta}{\partial t} = -\frac{eE_0}{mc\gamma} + c \nabla \gamma |\beta|^2 \, . \tag{10.140}$$

Notice also that

$$(1 + \gamma^2 \tilde{\beta}^2) = 1 + \frac{\tilde{\beta}^2}{1 - \beta_b^2 - \tilde{\beta}^2} = \frac{1 - \beta_b^2}{1 - \beta_b^2 - \tilde{\beta}^2} \, . \tag{10.141}$$

Replacing this in the momentum equation (10.139), we obtain

$$\frac{\partial \tilde{\beta}}{\partial t} = -(1 - \gamma_b^2 \tilde{\beta}^2) \left( \frac{eE_0}{mc\gamma} - c\nabla\gamma|\beta|^2 \right) , \tag{10.142}$$

where we have used the unperturbed gamma-factor $\gamma_b = (1 - \beta_b^2)^{1/2}$. This equation can easily be solved if we replace $\gamma$ by $\gamma_b$ in the denominator of the electric force. This assumption is valid for $\tilde{\beta}^2 \ll \beta_b^2$. A solution of the form

$$\tilde{\beta}(z, t) = \sum_{2n+1} \tilde{\beta}_{2n+1} \exp[i(2n+1)k_0 z - i\omega_0 t) , \tag{10.143}$$

is then possible. The first term of this solution corresponds to the fundamental frequency $n = 0$, and evolves as $\exp(ik_0 z - i\omega_0 t)$. Noting that the ponderomotive term in equation (10.142) has no contributions to this frequency component, we get

$$\tilde{\beta} = -i\frac{a_0}{\gamma_b} \exp(ik_0 z - i\omega_0 t) , \tag{10.144}$$

where we have used the dimensionless field amplitude $a_0$. Adopting an iterative procedure, we use for the second-order term $\tilde{\beta}_3 \exp(i3k_0 z - i3\omega_0 t)$, and obtain

$$\tilde{\beta}_3 = -i\frac{a_0^3}{3\gamma_b} . \tag{10.145}$$

This procedure is valid to the lowest order of the parameter $a_0/\gamma_b$, and leads to the following amplitudes of the series in eq. (10.143)

$$\tilde{\beta}_{2n+1} = -i\frac{a_0^{2n+1}}{(2n+1)!!\gamma_b} , \tag{10.146}$$

with the notation $(2n+1)!! = 3 \times 5 \times ...(2n+1)$. This solution is simple but has divergence issues when $a_0 > 1$.

For this reason, we can adopt an alternative and more accurate method. For that purpose, we write the equation of motion (10.142) as

$$\frac{\partial \tilde{\beta}}{\partial t} = -\omega_0 \frac{a_0(z,t)}{\gamma_b} (1 - \gamma_0^2 \tilde{\beta}^2) . \tag{10.147}$$

When the laser is described by $a_0(z,t) = a_0 \cos(k_0 z - \omega_0 t)$, this can be exactly integrated, and the result is

$$\tilde{\beta}(z,t) = \frac{1}{\gamma_b} \tanh\left[a_0 \sin(k_0 z - \omega_0 t)\right] . \tag{10.148}$$

This solution can also be written as a Fourier series, with the generic form of equation (10.143), where the amplitudes $\tilde{\beta}_n$ now have to be determined by the inverse Fourier integrals. For large intensities of the incident radiation,

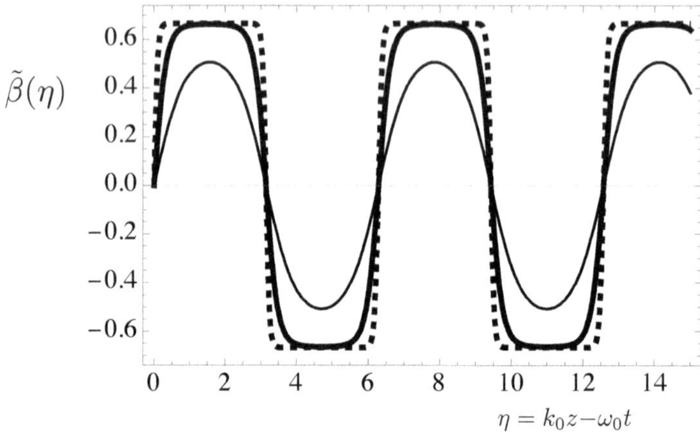


**FIGURE 10.5**
Solution of eq. (10.147): Perturbed velocity $\tilde{\beta}(\eta)$, as a function of $\eta = k_0z - \omega_0 t$, for different values of the incident field amplitude: $a_0 = 1$ (thin line), $a_0 = 3$ (in bold) and $a_0 = 10$ (dashed line). The value of $\gamma_b = 1.5$ is used.

$a_0 \gg 1$, this new solution tends to a square function, with a large amount of harmonics $\omega_n$, as illustrated in Figure 10.5.

Let us now consider the scattered field. Using eq. (10.137), valid in the one-dimensional approximation where the transverse beam structure is ignored, we get

$$\left( \frac{\partial^2}{\partial z^2} - \frac{1}{c^2} \frac{\partial^2}{\partial t^2} \right) A(x, t) = e n_b \mu_0 c f(z + v_b t) \tilde{\beta} \,, \tag{10.149}$$

where $\tilde{\beta} \equiv \tilde{\beta}(z, t)$. Using the temporal Fourier transformation, of the scattered field, as

$$A(z, t) = \int A_\omega(z) e^{-i\omega t} \frac{d\omega}{2\pi} \,, \tag{10.150}$$

we get

$$\left( \frac{\partial^2}{\partial z^2} + \frac{\omega^2}{c^2} \right) A_\omega(z) = \frac{\omega_b^2}{emc} \sum_n \int f(z + v_b t) \tilde{\beta}_n(z) e^{i\varphi_{2n+1} + i\omega t} dt \,, \tag{10.151}$$

where $\omega_b$ is the beam plasma frequency, and the phase $\varphi_n$ is determined by

$$\varphi_{2n+1} \equiv k_n z - \omega_{2n+1} t = (2n + 2)(k_0 z - \omega_0 t) \,. \tag{10.152}$$

For each term of the sum in eq. (10.151), we have to solve an integral of the form

$$\mathcal{I}_n(z) = -\frac{i}{\Omega_{2n+1}} \int f'(z + \xi) e^{i k_{2n+1} z - i\Omega_{2n+1} \eta / v_b} d\xi \,, \tag{10.153}$$

where $\Omega_n = \omega_n - \omega$, and $f'$ is the derivative of $f$ with respect to the new variable $\xi = v_b t$. Replacing this in eq, (10.151), we get

$$\left(\frac{\partial^2}{\partial z^2} + \frac{\omega^2}{c^2}\right) A_\omega(z) = \frac{\omega_b^2}{emc} \sum_n \tilde{\beta}_n(z) I_n(z) . \tag{10.154}$$

For backward radiation, which propagates in the negative $z$-direction, the solution is

$$A_\omega(z) = i \sum_n \frac{a_n}{2k} e^{-ikz} \int_{-\infty}^{z} e^{ikx'} \mathcal{I}_n(z') dz' , \tag{10.155}$$

where $k = \omega/c$, and

$$a_n = \frac{\omega_b^2}{emc} \tilde{\beta}_n . \tag{10.156}$$

Let us now consider the influence of beam shape. For a long beam with a sharp front, we can use $f(z + v_b t) = H(z + v_b t)$, where $H(z)$ is the Heaviside step function. This gives $f'(\xi + z) = \delta(\xi + z)$ and the integral $\mathcal{I}_n(z)$ reduces to

$$\mathcal{I}_n(z) = -\frac{i}{\Omega_{2n+1}} \exp\left[i\left(k_{2n+1} + \frac{\Omega_{2n+1}}{v_b}\right) z\right] . \tag{10.157}$$

Replacing this in eq. (10.155), we obtain

$$A_\omega(z) = \sum_n \frac{\pi a_n}{k\Omega_{2n+1}} e^{-ikz} \delta(k + k_{2n+1} + \Omega_{2n+1}/v_b) . \tag{10.158}$$

The frequency of the scattered radiation is determined by the argument of the delta function, which implies that

$$\frac{\omega}{c} = \frac{\omega}{u} - (2n+1)\left(\frac{\omega_0}{v_b} + \frac{\omega_0}{c}\right) . \tag{10.159}$$

This can be rearranged, for $\omega = \omega_{2n+1}$, to give

$$\omega_{2n+1} = \gamma_b^2 (1 + \beta_b)^2 (2n+1) \omega_0 . \tag{10.160}$$

We can see that a double-Doppler shift occurs for all the harmonics, up-shifting them by a factor of $4\gamma_0^2$. This is a characteristic feature of back-reflection by a relativistic mirror, but now for all the odd harmonics of the scattered spectrum. This can lead to a very large frequency shift.

However, the electron beams are very imperfect mirrors, where total reflection is almost never possible. An estimate of the scattered intensity can be made using the energy density for each harmonic, as

$$W_n = |E_\omega(2n+1)|^2 = \omega_{2n+1}^2 \left| \frac{\pi a_n}{k\Omega_{2n+1}} \right|^2 \tag{10.161}$$

This can also be written as

$$W_n = \omega_{2n+1}^2 \frac{\pi^2}{k^2 \Omega_{2n+1}^2} \left(\frac{\omega_b^2}{emc}\right)^2 \left|\tilde{\beta}_{2n+1}\right|^2 \tag{10.162}$$

We notice that $k = \omega/c$ and

$$\Omega_{2n+1} = (2n+1)\omega_0 - \omega(n) = (2n+1)\frac{2\beta}{1-\beta}\omega_0 \,. \tag{10.163}$$

Using the expression for $\tilde{\beta}_n$ valid for $a_0 >> 1$, we can then write

$$\frac{W_{2n+1}}{W_0} = \frac{\omega_b^4}{\omega_0^4}\frac{(1-\beta_b)^2}{\beta_b^2\gamma_b^2}\frac{1}{(2n+1)^2\gamma_b^2} \,. \tag{10.164}$$

This is proportional to the ratio $(\omega_b/\omega_0)^4$ and to the factor $\gamma_b^{-8}$. However, in terms of instantaneous power the results are more promising. First, the reduction in pulse duration is of the same order of the frequency up-shift, which increases the backscattered power by a factor of $\gamma_b^2$. Second, the relativistic focusing in the backward direction is of order $\gamma_b^4$, due to the fact that only the perpendicular component is up-shifted. This means that the scattered beam divergence is reduced by a factor of $\gamma_b^2$ in each transverse direction. As a result, the backscattered power $P_n$ is of order

$$P_n \simeq \left(\frac{\omega_b}{\omega_0}\right)^4 \frac{P_0}{(2n+1)^2} \,, \tag{10.165}$$

where $P_0$ is incident laser pulse power.

Let us finally consider a more general kind of electron beam shape, described by a Fourier spectrum of the form

$$f(z + v_b t) = \int F(p) \exp\left[ip(z + v_b t)\right] \frac{dp}{2\pi} \,, \tag{10.166}$$

where the Fourier amplitudes satisfy the condition $0 \leq |F(p)| \leq 1$. In this case, we have

$$\mathcal{I}_n(z) = F(\Omega/v_b) \exp\left[i\left(k_{2n+1} + \frac{\Omega_{2n+1}}{v_b}\right)\right] \,, \tag{10.167}$$

where $F(\Omega/v_b) \equiv F(p = \Omega/v_b)$. This leads to the following scattered wave amplitude

$$A_\omega(z) = \sum_n \frac{\pi a_{2n+1}}{k v_b} F(\Omega/v_b) e^{-ikz} \delta(k + k_{2n+1} + \Omega_{2n+1}/v_b) \,. \tag{10.168}$$

The backscattered power now scales as

$$P_n \simeq F(\Omega/v_b) \left(\frac{\omega_b}{\omega_0}\right)^4 \frac{P_0 \gamma_b^4}{(2n+1)^2} \,. \tag{10.169}$$

This could lead to a significant enhancement of backscattered power with respect to value given by eq. (10.165). The difference is that, for an appropriate choice of beam modulation, the entire beam can contribute to scattering, not just its front.

## 10.7 Relativistic Bubbles

In contrast with the previous section, where we discussed scattered by electron beams moving in vacuum, we now consider scattering by relativistic bubbles moving in a plasma. These bubbles could be associated with plasma voids, such as those resulting from laser wakefields. This scattering process, not only leads to large frequency shifts, as in the previous case, but could also be useful to plasma diagnostics.

Let us then assume a relativistic bubble, with an arbitrary shape, moving in a plasma with velocity $\mathbf{u}$. The electron plasma density can be described by

$$n(\mathbf{vr}, t) = n_0 - \delta n \, F(\mathbf{r} - \mathbf{u}t) , \qquad (10.170)$$

where $n_0$ is the equilibrium density of the background plasma, $\delta n$ is the density perturbation associated with the bubble, and $F(\mathbf{r} - \mathbf{u}t)$ is the form function. As in the previous section, we consider an incident electromagnetic wave, described by the vector potential

$$\mathbf{A}_{inc}(\mathbf{r}, t) = \mathbf{A}_0 \exp(i\mathbf{k}_0 \cdot \mathbf{r} - i\omega_0 t) + c.c.. \qquad (10.171)$$

The nonlinear current, resulting from by the interaction of the incident wave with the plasma bubble, can be described by

$$\mathbf{J}_s(\mathbf{r}, t) = -\frac{e^2 n_0}{m} \mathbf{A}_s + \frac{e^2 \delta n}{m} F(\mathbf{r} - \mathbf{u}t) \mathbf{A}_{inc} , \qquad (10.172)$$

where $\mathbf{A}_s$ is the scattered wave potential. This field should satisfy the wave equation (10.137), and its Fourier component $\mathbf{A}_\omega$, should be determined by

$$\left( \nabla^2 + \frac{\omega^2}{c^2} \right) \mathbf{A}_\omega = -\mu_0 \mathbf{J}_\omega , \qquad (10.173)$$

where the source current $\mathbf{J}_\omega$ is

$$\mathbf{J}_\omega \mathbf{r}) = \mathbf{A}_0 \frac{\delta n}{n_0} \frac{\omega_p^2}{\mu_0 c^2} \int F(\mathbf{r} - \mathbf{u}t) e^{i\omega t} \left( e^{i\mathbf{k}_0 \cdot \mathbf{r} - i\omega_0 t} + c.c. \right) dt , \qquad (10.174)$$

and $\omega_p$ is the equilibrium plasma frequency.

Let us now assume that the bubble moves along the $z$-axis, such that $\mathbf{u} = u\mathbf{e}_z$, with a shape that remains constant during the scattering process. This allows us to write the form function as

$$F(\mathbf{r} - \mathbf{u}t) = G(\mathbf{r}_\perp) F_z(\eta) , \quad \eta = z - ut , \qquad (10.175)$$

where $G(\mathbf{r}_\perp)$ and $F_z(\eta)$ describe the perpendicular and parallel bubble shapes. We can describe the parallel shape function by the Fourier integral

$$F_\parallel(\eta) = \int_{-\infty}^{\infty} F_\parallel(q) \exp(iq\eta) \frac{dq}{2\pi} . \qquad (10.176)$$

As a typical example, we can assume a Gaussian shape of width $\sigma$, such that

$$F_z(\eta) = \exp\left(-\frac{\eta^2}{2\sigma^2}\right), \quad F_z(q) = 2\pi\sigma \exp\left(-\frac{1}{2}q^2\sigma^2\right). \tag{10.177}$$

Another typical example is that of a periodic bubble structure, similar to a long wakefield, as described by

$$F_z(\eta) = [1 - H(|\eta| - \pi\sigma/2)] \cos(\eta/\sigma) \tag{10.178}$$

where $H(x)$ is the Heaviside function. The corresponding Fourier spectrum is given by

$$F_z(q) = 2\int_0^{\pi/2} \cos z \, e^{iq\sigma z} dz, \tag{10.179}$$

Replacing eq.(10.176) in the nonlinear source current of eq. (10.174), we obtain

$$J_\omega = \mathbf{A}_0 \frac{\delta n}{n_0} \frac{\omega_p^2}{\mu_0 c^2} \left[\mathcal{J}_+(\mathbf{r}_\perp) e^{i(q_+ \mathbf{e}_x + \mathbf{k}_0)\cdot \mathbf{r}} + \mathcal{J}_-(\mathbf{r}_\perp) e^{i(q_- \mathbf{e}_x - \mathbf{k}_0)\cdot \mathbf{r}}\right], \tag{10.180}$$

where we have use the auxiliary quantities

$$\mathcal{J}_\pm(\mathbf{r}_\perp) = (\mathbf{e}_0 \cdot \mathbf{e}_\omega) \frac{f(\mathbf{r}_\perp)}{u} F_\parallel(q_\pm), \quad q_\pm = \frac{1}{u}(\omega \mp \omega_0), \tag{10.181}$$

with $\mathbf{e}_0 = \mathbf{A}_0/A_0$ and $\mathbf{e}_\omega = \mathbf{A}_\omega/A_\omega$. This will allow us to determine the intensity and spectrum of the scattered radiation.

Let us now consider the scattered wave signal, propagating in a given direction with wavevector $\mathbf{k}$. Solving the wave equation (10.173), and noting that $(\mathbf{k} \cdot \mathbf{e}_\omega) = 0$, we obtain the scattered field observed at a given position $\mathbf{r}$, as

$$A_\omega(\mathbf{r}) = -\mathbf{A}_0 \frac{\delta n}{n_0} \frac{\omega_p^2}{2kc^2} e^{i\mathbf{k}\cdot\mathbf{r}} \int^{\mathbf{r}} d\mathbf{r}' \int \frac{d\mathbf{q}_\perp}{(2\pi)^2} \times \\ \times e^{-i\mathbf{k}\cdot\mathbf{r}'} \sum_\pm \left[\mathcal{J}_\pm(\mathbf{q}_\perp) e^{i(\mathbf{q}_p m \pm \mathbf{k}_0)\cdot bf r'}\right], \tag{10.182}$$

where $\mathcal{J}_\pm(\mathbf{q}_\perp)$ are the Fourier transforms of the quantities (10.181), as determined by

$$\mathcal{J}_\pm(\mathbf{q}_\perp) = \int \mathcal{J}_\pm(\mathbf{r}_\perp) e^{-i\mathbf{q}_\perp\cdot\mathbf{r}_\perp} d\mathbf{r}_\perp. \tag{10.183}$$

If we take the limit $|\mathbf{r}| \to \infty$ the above solution reduces to

$$A_\omega(\mathbf{r}) \simeq -\mathbf{A}_0 \frac{\delta n}{n_0} \frac{\pi\omega_p^2}{kc^2} e^{i\mathbf{k}\cdot\mathbf{r}} \sum_\pm \mathcal{J}_\pm(\mathbf{q}_\perp) \delta(\mathbf{q}_\pm - (\mathbf{k} - \mathbf{k}_0)), \tag{10.184}$$

where we have used $\mathbf{q}_\pm = q_\pm \mathbf{e}_x + \mathbf{q}_\perp$. We can see that the scattered radiation propagates along the two directions defined by

$$\mathbf{k} = \mathbf{q}_\pm \pm \mathbf{k}_0 \equiv \frac{1}{u}(\omega \mp \omega_0)\mathbf{e}_x + \mathbf{q}_\perp \pm \mathbf{k}_0. \tag{10.185}$$

The frequency $\omega$ of the scattered radiation can also be derived, using the plasma dispersion relation. This allows us to write

$$\frac{\omega^2}{\gamma_b^2} \mp 2\omega\omega_0 \left(1 - \frac{\mathbf{k}_0 \cdot \mathbf{u}}{\omega_0}\right) + \omega_0^2 \left(\frac{u^2}{c^2} - 2\frac{\mathbf{k}_0 \cdot \mathbf{u}}{\omega_0}\right) + Q_\perp^2 u^2 = 0\,, \tag{10.186}$$

where we have defined $Q_\perp^2 = q_\perp^2 \pm 2(\mathbf{q}_\perp \cdot \mathbf{k}_{0\perp}^2)$, and introduced the bubble gamma-factor $\gamma_b = (1 - u^2/c^2)^{-1/2}$. Solving this equation, we get

$$\omega = \pm\omega_0\gamma_b^2 \left(1 - \frac{\mathbf{k}_0 \cdot \mathbf{u}}{\omega_0}\right) \{1 \pm \xi\}\,. \tag{10.187}$$

The two $\pm$ signs appearing in this expression are independent, and the quantity $\xi$ is determined by

$$\xi^2 = 1 - \frac{[(1 + Q_\perp^2 c^2/\omega_0^2)u^2/c^2 - 2\mathbf{k}_0 \cdot \mathbf{u}/\omega_0]}{(1 - \mathbf{k}_0 \cdot \mathbf{u}/\omega_0)^2\gamma_b^2}\,. \tag{10.188}$$

This result shows that the scattered frequency $\omega$ strongly depends on the scattering direction, determined by $\mathbf{k}$. Large frequency shifts can eventually occur, due to the multiplying factor $\gamma_b^2$ in eq. (10.187). This is reminiscent of the double-Doppler shift of the relativistic mirrors. To understand this more clearly, we define the scattering angle $\theta_0$ as the angle between the bubble velocity $\mathbf{u}$, and the direction of propagation $\mathbf{k}_0/k_0$ of the incident wave. It is given by

$$\cos\theta_0 = -\frac{\mathbf{k}_0 \cdot \mathbf{u}}{\beta\sqrt{\omega_0^2 - \omega_p^2}}\,, \tag{10.189}$$

with $\beta = u/c$. Replacing this in eq. (10.187), we obtain

$$\omega = \omega_0\gamma_b^2 \left(1 + \beta\cos\theta_0\sqrt{1 - \frac{\omega_p^2}{\omega_0^2}}\right) \{1 \pm \xi\}\,. \tag{10.190}$$

An useful approximation for the quantity $\xi$, valid in the high-frequency limit, such that $\omega_0^2 \gg Q_\perp^2 c^2, \omega_p^2$, is given by

$$\xi^2 = 1 - \frac{\beta^2 + 2\beta\cos\theta_0}{\gamma_b^2(1 + \beta\cos\theta_0)^2} \tag{10.191}$$

The resulting frequency shifts $\omega/\omega_0$ are illustrated in Figure 10.6, where only the upper frequency branch is represented.

Finally, let us discuss the intensity of the scattered radiation. Using eq. (10.184), we obtain the intensity ratio

$$\frac{|A_\omega|^2}{|A_0|^2} = \left(\frac{\delta n}{n_0}\frac{\pi\omega_p^2}{kuc^2}\right)^2 |(\mathbf{e}_0 \cdot \mathbf{e}_\omega)|^2 |F(\mathbf{q}_\perp, q_\pm)|^2\,. \tag{10.192}$$

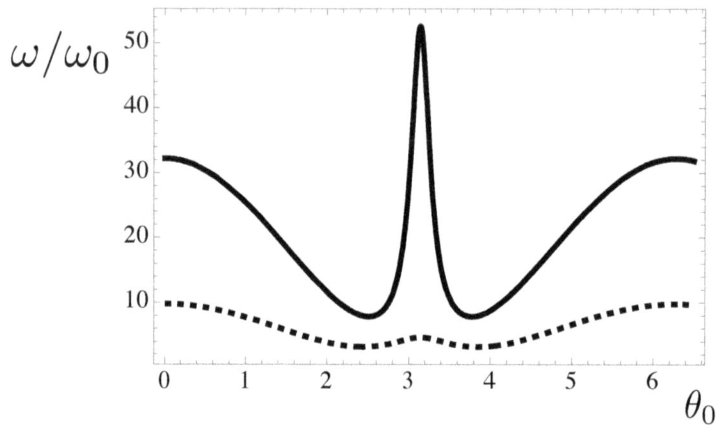


**FIGURE 10.6**
Scattering by relativistic bubbles: Frequency upshift $\omega/\omega_0$, as a function of the scattering angle $\theta_0$ (in radians), for $\gamma_b = 3$, and $\gamma_b = 1.7$ (dashed), with $\omega_p^2/\omega_0^2 = 0.3$.

For a Gaussian bubble, we have

$$F(\mathbf{q}_\perp, q_\pm) = \exp\left(-\frac{\eta^2}{2\sigma^2} - \frac{r_\perp^2}{2\sigma_\perp^2}\right) . \tag{10.193}$$

Taking $(\mathbf{e}_0 \cdot \mathbf{e}_\omega) = 1$, this leads to

$$\frac{|A_\omega|^2}{|A_0|^2} = \left(\frac{\delta n}{n_0}\right)^2 \frac{2^3\pi^5\omega_p^4}{k^2u^2c^4}\sigma^2\sigma_\perp^2 \exp\left[-q_\pm^2\sigma^2 - (\mathbf{k} \mp \mathbf{k}_0)_\perp^2\sigma_\perp^2\right] . \tag{10.194}$$

This result shows that the scattered intensity is maximum for $|q_\pm| \leq 1/\sigma$ and $|q_\perp| \leq 1/\sigma_\perp$.

# 11

# *Photon Kinetics*

In this chapter we consider the main properties of photon kinetics, and explore the analogues between photons and electrons, when they interact with waves and nonlinear structures in a plasma. The main source of this analogy is the existence of an effective photon charge in a plasma. Photon trapping and photon instabilities of electron plasma waves can be described in a way that resembles electron trapping and electron beam-plasma instabilities, already discussed in this book.

A first approach to photon kinetics can be found in [194], and more recent work includes the discussion of photon Landau damping and wave-kinetics, where the role of photons can be played by other quasi-particles existing in the medium. This opens the way to a global description of wave turbulence.

In this chapter, we start with a discussion of single particle effects, based on the concepts of geometric optics. This allows us to describe electromagnetic radiation (for instance, a laser pulse) as a gas of wave-packets, which follow regular trajectories. Such a description can be seen as the analogue of relativistic classical dynamics and statistical mechanics. A photon kinetic equation of the Klimontovich type can be derived, which by a statistical average can be transformed into a Vlasov equation for photons. This leads to the concept of photon Landau damping, first established in 1997 [195]. In order to reinforce the analogies between photons and relativistic charged particles, the concepts of photon effective charge and mass can be explored and used to describe intense laser-plasma interaction processes [196].

But, electromagnetic radiation is made of waves, which show interference and diffraction effects, which cannot be accounted for in geometric optics. We therefore need to consider a more consistent description based on wave optics. This is provided by a more general photon-kinetic equation, where such effects are included. This improved description of electromagnetic waves lead us to realise that photon-electron interactions in a plasma, because of energy and momentum conservation, lead to recoil effects that are usually associated with quantum physics [197].

Such formal analogies can be pushed forward, with photons replaced by quasi-particles in the plasma medium, such as plasmons, phonons or driftons. This means that photon-kinetics can be extended to a more general wave-kinetic theory, allowing a description of plasma turbulence in terms of quasi-particles. This is particularly useful to describe plasma turbulence. A generic approach to wave-kinetics will be given in this chapter, which includes the

DOI: 10.1201/9781003514701-11

description of a quasi-linear theory of turbulence based on quasi-particles. This is important to describe several phenomena, such as anomalous transport.

Finally, we include a brief description Bose-Einstein condensation of photons in a plasma. This effect is a collective photon processes, provided by electron-photon coupling via Compton scattering. Photon condensation was first predicted by Zel'dovich and Levich in 1969, in the frame of Astrophysics [198]. In this pioneering work photon mass effects were ignored, and the electromagnetic spectrum was predicted to condensate at zero frequency. This, of course, is not physically valid and has been improved in recent theoretical models, where plasma frequency (and therefore photon mass) is retained [199]. Furthermore, photon condensation can also be considered in laboratory plasmas [200].

## 11.1 Geometric Optics

### i) Photon Canonical Equations

Let us start with the simple case of an electromagnetic wavepacket, moving in a varying medium, where the refractive index varies on a time scale much longer than the wave period. It is known that, in this case, electromagnetic wave propagation can be described by geometric optics [201, 202]. In this approximation, the wavepacket can be seen as a quasi-particle, with trajectories described by ray-equations. It is interesting to note that these ray equations are similar to particle equations of motion, as described by Classical Mechanics. In particular, these equations can be written in a canonical form, as

$$\frac{d\mathbf{r}}{dt} = \frac{\partial H}{\partial \mathbf{k}} \,, \quad \frac{d\mathbf{k}}{dt} = - \frac{\partial H}{\partial \mathbf{r}} \,, \tag{11.1}$$

where $\mathbf{r}$ determines the wavepacket position, and $\mathbf{k}$ its momentum (or wavevector). The wavepacket is a kind of classical photon, and we can identify these equations as the photon equations of motion. A more sophisticated description of electromagnetic wave propagation will emerge later in this chapter.

The Hamiltonian $H \equiv H(\mathbf{r}, \mathbf{k}, t)$ appearing above is equal to the photon frequency, or the wavepacket frequency. For transverse waves propagating in a non-stationary isotropic plasma, this frequency depends to the exact location, and on time, $\omega = H(\mathbf{r}, \mathbf{k}, t)$, and can be written as

$$\omega = H(\mathbf{r}, \mathbf{k}, t) = \sqrt{k^2c^2 + \omega_p^2(\mathbf{r}, t)/\gamma_a} \,, \tag{11.2}$$

where $\omega_p$ is the space and time-varying plasma frequency, and $\gamma_a$ is the relativistic gamma-factor associated with the amplitude of the wavepacket field $a$. In the usual formulation, this factor is absent, $\gamma_a = 1$. The similarity of this expression with the Hamiltonian of a relativistic particle with finite rest mass (for instance, an electron) is obvious, and was first noticed by Anderson in 1963 [203]. We can therefore rewrite it as

$$\omega = m_{\mathbf{k}} \gamma_{\mathbf{k}} c^2 = \gamma_{\mathbf{k}} \frac{\omega_p}{\sqrt{\gamma_a}} , \tag{11.3}$$

where $m_{\mathbf{k}}$ the *photon effective mass*, and $\gamma_{\mathbf{k}}$ is the *photon gamma-factor*, defined as

$$\gamma_{\mathbf{k}} = \sqrt{1 + \frac{k^2 c^2}{\omega_p^2} \gamma_a} . \tag{11.4}$$

We can see that these effective quantities depend on the plasma frequency, as well as on the intensity of the considered wavepacket. Notice that this quantity diverges and loses its meaning in vacuum, where $m_{\mathbf{k}}$ reduces to zero, and $\gamma_{\mathbf{k}}$ diverges. However, in a plasma, this can be significant and eventually of the order of the electron gamma-factor for large intensities, $\gamma_a \gg 1$.

Using an appropriate Legendre transformation, the photon Hamiltonian $H(\mathbf{r}, \mathbf{k}, t)$ can also be derived from a Lagrangian $\mathcal{L}(\mathbf{r}, \mathbf{v}, t)$, defined as

$$\mathcal{L}(\mathbf{r}, \mathbf{v}, t) = -m_{\mathbf{k}} \frac{c^2}{\gamma_{\mathbf{k}}} = -\frac{\omega_p}{\sqrt{\gamma_a}} \left( 1 - \frac{v_{\mathbf{k}}^2}{c^2} \right)^{1/2} . \tag{11.5}$$

The quantity $\mathbf{v}_{\mathbf{k}} = kc^2/\omega$ is obviously the photon group velocity. To complete the basic description of ray-equations in a plasma, it is now useful to write them in covariant form. For that purpose, we define the photon 4-vector position and momentum, which are similar to those of a relativistic particle

$$x^\mu \equiv (ct, \mathbf{r}) , \quad k^\mu \equiv (\omega/c, \mathbf{k}) , \tag{11.6}$$

with the index $\mu = (0, 1, 2, 3)$. The module square of the 4-momentum is given by

$$k_\mu k^\mu = k^2 - \frac{\omega^2}{c^2} = -m_{\mathbf{k}}^2 c^2 . \tag{11.7}$$

Following the procedure used in Section (2.2.3) for material particles, we can similarly define a covariant Hamiltonian for electromagnetic wavepackets (photons), as

$$\mathcal{H}(x^\mu, k^\mu) = \frac{k_\mu k^\mu}{2m_{\mathbf{k}}} + \frac{1}{2} m_{\mathbf{k}} c^2 . \tag{11.8}$$

This quantity is always constant and equal to zero, $\mathcal{H}(x^\mu, k^\mu) = 0$. The photon ray-equation in covariant form can then be written as

$$\frac{dx^\mu}{d\tau} = \frac{\partial \mathcal{H}}{\partial k_\mu} , \quad \frac{dk_\mu}{d\tau} = -\frac{\partial \mathcal{H}}{\partial x^\mu} , \tag{11.9}$$

where $\tau$ is the photon proper time, as defined in the plasma. Notice that such a definition is meaningless in vacuum.

### ii) Photon Trapping

It is now useful to illustrate the photon ray-equations using a simple example. For that purpose, we consider the motion of a photon in the field of an electron plasma wave. This will lead to the concept of *photon trapping*. We assume an electron plasma wave with frequency $\omega'$ and wavevector $\mathbf{k}'$, propagating along some direction $\mathbf{e}_z$. For a photon moving in the presence of this wave, we have

$$\omega_p^2(\mathbf{r}, t) = \omega_{p0}^2 \left[1 + \epsilon \cos(\mathbf{k}' \cdot \mathbf{r} - \omega' t)\right] . \tag{11.10}$$

where the constant $\omega_{p0}$ represents the background plasma frequency, and $\epsilon = \tilde{n}/n_0$ determines the plasma wave amplitude. The electron plasma wave satisfies the usual dispersion relation $\omega'^2 = k'^2 c^2 + \omega_{p0}^2$, where it is assumed that the size of the electromagnetic wavepacket is small compared with the plasma wavelength. Therefore, the nonlinear correction associated with $\gamma_a$ is only valid locally and has no influence on the electrostatic wave dispersion. We consider a simple one-dimensional problem where $\mathbf{k} = k\mathbf{e}_z$, and propagation perpendicular to $\mathbf{e}_z$ is ignored.

In order to describe the photon evolution, it is useful to introduce a canonical transformation from the canonical variables $(z, k)$ into a new pair of variables $(q, p)$, with $q = z - (\omega'/k')t$, and $p = k$, such that the electron plasma frequency becomes $\omega_p^2(q)$. The photon ray-equations (11.1) reduce to

$$\frac{dq}{dt} = \frac{\partial H_p}{\partial p} , \quad \frac{dp}{dt} = -\frac{\partial H_p}{\partial q} , \tag{11.11}$$

The new Hamiltonian $H_p \equiv H_p(q, p)$ is determined by

$$H_p = \omega(q, p) - p\frac{\omega'}{k'} = \sqrt{p^2 c^2 + \frac{1}{\gamma_a}\omega_p^2(q)} - p\frac{\omega'}{k'} . \tag{11.12}$$

with

$$\omega_p^2(q) = \omega_{p0}^2 \left[1 + \epsilon \cos(k' q)\right] . \tag{11.13}$$

We notice that the new Hamiltonian is a constant of motion, $H_p = const.$, and that no Lorentz transformation was involved, just a canonical transformation. The above equations of motion can then be explicitly written as

$$\frac{dq}{dt} = \frac{c^2 p}{\omega(q, p)} - \frac{\omega'}{k'} , \quad \frac{dp}{dt} = \epsilon \frac{\omega_{p0}^2 k'}{\gamma_a^{1/2} \omega(q, p)} \sin(k' q) . \tag{11.14}$$

Of particular interest are the fixed points determined by the condition $d/dt = 0$. We can easily realize that an elliptic (stable) fixed-point exists at $q = q_e$ and $p = p_e$ such that

$$q_e = \frac{\pi}{k'} , \quad p_e = \gamma' \beta' \frac{\omega_{p0}}{\gamma_a^{1/2} c} \sqrt{1 - \epsilon} , \tag{11.15}$$

where the relative velocity of the electron plasma wave $\beta'$ and the corresponding gamma-factor $\gamma'$, are defined by

$$\beta' = \frac{\omega'}{k'c}, \quad \gamma' = \frac{1}{1 - \beta'^2}. \tag{11.16}$$

For deeply trapped photon trajectories moving near this fixed-point, we can use the variable $\tilde{q} = q - q_e$. And, linearizing the equations of motion (11.14) around this fixed-point, we are reduced to a linear oscillator, as

$$\frac{d^2\tilde{q}}{dt^2} + \omega_b^2 \tilde{q} = 0, \quad \omega_b = \frac{ck'}{\gamma'} \sqrt{\frac{\epsilon}{2(1-\epsilon)}}. \tag{11.17}$$

The frequency of this linear oscillator can be called the *photon bounce frequency* and determines the oscillation of trapped photons inside the electron plasma waves. The similarity of this behavior with that of trapped electrons is remarkable.

Another important aspect of photon ray-equations in an electron plasma wave is the existence of separatrix curves, which are special kinds of trajectories that delimit the trapped photon region. A simple analysis of these equations shows that these curves are determined by

$$p(q) = \gamma' \omega_{p0} \gamma^{1/2} c \sqrt{1+\epsilon} \left[ \beta' \pm \sqrt{1 - \frac{1 + \epsilon(k'q)}{1+\epsilon}} \right]. \tag{11.18}$$

This is illustrated in Figure 11.1, where the separatrix curves and deeply trapped trajectories are represented.

### iii) Photon Cyclotron Motion

Another useful example of the use of the photon ray-equations is related it the existence of cyclotron-like photon trajectories in plasma channels. These channels can pre-exist in the background plasma, or result from the action of intense laser pulses that locally change the electron density profile by pushing out the electrons and modify the relativistic factor $\gamma_a$. For this purpose, we consider a dispersion relation

$$\omega = \sqrt{k^2c^2 + \omega_{p0}^2 \, g(\mathbf{r}, t)}, \quad g(\mathbf{r}, t) = \frac{n}{n_0} \frac{1}{\gamma_a}. \tag{11.19}$$

Here, the quantity $n \equiv n(\mathbf{r}, t)$ is the plasma density locally perturbed by a plasma channel, to be specified latter, $n_0$ is the equilibrium density and $\omega_{p0}$ the equilibrium plasma frequency.

Let us focus on a density channel, with cylindrical symmetry. In this case, the ray-equations can be decoupled for axial and perpendicular motion. We have, for the axial photon motion

$$\frac{dz}{dt} = v_z \equiv \frac{k_z c^2}{\omega}, \quad \frac{dk_z}{dt} = F_z \equiv -\frac{\omega_{p0}^2}{2\omega} \frac{\partial g}{\partial z}. \tag{11.20}$$

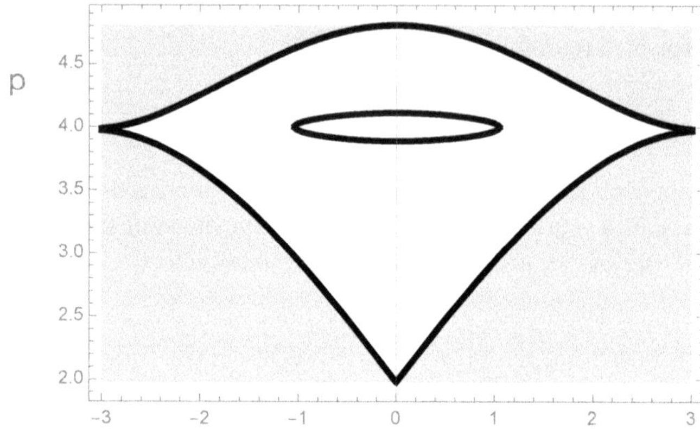


**FIGURE 11.1**
Trapped photon trajectories in the field of an electron plasma wave: deeply trapped trajectories and separatrix curves. The untrapped photon region is shaded.

And, for the perpendicular motion

$$\frac{d\mathbf{r}_\perp}{dt} = \mathbf{v}_\perp \equiv \frac{\mathbf{k}_\perp c^2}{\omega} \,, \quad \frac{d\mathbf{k}_\perp}{dt} = \mathbf{F}_\perp \equiv -\frac{\omega_{p0}^2}{2\omega}\frac{\partial g}{\partial \mathbf{r}_\perp} \,. \tag{11.21}$$

Let us assume a generic cylindrical channel, of the form $g(\mathbf{r}, t) = f_\perp(r) f_z(z, t)$. For a pre-formed and static channel, we can simply use $f_z(z, t) = 1$. In contrast, if the channel is created by some intense laser pulse, we typically have $f_z(z, t) = f(\eta) = \exp(-z^2/2\sigma_z^2)$, with $\eta = (z - v_g t)$, where $v_g$ is the group velocity of the laser pulse and $\sigma_z$ its length.

As for the radial profile function $f_\perp$, typical examples are Gaussian profiles, $f_G = 1 - \alpha_G \exp(-r^2/2\sigma_\perp^2)$, where $\sigma_\perp$ is the channel radius, or parabolic profiles, $f_P(r) = 1 + \alpha_P r^2$. The constants $\alpha_G$ and $\alpha_P$ determine the channel depth. These two profiles are very similar near the axis, for $r \ll \sigma_\perp$, if we choose $\alpha_P \simeq \alpha_G / 2\sigma_\perp^2$. With such profiles, the axial and perpendicular forces are

$$F_z = -\frac{\omega_{p0}^2}{2\omega} f_\perp(r) \frac{\partial f_z}{\partial z} \,, \quad \mathbf{F}_\perp = -\frac{\omega_{p0}^2}{2\omega} f_z(z, t) \frac{\partial f_\perp}{\partial r} \mathbf{e}_r \,, \tag{11.22}$$

where $\mathbf{e}_r$ is the radial unit vector. Using the two examples of radial profiles, we can write $(\partial f_\perp / \partial r) = \alpha r$, with $\alpha = 2\alpha_P = \alpha_G / \sigma_\perp^2$. In this case the $x$ and $y$ components of the perpendicular equations of motion can be decoupled, such that

$$\frac{dr_j}{dt} = \frac{k_j c^2}{\omega}, \frac{dk_j}{dt} = -\frac{\omega_{p0}^2}{2\omega} f_z(z, t) \alpha r_j \,, \tag{11.23}$$

for $j = x, y$, and $r_x = x$, $r_y = y$. This leads to the following equation for a damped harmonic oscillator

$$\frac{d^2 r_j}{dt^2} + \omega_{rot}^2 r_j + 2\Gamma \frac{dr_j}{dt} = 0\,, \tag{11.24}$$

where we have defined a rotation frequency $\omega_{rot}$ and a damping factor $\Gamma$, such that

$$\omega_{rot} = \frac{\omega_{p0}^2}{2\omega}\sqrt{\frac{\alpha}{2} f_z(z,t)}\,, \quad \Gamma = \frac{\omega_{p0}^2}{4\omega} f_\perp \frac{\partial f_z}{\partial t}\,. \tag{11.25}$$

Using these equations, we can easily derive first-order solutions, if we make the following assumptions: First, we assume that the plasma is strongly underdamped, such that we can use $\omega \simeq k_0 c$, where $k_0$ is a constant, in the denominators. Second, we assume a small damping factor $\Gamma \ll \omega_{rot}$, which is true in a preformed plasma channel (where $f_z = 1$) and also approximately valid for the relevant photon trajectories in a laser created channel. The solutions can then be written as

$$x(t) = r_c \sin(\omega_c t)\,, \quad k_x = k_x(0)\cos(\omega_c t)\,, \tag{11.26}$$

with similar equations for $y(t)$, with the photon radius $r_c = ck_x(0)/k_0\omega_c$ and gyration frequency $\omega_c \equiv \omega_{rot}(f_z = 1)$. This is a kind of photon cyclotron motion, which reveals qualitatively new features, confirmed by simulations.

This nearly circular photon motion results from what could be called photon magnetization, where the influence of a plasma channel can be described in terms of an effective magnetic field $\mathcal{B}$, which is parallel to the channel axis. In order to determine this field, we use the definitions of photon effective mass and charge, $m_{ph}$ and $q_{ph}$, as

$$m_{ph} = \frac{\hbar}{c^2}\omega_p\,, \quad q_{ph} = -\left(\frac{e}{m}\right)\frac{\hbar}{4}\frac{\omega_p^2}{v_k^2 \omega}\,, \tag{11.27}$$

where $e$ and $m$ are the electron charge and mass. As we have seen, the effective photon mass, or photon inertia [203], is determined by the plasma frequency, which in the present context is equal to $\omega_p = \omega_{p0}\sqrt{g}$, as shown by the dispersion relation (11.19). It is therefore a linear property of the photon in a plasma. On the other hand, the effective charge is a nonlinear property and results from the ponderomotive force. From here, we get the simple ratio

$$\left(\frac{q_{ph}}{m_{ph}}\right) = -\left(\frac{e}{m}\right)\frac{c^2}{v_k^2}\frac{\omega_p}{\omega}\,. \tag{11.28}$$

We can then identify the above photon gyro-frequency $\omega_c = |q_{ph}|\mathcal{B}/m_{ph}$. This leads to the follow expression for the effective magnetic field

$$\mathcal{B} = \sqrt{\frac{\alpha}{2}}\left(\frac{m}{e}\right)\frac{v_k^2}{c^2}\frac{\omega}{k_0}\mathbf{e}_z\,. \tag{11.29}$$

Such a definition demonstrates the existence of photon magnetization, as due to channel guiding propagation. A similar approach can be used for the axial motion, where we define an effective electric field $\mathcal{E}$, of the form

$$\mathcal{E} = \left(\frac{m}{e}\right) \frac{v_k^2 \omega_p}{2\omega} \frac{\partial \ln g}{\partial z} \mathbf{e}_z \,. \tag{11.30}$$

This means that the ray-equations of the guided motion can be equivalently written as $m_{ph} d\mathbf{v}_k/dt = \mathcal{F}$, with an effective Lorentz force given by

$$\mathcal{F} = q_{ph} \left(\mathcal{E} + \mathbf{v}_k \times \mathcal{B}\right) \,, \tag{11.31}$$

where the effective electromagnetic field is determined by eqs. (11.29) and (11.30).

The existence of such striking analogies between photons and changed particles in a magnetic field, should not make us forget the obvious qualitative differences. First, there is only one *magnetic field line*, which coincides with the axis of the plasma channel, in contrast with the infinite number of field lines, for charged particles in real magnetic fields. Second, the photon motion can also be elliptical, with the same orbital frequency, depending on the dephasing between $x(t)$ and $y(t)$. This contrasts with the purely circular motion of magnetized charged particles. Actually, we could replace the effective magnetic field by a potential well, trapping the photon in the transverse direction.

A most interesting situation occurs when the plasma channel is created by a strong laser pulse, around which another pulse oscillates. We then notice the occurrence of an atom-of-light. This is not a consequence of the ray-tracing approximation, and is supported by full wave simulations, as illustrated in Figure 11.2. For pulses of identical intensity, they would both rotate around each other, leading to what has been called *braided light* [204].

It is now useful to consider the conditions for the existence of these atoms-of-light in a steady state. First of all, a stable atom can only persist in time if both pulses propagate with the same axial velocity. In order to discuss this issue, we assume that the driving pulse, with group velocity $\mathbf{v}_1 = c^2\mathbf{k}_1/\omega_1$ only propagates in the axial direction, $\mathbf{k}_1 = k_1 \mathbf{e}_z$. On the other hand, the satellite pulse moves with a finite velocity in the perpendicular plane $\mathbf{v}_2 = c^2\mathbf{k}_2/\omega_2$, with $\mathbf{k}_2 = k_{2z}\mathbf{e}_z + \mathbf{k}_{2\perp}$. The existence of a stable atom therefore implies the equality $v_1 = v_{2z}$, which leads to the following relation between the frequencies of the two pulses

$$\left(\omega_2^2 - \omega_1^2\right) \omega_p^2 = \omega_1^2 k_{2\perp}^2 c^2 \,. \tag{11.32}$$

Assuming that $\omega_2 = \omega_1 + \delta\omega$, with $\delta\omega \ll \omega_1$, we get the frequency difference between the two pulses as

$$\delta\omega = \omega_1 \frac{k_{2\perp}^2 c^2}{\omega_p^2} \,. \tag{11.33}$$

This is a necessary but not a sufficient condition for the existence of a stable

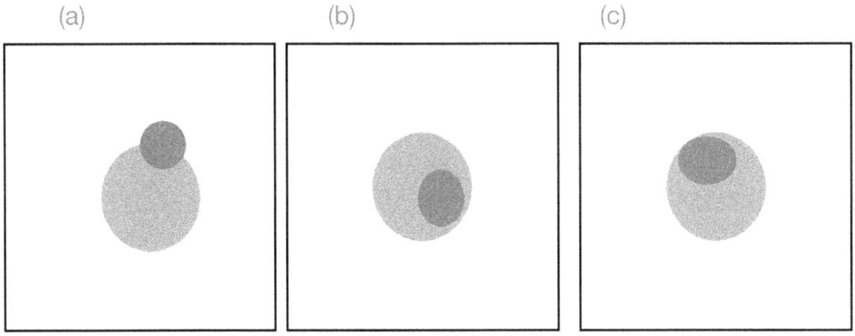


**FIGURE 11.2**
Atom of light, with two laser pulses in a plasma: the probe pulse is attracted and rotates around the strong one. (Adapted from pic simulations with the OSIRIS code): (a) initial conditions, $t = 0$; (b) $t \simeq \pi/\omega_c$, (c) $t = 2\pi/\omega_c$. [Courtesy J. Vieira]

atomic structure. Due to the wave nature of the photon, we need to impose that the corresponding field is single-valued. This can be satisfied by the quantization rule

$$\int_C \mathbf{k} \cdot \mathbf{r}_\perp = 2\pi\nu \,, \quad \mathbf{r}_\perp = \frac{c^2 k_{2\perp}}{\omega_2 \omega_c} \,, \tag{11.34}$$

where $\nu = 1, 2, 3, ..$ is an integer, and $C$ is a closed contour in the perpendicular plane. This leads to a quantization of the possible values of photon transverse momentum inside the atom-of-light. As a consequence, the frequency or energy of the orbiting pulse is also quantized, as

$$\delta\omega_\nu \simeq \omega_c \frac{\omega_1^2}{\omega_p^2} \nu \,. \tag{11.35}$$

This defines the possible energy states of the atomic light structure. We have used a quasi-classical description of the orbital photon motion, because geometric optics is valid in our model. But we could also use a more exact description of this atom-of-light, based on a full wave model.

## 11.2 Photon-Kinetic Theory

Let us now consider a more complete description of the electromagnetic pulse propagating in a plasma. This new approach is based on a photon-kinetic equation. There are two ways of deriving this equation, one is heuristic and based on the above photon ray-equations. The other is more rigorous and based on the exact wave equation. The theoretical procedure is somewhat complicated but the final result will be simple and physical appealing.

Let us start with the heuristic approach, and consider a large number of photon trajectories, determined by $N \gg 1$ different solutions of the ray-equations (11.1), as $(\mathbf{r} = \mathbf{r}_j(t), (\mathbf{k} = \mathbf{k}_j(t))$, for $j = 1, 2, ...N$. In the photon phase-space $(\mathbf{r}, \mathbf{k})$, this bunch of trajectories evolves in time, and can be described by a micro-density distribution $\rho(\mathbf{r}, \mathbf{k}, t)$, defined as

$$\rho(\mathbf{r}, \mathbf{k}, t) = \sum_{j=1}^{N} \delta(\mathbf{r} - \mathbf{r}_j(t))\, \delta(\mathbf{k} - \mathbf{k}_j(t))\,. \tag{11.36}$$

By definition, this quantity satisfies the conservation equation

$$\frac{d}{dt}\rho(\mathbf{r}, \mathbf{k}, t) = 0\,, \quad \frac{d}{dt} \equiv \frac{\partial}{\partial t} + \mathbf{v_k} \cdot \nabla + \mathbf{f_k} \cdot \frac{\partial}{\partial \mathbf{k}}\,. \tag{11.37}$$

The photon group velocity $\mathbf{v_k}$ and the force $\mathbf{f_k}$, are determined by

$$\mathbf{v_k} = \frac{\partial H}{\partial k}\,, \quad \mathbf{f_k} = -\frac{\partial H}{\partial r}\,. \tag{11.38}$$

We notice the formal analogy with the Klimontovich equation discussed in Section 3.2. The micro-density $\rho(\mathbf{r}, \mathbf{k}, t)$ is a singular function, and it is useful to replace it by a statistical average $W(\mathbf{r}, \mathbf{k}, t) = \langle \rho(\mathbf{r}, \mathbf{k}, t) \rangle$, where the singularities disappear. Following the procedure outlined in Chapter 3, and ignoring the fluctuations around the mean value, $\delta\rho = \rho - W$, we can then derive a Vlasov equation for the averaged distribution, of the form

$$\left( \frac{\partial}{\partial t} + \mathbf{v_k} \cdot \nabla + \mathbf{f_k} \cdot \frac{\partial}{\partial \mathbf{k}} \right) W(\mathbf{r}, \mathbf{k}, t) = 0\,, \tag{11.39}$$

Thisis sometimes called the *photon-kinetic equation*. It states the conservation of the number of rays in the photon phase-space $(\mathbf{r}, \mathbf{k})$. This result is, of course, valid in the geometric optics approximation. But, it can also be justified by a proper wave description of electromagnetic radiation. For that purpose, we start from the standard wave equation for the vector potential $\mathbf{a}(\mathbf{r}, t)$, as

$$\left[ \nabla^2 - \frac{1}{c^2}\frac{\partial^2}{\partial t^2} - \frac{\omega_p^2}{\gamma_a c^2} \right] \mathbf{a} = 0\,, \tag{11.40}$$

where $\mathbf{a} = e\mathbf{A}/mc$ is the normalized field, and the electron plasma frequency $\omega_p$ is assumed to vary in space and time, due to the eventual existence of density perturbations, $\tilde{n}(\mathbf{r}, t)$. It is now useful to introduce the field auto-correlation function, defined by

$$K(\mathbf{r}, \mathbf{s}, t, \tau) = \mathbf{a}(\mathbf{r} + \mathbf{s}/2, t + \tau/2) \cdot \mathbf{a}^*(\mathbf{r} - \mathbf{s}/2, t - \tau/2)\,. \tag{11.41}$$

Using a Fourier transformation of this quantity, we define the Wigner function for the electromagnetic field as

$$W(\omega, \mathbf{k}; \mathbf{r}, t) = \int d\vec{s} \int d\tau \; K(\mathbf{r}, \vec{s}, t, \tau) \exp(-i\mathbf{k} \cdot \vec{s} + i\omega\tau) \,. \tag{11.42}$$

This is similar, but not identical, to the original definition of Wigner function, which was derived for quantum particles [205]. This quantity can be called quasi-probability, for reason that will become apparent later. At this point, it should be noticed that

$$|\mathbf{a}(\mathbf{r}, t)|^2 = \int W(\omega, \mathbf{k}; \mathbf{r}, t) \frac{d\omega}{2\pi} \frac{d\mathbf{k}}{(2\pi)^3} \,. \tag{11.43}$$

This allows us to write the gamma factor $\gamma_a$ in terms of quasi-probabilities. Taking the so-called *Wigner-Moyal procedure* [206, 207, 208], and using a straightforward generalization of this procedure to the electromagnetic field [209, 210], we arrive at an evolution equation for the Wigner function $W(\omega, \mathbf{k}; \mathbf{r}, t)$, of the form

$$i \left( \frac{\partial}{\partial t} + \frac{c^2 \mathbf{k}}{\omega} \cdot \nabla \right) W = \mathcal{C}_{int} \,, \tag{11.44}$$

where the quantity $\mathcal{C}_{int}$ represents the interaction of photons with the electrons in the plasma, and is defined by the integral

$$\mathcal{C}_{int} = \frac{\omega_{p0}^2}{n_0} \int \int n(\Omega, \mathbf{q}) \, [W_- - W_+] \exp(i\mathbf{q} \cdot \mathbf{r} - i\Omega t) \frac{d\Omega}{2\pi} \frac{d\mathbf{q}}{(2\pi)^3} \,, \tag{11.45}$$

In this derivation we usually ignore the gamma factor $\gamma_a$, but relativistic corrections can be introduced when necessary. Here, we have used the plasma frequency, $\omega_{p0} = (e^2 n_0 / \epsilon_0 m)^{1/2}$, where $n_0$ is the equilibrium density. The quantities $W_\pm$ used above are defined as

$$W_\pm = W(\omega \pm \Omega/2, \mathbf{k} \pm \mathbf{q}/2) \,. \tag{11.46}$$

They correspond to displaced Wigner functions. We have also used the Fourier spectrum of density fluctuations, such that

$$n(\mathbf{r}, t) = \int \frac{d\mathbf{q}}{(2\pi)^3} \int \frac{d\Omega}{2\pi} \, n(\Omega, \mathbf{q}) \exp(i\mathbf{q} \cdot \mathbf{r} - i\Omega t) \,, \tag{11.47}$$

where $n(\Omega, \mathbf{q})$ are the Fourier components of the electron density fluctuations $n(\mathbf{r}, t)$.

Equation (11.44) can be seen as a complicated version of the photon-kinetic equation (11.39). But it can be simplified, using simple and plausible arguments. First, we note that the photon frequencies $\omega$ and wavevectors $\mathbf{k}$ are not independent, because they locally satisfy a dispersion relation, such that $\omega = \omega_{\mathbf{k}}$ is a function of $\mathbf{k}$. This means that we can use a reduced quasi-distribution, as

$$W(\omega, \mathbf{k}, \mathbf{r}, t) = W_{\mathbf{k}}(\mathbf{r}, t)\, \delta(\omega - \omega_{\mathbf{k}})\,. \tag{11.48}$$

where $\omega_k$ can be determined by the dispersion relation

$$\omega^2 = k^2c^2 + \frac{1}{\gamma(W)}\omega_p^2(\mathbf{r}, t)\,, \quad \gamma(W) = \left[1 + \int W_{\mathbf{k}}(\mathbf{r}, t)\frac{d\mathbf{k}}{(2\pi)^3}\right]^{1/2}\,. \tag{11.49}$$

Here, $\gamma(W)$ is the electron gamma-factor, written in terms of the Wigner function. The photon-kinetic equation then takes the form

$$i\left(\frac{\partial}{\partial t} + \mathbf{v}_{\mathbf{k}} \cdot \nabla\right) W_{\mathbf{k}} = \mathcal{C}_{\mathbf{k}}\,, \tag{11.50}$$

where the interaction integral becomes

$$\mathcal{C}_{\mathbf{k}} = \frac{\omega_{p0}^2}{n_0} \int n(\mathbf{q}, t)\, [W_{\mathbf{k}-} - W_{\mathbf{k}+}] \exp(i\mathbf{q} \cdot \mathbf{r} - i\Omega t)\frac{d\mathbf{q}}{(2\pi)^3}\,, \tag{11.51}$$

where now $W_{\mathbf{k}\pm}$ are defined by

$$W_{\mathbf{k}\pm} \equiv W_{\mathbf{k}\pm\mathbf{q}/2}\,. \tag{11.52}$$

A further simplification occurs when plasma oscillations reduce to a single Fourier component, representing an electrostatic wave. In this case, we use $n(\mathbf{r}, t) = n_0 + \tilde{n}\cos(\mathbf{q} \cdot \mathbf{r} - \Omega t)$, where $\tilde{n}$ is the perturbation amplitude. The photon kinetic equation (11.50) becomes

$$i\left(\frac{\partial}{\partial t} + \mathbf{v}_{\mathbf{k}} \cdot \nabla\right) W_{\mathbf{k}} = \omega_{p0}^2 \frac{\tilde{n}(t)}{n_0}\, [W_{\mathbf{k}-} - W_{\mathbf{k}+}] \exp(i\mathbf{q} \cdot \mathbf{r})\,. \tag{11.53}$$

As a final step toward simplification, let us now consider the quasi-classical (or geometric optics limit). This is valid when the scale of the electron density perturbations is much larger than the photon wavenumbers, or $|\mathbf{q}| \ll |\mathbf{k}|$. This means that we develop $W_{\mathbf{k}\pm}$ around $W_{\mathbf{k}}$, leading to

$$W_{\mathbf{k}\pm} \simeq W_{\mathbf{k}} \pm \frac{\mathbf{q}}{2} \cdot \frac{\partial}{\partial \mathbf{k}} W_{\mathbf{k}} + \frac{1}{2}\left(\frac{\mathbf{q}}{2} \cdot \frac{\partial}{\partial \mathbf{k}}\right)^2 W_{\mathbf{k}} + ... \tag{11.54}$$

Replacing this in eq. (11.53), we obtain

$$\left(\frac{\partial}{\partial t} + \mathbf{v}_{\mathbf{k}} \cdot \nabla + \mathbf{f}_k \cdot \frac{\partial}{\partial \mathbf{k}}\right) W_{\mathbf{k}} = 0\,. \tag{11.55}$$

where $\mathbf{f}_{\mathbf{k}} \propto \tilde{\nabla} n(\mathbf{r}, t)$ is the force acting of photons with wavevector $\mathbf{k}$, due to the existence of a density perturbation. We can see that, in this limit, we are reduced to the photon Vlasov equation (11.39), thus confirming the derivation based on the Klimontovich type of micro-density (11.36).

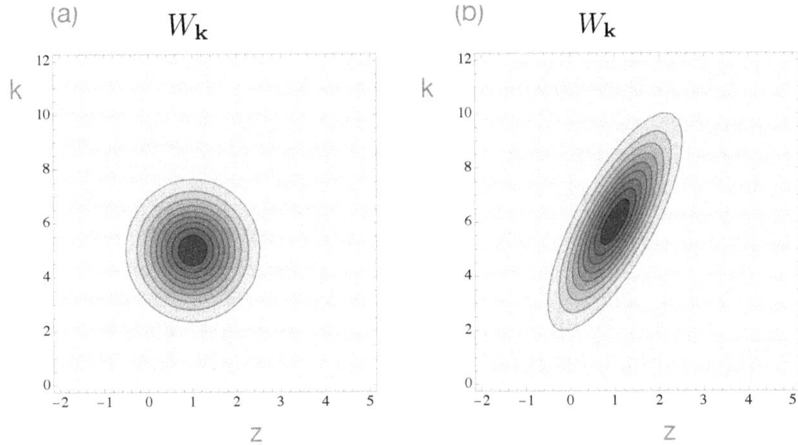


**FIGURE 11.3**
Typical examples of Wigner distributions in the two-dimensional phase space $(z, k)$: (a) Gaussian Pulse and (b) Chirped Pulse.

Our discussion on the photon-kinetic equation needs to be completed by establishing a relation between the photon quasi-distribution $W_{\mathbf{k}}$ and the photon number density $N_{\mathbf{k}}$, sometimes also known as the *photon occupation number*. This can be done using the electromagnetic wave energy. We know that, the energy density $U_{\mathbf{k}}$ in a plasma is given by [212]

$$U_{\mathbf{k}} = \frac{\epsilon_0}{4} \left( \frac{\partial \omega \mathcal{D}}{\partial \omega} \right)_{\mathbf{k}} |E_{\mathbf{k}}|^2 , \tag{11.56}$$

where $\mathcal{D} \equiv \mathcal{D}(\omega, \mathbf{k})$ is the dispersion relation, as defined in Chapter 4. On the other hand, we can define the photon number density $N_{\mathbf{k}}$ by the relation $U_{\mathbf{k}} = 2\hbar\omega_{\mathbf{k}} N_{\mathbf{k}}$, where the factor of 2 pertains to two independent polarization states. Noting that $\partial D/\partial \omega) \sim 2/\omega$, and using the definition of the Wigner function in terms of the normalized field amplitudes $a(\mathbf{r}, t)$, as defined by eq. (11.43), we finally obtain

$$N_{\mathbf{k}} = \frac{\epsilon_0}{4\hbar} \left( \frac{mc}{e} \right)^2 \omega_{\mathbf{k}} W_{\mathbf{k}} . \tag{11.57}$$

This shows that the Wigner function determines the photon occupation number and the photon-kinetic equation discussed above describes their spacetime evolution. This is valid in the classical limit, where this function is always positive.Two typical cases of electromagnetic wavepackets, which correspond to Gaussian pulses and to chirped pulses, are illustrated in Figure 11.3.

## 11.3 Photon Landau Damping

The photon-kinetic equations derived in the previous Sections allow us to describe an electromagnetic wavepacket, for instance a laser pulse, as equivalent to a gas of photons evolving in space and time inside the plasma. Not only they give a new perspective on the electromagnetic wave phenomena, but they also put a new emphasis on field correlations. This helps to reveal new phenomena, such as the photon Landau damping to be discussed next. For that purpose, we need to couple the photon-kinetic equation with the wave equation describing the electron plasma waves. This is achieved by the ponderomotive force.

Let us then focus on slowly varying processes, much slower than the field oscillations of the electromagnetic waves. We can describe the electron density $n$, and the electron velocity $\mathbf{u} = \gamma_a \mathbf{v}$, where $\gamma_a = \sqrt{1 + a^2}$ is the electron relativistic factor associated with the electromagnetic spectrum of amplitude $a(\mathbf{r}, t)$, by the electron fluid equations

$$\frac{\partial n}{\partial t} + \nabla \cdot \left\langle \frac{n\mathbf{u}}{\gamma_a} \right\rangle = 0 \,, \tag{11.58}$$

and

$$\frac{\partial \mathbf{u}}{\partial t} + \frac{c}{mn\gamma_s} \nabla \mathcal{P}_e = c^2 \nabla \Phi - c \left\langle \frac{1}{\gamma_a} \left[ \mathbf{u} \cdot \nabla \mathbf{u} + \mathbf{u} \times (\nabla \times \mathbf{a}) \right] \right\rangle \,. \tag{11.59}$$

Here, $\mathcal{P}_e$ is the electron pressure, $\langle\rangle$ denotes an average over the fast time scales, and $\Phi = eV/mc^2$ represents the normalized scalar potential, as described by the Poisson's equation

$$\nabla^2 \Phi = - \frac{e^2 n}{\epsilon_0 mc^2} \,. \tag{11.60}$$

The term on the right hand side of eq. (11.59) represents the ponderomotive force due to the presence of radiation From this, it is then possible to derive an equation describing the evolution of the electron density perturbations, $\tilde{n} = n - n_0$, where $n_0$ is the equilibrium value, of the form

$$\frac{\partial^2 \tilde{n}}{\partial t^2} - \frac{S_e^2}{\gamma_s} \nabla \tilde{n} + \frac{\omega_{p0}^2}{\gamma_s} \tilde{n} = \frac{c^2}{2} n_0 \nabla \cdot \left( \frac{1}{\gamma_s^2} \nabla |a|^2 \right) \,, \tag{11.61}$$

where $S_e = \sqrt{3T_e/m}$ is the electron thermal velocity. We can now simplify the problem and take the weakly relativistic limit, where $\gamma_s \simeq 1$, and rewrite this equation as

$$\left( \frac{\partial^2}{\partial t^2} - S_e^2 \nabla \right) \tilde{n} + \omega_{p0}^2 \tilde{n} = \frac{c^2 n_0}{2} \nabla^2 \int W_{\mathbf{k}} \frac{d\mathbf{k}}{(2\pi)^3} \,. \tag{11.62}$$

The ponderomotive force is now written in terms of the photon quasi-distribution $W_{\mathbf{k}}$ and, in order to solve the equation, we need to couple it to the photon-kinetic equation. We assume that the electron plasma wave is monochromatic, described by $\tilde{n}(\mathbf{r}, t) = \tilde{n}\exp(i\mathbf{q}\cdot\mathbf{r} - i\Omega t)$, as well as slow perturbation in the photon spectrum, described as $\tilde{W}_{\mathbf{k}}(\mathbf{r}t) = \tilde{W}\exp(i\mathbf{q}\cdot\mathbf{r} - i\Omega t)$. Linearizing eq. (11.53) with respect to these perturbations, we obtain

$$i\left(\frac{\partial}{\partial t} + \mathbf{v}_{\mathbf{k}}\cdot\nabla\right)\tilde{W}_{\mathbf{k}}(\mathbf{r}, t) = \omega_{p0}^2 \frac{\tilde{n}(t)}{n_0}\left[W_{0,\mathbf{k}-} - W_{0,\mathbf{k}+}\right]\exp(i\mathbf{q}\cdot\mathbf{r})\,. \tag{11.63}$$

where $W_{0,\mathbf{k}}$ is the equilibrium spectrum. We can see that the amplitude of the perturbations $\tilde{n}$ and $\tilde{W}$ are linked by ponderomotive force effects, as described by the above two equations. We then get

$$\tilde{W}_{\mathbf{k}} = \frac{\tilde{n}q^2}{n_0}\frac{\omega_{p0}^2}{(\Omega - \mathbf{q}\cdot\mathbf{v}_{\mathbf{k}})}\left[W_{0,\mathbf{k}-} - W_{0,\mathbf{k}+}\right]. \tag{11.64}$$

On the other hand, using eq. (11.62) we have

$$\tilde{n} = \frac{1}{2}\frac{c^2 n_0}{(\Omega^2 - S_e^2 q^2 - \omega_{p0}^2)}\int \tilde{W}_{\mathbf{k}}\frac{d\mathbf{k}}{(2\pi)^3}\,. \tag{11.65}$$

From these two equations we then obtain the dispersion relation of electron plasma waves, with frequency $\Omega$ and wavevector $\mathbf{q}$, in the presence of the electromagnetic spectrum, as

$$\Omega^2 = S_e^2 q^2 + \omega_{p0}^2\left[1 + \frac{q^2c^2}{2}\int \frac{W_0(\mathbf{k} - \mathbf{q/2}) - W_0(\mathbf{k} + \mathbf{q}/2)}{(\Omega - \mathbf{q}\cdot\mathbf{v}_{\mathbf{k}})}\frac{d\mathbf{k}}{(2\pi)^3}\right]. \tag{11.66}$$

Or, equivalently

$$\Omega^2 = S_e^2 q^2 + \omega_{p0}^2\left[1 + \frac{q^2c^2}{2}\int W_0\left(\frac{1}{(\Omega - \mathbf{q}\cdot\mathbf{v}_+)} - \frac{1}{(\Omega - \mathbf{q}\cdot\mathbf{v}_-)}\right)\frac{d\mathbf{k}}{(2\pi)^3}\right]. \tag{11.67}$$

where we have defined the displaced photon group velocities, $\mathbf{v}_\pm = \mathbf{v}(\mathbf{k}\pm\mathbf{q}/2)$. This dispersion relation can be written in the standard form

$$\epsilon(\Omega, \mathbf{q}) \equiv 1 + \chi_e(\Omega, \mathbf{q}) + \chi_{ph}(\Omega, \mathbf{q}) = 0\,. \tag{11.68}$$

The electron susceptibility $\chi_e(\Omega, \mathbf{q})$ is that of a cold plasma

$$\chi_e(\Omega, \mathbf{q}) = -\frac{\omega_{p0}^2}{\Omega^2}\left(1 + \lambda_D^2 q^2\right)\,, \tag{11.69}$$

where $\lambda_D = S_e/\omega_{p0}$ is the electron Debye length, and the additional term $\chi_{ph}(\Omega, \mathbf{q})$ is the photon susceptibility, defined by

$$\chi_{ph}(\Omega, \mathbf{q}) = -\frac{\omega_{p0}^2}{\Omega^2}\frac{1}{2}\int W_0\left[\frac{1}{(\Omega - \mathbf{q}\cdot\mathbf{v}_+)} - \frac{1}{(\Omega - \mathbf{q}\cdot\mathbf{v}_-)}\right]\frac{d\mathbf{k}}{(2\pi)^3}\,. \tag{11.70}$$

The remarkable thing about this dispersion relation is that the photons contribute to dispersion as an additional new plasma kind of particles. In particular, they can contribute to the plasma wave damping in a way that is similar to electron Landau damping. This is relevant to the case of relativistic electron plasma waves, such that $\Omega/q \gg S_e$, where resonant electrons are absent, but the photons (or equivalently, electromagnetic radiation) can be resonant with the electrostatic wave, and contribute to damping.

In order to discuss this damping process, it is now useful to consider parallel and perpendicular wavevector and group velocity components such that

$$\mathbf{k} = p\,\mathbf{e}_z + \mathbf{k}_\perp \,, \quad \mathbf{v_k} = v\,\mathbf{e}_z + \mathbf{v}_\perp \,, \tag{11.71}$$

where $p$ and $v$ are the parallel momentum and group velocity, with respect to the direction of plasma wave propagation, assumed along the $z$-axis, $\mathbf{q} = q\mathbf{e}_z$. Integrating over the perpendicular momentum $\mathbf{k}_\perp$ the photon susceptibility becomes

$$\chi_{ph}(\Omega, \mathbf{q}) = -\frac{\omega_{p0}^2}{\Omega^2}\frac{q^2c^2}{2}\int G(p)\left[\frac{1}{(p-p_{0+})} - \frac{1}{(p-p_{0-})}\right]\frac{dp}{2\pi} \,. \tag{11.72}$$

Here, the new function, $G(p)$, is the equilibrium distribution in the parallel direction, defined by

$$G(p) = \int \frac{W_0(p, \mathbf{k}_\perp)}{(\partial v/\partial p)_0}\frac{d\mathbf{k}_\perp}{(2\pi)^2} \,. \tag{11.73}$$

where the derivative in the denominator is taken at the resonant value $p = p_0$, defined by the equality $v(p_0) = \Omega/q$. We have also used $p_{0\pm} = p_0 \pm q/2$. If we just take the principal value of the integral in eq. (11.72), we obtain the photon contribution to the plasma wave dispersion. But, if we retain the contribution of the poles at $p = p_{0\pm}$, we obtain a damping rate, which is given by

$$\Gamma_{ph} = -\omega_{p0}\frac{\pi q c^2}{4}\left[G(p_0 - q/2) - G(p_0 + q/2)\right] \,. \tag{11.74}$$

This is *photon Landau damping*. In order to extend the analogy with the traditional electron Landau damping, it is useful to take the limit of geometric optics, where $q \ll p$, and use the expansion

$$G(p_0 \pm q/2) \simeq G(p_0) \pm \frac{q}{2}\left(\frac{\partial G}{\partial p}\right)_0 \,, \tag{11.75}$$

The damping rate is then reduced to

$$\Gamma_{ph} = -\omega_{p0}\frac{\pi q^2 c^2}{8}\left(\frac{\partial G}{\partial p}\right)_0 \,, \tag{11.76}$$

This expression is formally identical to the electron Landau damping and

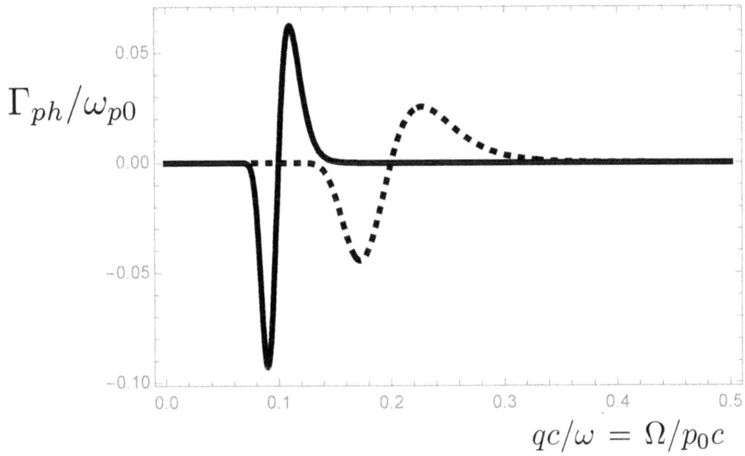


**FIGURE 11.4**
Photon Landau damping of electron plasma waves, $\Gamma_{ph}/\omega_{p0}$, with relativistic phase velocities $\Omega/q \simeq c$, as a function of the normalized electron plasma wavenumber $qc/\omega = \Omega/p_0c$. We assume a Gaussian photon beam, $G(p_0) = G_0 \exp[-(p_0 - \bar{p})^2/\sigma^2]$, for two different spectra, $\bar{p} = 3$ (dashed curve) and $\bar{p} = 10$ (bold curve).

shows that the plasma wave is damped by photons that are resonant with its phase velocity. This is particularly relevant to the energy dissipation of laser pulses by resonant wakefields.

Photon Landau damping was first derived in 1997 [195], using the Vlasov equation for photons (11.39), and the full-wave result given by eq. (11.74) was derived a few years later [197]. This exact result incorporates the wave nature of the electromagnetic radiation and shows that photon recoil occurs in the emission or absorption of plasmons. The recoil effect corresponds to a discrete increase or decrease of the photon momentum, typical of a quantum effect, which is here associated with the undulatory nature of the photon.

The photon Landau damping is illustrated in Figure 11.4, for two different spectra. We notice that there is always a region of damping, followed by an amplification region. This shows that a photon beam can excite electron oscillations, as previously discussed in Section 6.5. As noticed before, the instability also occurs for a cold photon beam, which corresponds to monochromatic radiation, as

$$W_{\mathbf{k}} = (2\pi)^3 W_0 \delta(\mathbf{k} - \mathbf{k_0}) \,, \tag{11.77}$$

In this case, integration of eq. (11.70) leads to a photon susceptibility of the form

$$\chi_{ph}(\Omega, \mathbf{q}) = \frac{\omega_{p0}^2}{\Omega^2} \frac{q^2c^2}{2} \frac{W_0 \mathbf{q} \cdot (\mathbf{v}_+ - \mathbf{v}_-)}{(\Omega - \mathbf{q} \cdot \mathbf{v}_+)(\Omega - \mathbf{q} \cdot \mathbf{v}_-)} \tag{11.78}$$

In the geometric optics limit, such that $|q| \ll k|$, this reduces to

$$\chi_{ph}(\Omega, \mathbf{q}) = \frac{\omega_{p0}^2}{\Omega^2} \frac{q^4 c^4}{2} \frac{W_0}{\omega_0(\Omega - \mathbf{q} \cdot \mathbf{v}_0)^2} \tag{11.79}$$

where $\omega_0 = \omega(\mathbf{k}_0)$, and $\mathbf{v}_0 = \mathbf{v}(\mathbf{k}_0)$. Using this in the dispersion relation (11.68), we can easily show that the electron plasma waves are unstable, with a maximum growth rate

$$\Gamma = \omega_{p0} \frac{\sqrt{3}}{2} \left( W_0 \omega_{p0} q^4 c^4 \right)^{1/3} , \tag{11.80}$$

that corresponds to the double resonant condition

$$\Omega = q v_0 \simeq \omega_{p0} \,. \tag{11.81}$$

Here, again, we observe a beam instability with growth rate proportional to the cubic root of the beam density $W_0$, which is a typical feature of cold beam instabilities.

## 11.4 Photon Fluid Equations

Let us now focus on the fluid equations that can be derived from the photon-kinetic theory when we take averages over the electromagnetic wave spectrum. This leads to photon conservation laws that reveal new properties of the electromagnetic waves, in particular, the existence of Cherenkov emission of photon oscillations, as demonstrated next.

For the present purpose, we start from the Vlasov equation (11.55), written in terms of the photon number density $N_{\mathbf{k}}$, as

$$\left( \frac{\partial}{\partial t} + \mathbf{v}_{\mathbf{k}} \cdot \nabla + \mathbf{f}_{\mathbf{k}} \cdot \frac{\partial}{\partial \mathbf{k}} \right) N_{\mathbf{k}} = 0 \,, \tag{11.82}$$

with

$$\mathbf{v}_{\mathbf{k}} = \frac{\partial \omega_{\mathbf{k}}}{\partial \mathbf{k}} \,, \quad \mathbf{f}_{\mathbf{k}} = -\nabla \omega_{\mathbf{k}} \,. \tag{11.83}$$

It we integrate over the $\mathbf{k}$ spectrum, we obtain a continuity equation, of the form

$$\frac{\partial n_{ph}}{\partial t} + \nabla \cdot (\mathbf{v} n_{ph}) = 0 \,. \tag{11.84}$$

The mean density $n_{ph}$ and mean velocity $\mathbf{v}$ are determined by

$$n_{ph} = 2 \int N_{\mathbf{k}} \frac{d\mathbf{k}}{(2\pi)^3} \,, \quad \mathbf{v} = 2 \int \mathbf{v}_{\mathbf{k}} N_{\mathbf{k}} \frac{d\mathbf{k}}{(2\pi)^3} \,, \tag{11.85}$$

where the factor of 2 is introduced to account for the two possible polarization states. Notice that the integration of the second term in (11.82) gives

$$\int \mathbf{v_k} \cdot \nabla N_{\mathbf{k}} \frac{d\mathbf{k}}{(2\pi)^3} = \nabla \cdot (\mathbf{v} n_{ph}) - \int N_{\mathbf{k}} \left(\nabla \cdot \mathbf{v_k}\right) \frac{d\mathbf{k}}{(2\pi)^3} , \tag{11.86}$$

while integration of the third term leads to

$$\int \mathbf{f_k} \cdot \frac{\partial N_{\mathbf{k}}}{\partial \mathbf{k}} \frac{d\mathbf{k}}{(2\pi)^3} = \int N_{\mathbf{k}} \left(\nabla \cdot \mathbf{v_k}\right) \frac{d\mathbf{k}}{(2\pi)^3} , \tag{11.87}$$

The photon continuity equation (11.84) is, of course, only valid in the absence of radiation emission and absorption. The second conservation equation can be obtained if we multiply the kinetic equation (11.82) by the group velocity $\mathbf{v_k} = \mathbf{k}c^2/\omega$, before integrating over the spectrum. The result is

$$\frac{\partial \mathbf{v}}{\partial t} + \mathbf{v} \cdot \nabla \mathbf{v} = -\frac{1}{n_{ph}} \nabla \mathcal{P}_{ph} + \mathcal{F}_{ph} \, . \tag{11.88}$$

This is the momentum conservation equation, where $\mathcal{P}_{ph}$ represents the photon pressure, defined by

$$\mathcal{P}_{ph} = \frac{2}{3} \int N_{\mathbf{k}} (\mathbf{v_k} - \mathbf{v})^2 \frac{d\mathbf{k}}{(2\pi)^3} = n_{ph} T_{ph} \tag{11.89}$$

where the second equality defines $T_{ph}$, an effective temperature of the photon gas. The quantity $\mathcal{F}_{ph}$ is the mean force acting on the photon gas, and is determined by

$$\mathcal{F}_{ph} = \frac{2}{n_{ph}} \int N_{\mathbf{k}} \left[ \frac{\partial \mathbf{v_k}}{\partial t} + \nabla \cdot (\mathbf{v_k} \mathbf{v_k}) + \frac{\partial}{\partial \mathbf{k}} \cdot (\mathbf{f_k} \mathbf{v_k}) \right] \frac{d\mathbf{k}}{(2\pi)^3} \, . \tag{11.90}$$

We can see that this force is due to the space-time variations of the medium. For isotropic relativistic plasmas, this can conveniently be written as

$$\mathcal{F}_{ph} = -\frac{1}{n_{ph}} \int \frac{N_{\mathbf{k}}}{\omega_{\mathbf{k}}^2} \frac{d\mathbf{k}}{(2\pi)^3} \left( \nabla \omega_p^2 + \frac{\mathbf{v}}{c^2} \frac{\partial}{\partial t} \right) \frac{\omega_p^2}{\gamma_a} \, . \tag{11.91}$$

In a similar way, we can establish the energy conservation equation, if we multiply eq. (11.82) by the photon energy $\hbar \omega_{\mathbf{k}}$, and then integrate in $\mathbf{k}$. The result is

$$\frac{\partial}{\partial t} U_{ph} + \nabla \cdot (\mathbf{v} U_{ph}) = 2 \int N_{\mathbf{k}} \frac{\partial \omega_{\mathbf{k}}}{\partial t} \frac{d\mathbf{k}}{(2\pi)^3} , \tag{11.92}$$

where the mean energy density $U_{ph}$ is defined by

$$U_{ph} = \frac{2\hbar}{n_{ph}} \int \omega_{\mathbf{k}} N_{\mathbf{k}} \frac{d\mathbf{k}}{(2\pi)^3} \, . \tag{11.93}$$

In eq. (11.91) we can consistently assume that the electron relativistic factor is determined by $\gamma_a = \sqrt{1 + U_{ph}}$.

These photon fluid equations are usually less inspiring than the photon-kinetic equations discussed in previous sections. We can nevertheless use them to study slow oscillations of the electromagnetic radiation, coupled with the plasma. As a simple example, we consider an electromagnetic beam moving with mean velocity $\mathbf{v}_0$. In order to study eventual perturbations in the medium, evolving as $\exp(i\mathbf{q}\cdot\mathbf{r} - i\Omega t)$, linearizing the above conservation relations, and neglecting the electron density perturbations, we are able to derive a dispersion equation of the form

$$(\Omega - \mathbf{q}\cdot\mathbf{v}_0)^2 + \frac{\omega_p^2}{\gamma_0}\frac{\Omega_{ph}^2}{q^2c^2} = 0\,, \quad \Omega_{ph}^2 = \frac{q^2c^4}{4n_{ph0}}U_0\left(\gamma_0^2 - 1\right)\,, \tag{11.94}$$

where $\Omega_{ph} \propto U_0$ defines the photon plasma frequency. This equation can be solved as $\Omega = \Omega' + i\Omega''$, where the real and imaginary parts are given by

$$\Omega' = qv_0\cos\theta\,, \quad \Omega'' = \frac{\omega_p}{\sqrt{\gamma_0}}\frac{\Omega_{ph}}{qc}\,. \tag{11.95}$$

This result describes the excitation of low-frequency photon oscillations, with frequencies $\Omega' \ll \omega_{\mathbf{k}}$, inside a resonance cone. This is very similar to Cherenkov radiation by charged particles. However, in the present case, we obtain intensity oscillations inside the photon beam, produced by the photon beam itself [209].

## 11.5 Photon Condensation

We now consider the processes leading to thermal equilibrium in a system composed of a fully ionized plasma and a photon gas. As such, we are neglecting eventual absorption and emission of photons from neutral atoms and assume that photon thermalization results from electron-photon coupling, provided by Compton scattering. In order to describe this thermalization process, we therefore need to establish a new photon kinetic equation where Compton scattering is explicitly taken into account. For this purpose, we start from a simplified photon-kinetic equation, known as the *Kompaneets equation* [211]. This equation is usually written in the following way

$$\frac{\partial N}{\partial \tau} = \frac{1}{\kappa^2}\frac{\partial}{\partial \kappa}\left[\kappa^4\left(\frac{\partial N}{\partial \kappa} + N + N^2\right)\right]\,, \tag{11.96}$$

where $N \equiv N(\mathbf{k})$ is the photon occupation number for a given polarization state $\sigma$, the temporal variable $\tau$ needs to be defined and is sometimes called the Compton time, and $\kappa = \hbar\omega_{\mathbf{k}}/T_e$ is the photon energy, normalized to the electron temperature $T_e$ in energy units. We can see that the equilibrium solution of this equation, satisfying $\partial N/\partial\tau = 0$, is the Bose-Einstein distribution

$$N(\kappa) = \frac{1}{ze^{\kappa} - 1}\,, \quad z = e^{\kappa_0}\,, \tag{11.97}$$

where the constant $z$ is the *fugacity*. It reduces to a Planck distribution, if we take $z = 1$. Assuming a large number of photons $N \gg |\partial N/\partial \tau|$, eq. (11.96) can be reduced to

$$\frac{\partial f}{\partial \tau} = 2f \frac{\partial N}{\partial \kappa}, \quad f = \kappa^2 N . \tag{11.98}$$

Based on this simplified form of the Kompaneetz equation it is then possible to show that a shock front can occur ($N(\kappa) \to \infty, \kappa \to 0$) for times above some critical time $\tau \geq \tau_{cr}$. This shock front corresponds to an accumulation of photons at zero frequency, and can be identified as *photon condensation* [198]. This condensation process is only valid if we neglect the plasma frequency. But the existence of a frequency cutoff at $\omega_{\mathbf{k}} = \omega_p$ prevents the photon frequency to reduce to zero. This means that a consistent description of photon condensation in a plasma needs to include plasma dispersion effects.

Let us review the derivation of the Kompaneets equation, starting from a balance equation for the photon spectral distribution $N(\mathbf{k})$. We assume that electrons and photons exchange energy and momentum through the elementary process of Compton scattering. This means that a transition from the initial state $(\mathbf{p}, \mathbf{k}, N(\mathbf{k}))$, where $\mathbf{p}$ is the electron momentum, to a final state $(\mathbf{p}', \mathbf{k}', N(\mathbf{k}'))$, due to Compton scattering, implies a decrease of the photon number $N(\mathbf{k})$. Note that the number of electrons inside the elementary momentum volume $d\mathbf{p}$ around $\mathbf{p}$ is determined by $f(\mathbf{p})d\mathbf{p}$, where $f(\mathbf{p})$ is the electron distribution function. The total decrease of the number of photons per unit volume is then given by

$$-N(\mathbf{k})[1 + N(\mathbf{k}')] f(\mathbf{p}) d\mathbf{p} . \tag{11.99}$$

Here, the term in $N(\mathbf{k}')$ corresponds to the stimulated process. The Compton process implies a relation between the initial and final frequencies, of the form

$$\omega_{k'} = \frac{\omega_k}{1 + (1 - \cos\theta)\, \hbar\omega_k/(mc^2)}, \tag{11.100}$$

where $\theta$ is the scattering angle. Similarly, for the reverse case of a transition from $(\mathbf{p}', \mathbf{k}', N(\mathbf{k}'))$ to $(\mathbf{p}, \mathbf{k}, N(\mathbf{k}))$, we have an increase in the photon number of

$$N(\mathbf{k}')[1 + N(\mathbf{k})] f(\mathbf{p}') d\mathbf{p}' . \tag{11.101}$$

This leads to a detailed balance equation for the photon number, which is a *quantum Boltzmann equation* of the form

$$\frac{\partial N(\mathbf{k})}{\partial t} = \int d\mathbf{p} \int dw\, [f(\mathbf{p}')N(\mathbf{k}')\,(1 + N(\mathbf{k}) - f(\mathbf{p})N(\mathbf{k})\,(1 + N(\mathbf{k}'))] \,, \tag{11.102}$$

where $dw = cd\Omega(d\sigma/d\Omega)$, and the differential cross section for electron-photon collisions is determined by the Klein-Nishina formula

$$\frac{d\sigma}{d\Omega} = \frac{3}{16\pi} \left(\frac{\omega_{k'}}{\omega_k}\right)^2 \left[\frac{\omega_{k'}}{\omega_k} + \frac{\omega_k}{\omega_{k'}} - \sin^2\theta\right] \sigma_T . \tag{11.103}$$

Here, $\sigma_T = (8\pi/3)r_e^2$ is the Thomson cross section, and $r_e = e^2/(4\pi\epsilon_0 mc^2)$ is the classical electron radius. For a low-frequency spectrum, near $\omega_p$, this can be approximated by the non-relativistic expression

$$\frac{d\sigma}{d\Omega} = \frac{3}{16\pi}\left(1+\cos^2\theta\right)\sigma_T. \tag{11.104}$$

We now introduce some further approximations, and assume that each Compton collision process is associated with an energy exchange $\delta$, that is small compared with the electron thermal energy, as

$$\delta \equiv \hbar(\omega_{\mathbf{k}'} - \omega_{\mathbf{k}})/T_e \ll 1\,, \tag{11.105}$$

We can then expand the photon distribution to second-order with respect to this small quantity, as

$$N(\omega_{\mathbf{k}'}) \simeq N(\omega_{\mathbf{k}}) + \delta T_e \frac{\partial}{\partial\omega} N(\omega_{\mathbf{k}}) + \frac{\delta^2}{2} T_e^2 \frac{\partial^2}{\partial\omega^2} N(\omega_{\mathbf{k}}), \tag{11.106}$$

This allows us to reduce the integro-differential equation (11.102) to the following form

$$\frac{\partial N}{\partial t} = \left[N' + N(N+1)\right]\mathcal{I}_1(\mathbf{k}) + \left[\frac{1}{2}N'' + \left(N' + \frac{N}{2}\right)(1+N)\right]\mathcal{I}_2(\mathbf{k})\,, \tag{11.107}$$

where $N' = \partial N/\partial\kappa$ and $\kappa = \hbar\omega_k/T_e$ as defined above. The quantities $\mathcal{I}_j(\mathbf{k})$, for $j = 1, 2$, are determined by

$$\mathcal{I}_j(\mathbf{k}) = c\int d\mathbf{p}\int d\Omega \frac{d\sigma}{d\Omega}\delta^j f(\mathbf{p})\,. \tag{11.108}$$

for $j = 1, 2$. These integrals can be explicitly performed by noticing that, in the collision process, the energy exchange $\delta$ is approximately given by

$$\delta \simeq \frac{\hbar}{2m_e}\mathbf{p}\cdot(\mathbf{k} - \mathbf{k}')\,. \tag{11.109}$$

Using the photon dispersion relation $\omega_k^2 = \omega_p^2 + c^2k^2$, we can then integrate eqs. (11.108), and obtain

$$\mathcal{I}_1(\mathbf{k}) = \frac{\sigma_T n_0}{T_e}\frac{\hbar\sqrt{\omega_k^2 - \omega_p^2}}{m_e c}\left(4T_e - \hbar\sqrt{\omega_k^2 - \omega_p^2}\right)\,, \quad \mathcal{I}_2(\mathbf{k}) = 2\frac{T_e\sigma_T n_0}{m_e c}\left(\omega_k^2 - \omega_p^2\right). \tag{11.110}$$

where $n_0$ is the electron mean density. We are then able to derive a modified Kompaneets equation, of the form [199]

$$\frac{\partial N}{\partial \tau} = \frac{f(\kappa)}{\kappa^2}\frac{\partial}{\partial\kappa}\left[\kappa^4\left(\frac{\partial N}{\partial\kappa}f(\kappa) + N + N^2\right)\right]. \tag{11.111}$$

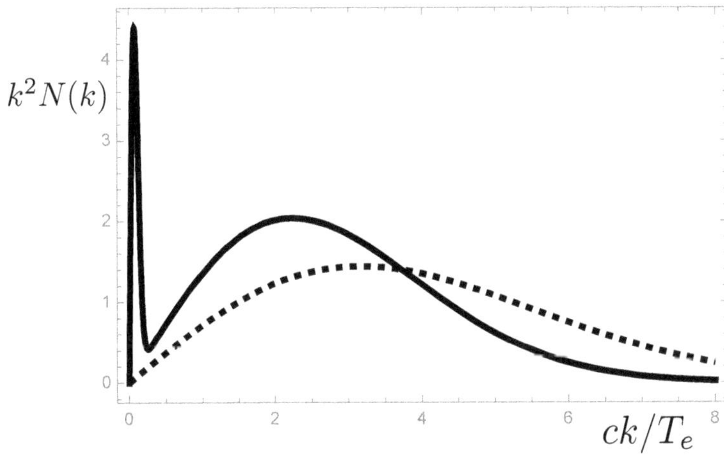


**FIGURE 11.5**
Photon condensation in a plasma: normalized function $k^2N(k)$, as a function of normalized wavenumber, $ck/T_e$, when the photon temperature is larger than the electron temperature, $T_{ph0} > T_e$. The initial photon spectrum is also shown (dashed curve).

Here we have used the Compton time $\tau$ and the Jacobian function $f(\kappa)$, as defined by

$$\tau = n_0 \sigma_T \frac{T_e}{mc^2} t, \quad f(\kappa) = \kappa \left[ \kappa^2 + \left( \frac{\hbar \omega_p}{T_e} \right)^2 \right]^{-1/2} . \tag{11.112}$$

Notice that, when the plasma frequency $\omega_p$ tends to zero, the function $f(\kappa)$ tends to 1. In that limit, eq. (11.111) reduces to the original equation derived by Kompaneets, stated as eq. (11.96) .

The modified eq. (11.111) can be used to study photon condensation in a plasma for a finite value of $\omega_p \neq 0$. This is illustrated in Figure 11.5, assuming that the initial the photon distribution is, $N_0 = 1/(\exp^{(\hbar\omega_k/T_{ph0})} - 1)$ where $T_{ph0}$ is the initial temperature of the photon gas. The numerical solutions show that, when the electromagnetic spectrum is cooler than the electron gas ($T_{ph0} < T_e$), no condensation is observed: the photon temperature will increase and tends to thermal equilibrium at $T_{ph} = T_e$. In contrast, for $T_{ph0} > T_e$, photon condensation can be attained. In this case, Compton scattering provides an accumulation of the photon spectrum near the lowest allowed frequency, at $\omega_{\mathbf{k}} \to \omega_p$, which corresponds to cutoff, $k = 0$. This accumulation is observed in the background of a thermal cloud of photons at the temperature $T_e$ (see Figure 11.5 for an illustration). This is in qualitative agreement with the well-known atomic theory of Bose-Einstein condensation [218]. In this integration, condensation becomes visible on a time-scale of the order of the Compton time, $\tau \simeq 1$.

## 11.6 Wave-Kinetics

The evolution of electromagnetic radiation in a plasma has been described as the evolution of a gas of relativistic particles, the photons. But, it is known that plasma supports a large variety of other waves, and the spectrum of all these waves can also be described as quasi-particles. They result from the correlation between fields and particles in the plasma. This suggests that the photon-kinetic theory described above can be generalized to other kind of waves in the medium, leading to a more general concept of *wave-kinetics.*

To discuss wave-kinetics we consider an unspecified kind of wave, with frequency $\omega$ and wavevector $\mathbf{k}$, obeying the dispersion relation

$$\mathcal{D}(\omega, \mathbf{k}) \equiv \mathbf{e}_{\mathbf{k}}^* \cdot \bar{\bar{\mathcal{D}}}(\omega, \mathbf{k}) \cdot \mathbf{e}_{\mathbf{k}} = 0\,, \tag{11.113}$$

where $\mathbf{e}_{\mathbf{k}}$ unit polarization vector, and $\bar{\bar{\mathcal{D}}}(\omega, \mathbf{k})$ is the dispersion tensor, as defined in Chapter 4, such that

$$\bar{\bar{\mathcal{D}}}(\omega, \mathbf{k}) = \left(\mathbf{NN} - N^2 \bar{\bar{1}} + \bar{\bar{\kappa}}\right)\,, \quad \bar{\bar{\kappa}} = \bar{\bar{1}} + \frac{i}{\omega \epsilon_0} \sum_\alpha \bar{\bar{\sigma}}_\alpha\,. \tag{11.114}$$

with the refractive index $\mathbf{N} = (c/\omega)\mathbf{k}$, and the dielectric tensor $\bar{\bar{\kappa}}$. The conductivity tensor for the different particle species $\alpha$, denoted by $\bar{\bar{\sigma}}_\alpha$, can be derived by some microscopic or fluid plasma model. Of course, this dispersion relation is valid for waves propagating in a static medium, but stays approximately valid for media evolving on time-scales much longer than $1/\omega$. Perturbations of the medium can be due to a variety of reasons, from density variations, $\delta n_\alpha$, external magnetic perturbations $\delta \mathbf{B}_0$, to temperature variations $\delta T$, and others. If we are dealing with a nonlinear dispersion relation, which depends on the wave intensity, we can also include $\delta |a|^2$, where $a \equiv a(\mathbf{r}, t)$ is the mode amplitude. As a result, the dispersion relation (11.113) is modified and has to be replaced by an expression of the form

$$[\mathcal{D}(\omega', \mathbf{k}') + \delta \mathcal{D}]\, a = 0\,, \tag{11.115}$$

where

$$\omega' = \omega + i\frac{\partial}{\partial t}\,, \quad \mathbf{k}' = \mathbf{k} - i\nabla\,, \tag{11.116}$$

and $\delta \mathcal{D}$ is a perturbed dispersion function, depending on the relevant fluctuations, as

$$\delta \mathcal{D}(\mathbf{r}, t) = \sum_\alpha \left(\frac{\partial \mathcal{D}}{\partial n_\alpha}\right) \delta n_\alpha + \frac{\partial \mathcal{D}}{\partial T} \delta T + \ldots. \tag{11.117}$$

We can expand (11.115) around the local linear dispersion (11.113), and obtain an evolution equation for the wave amplitude, valid for a slow time scale, as

$$\left( \frac{\partial}{\partial t} + \mathbf{v}_k \cdot \nabla - iU \right) a = 0 \,, \tag{11.118}$$

where the group velocity $\mathbf{v}_k$, and the potential $U(\mathbf{r}, t)$ associated with the perturbations of the medium, are defined by

$$\mathbf{v}_k = - \frac{(\partial \mathcal{D} / \partial \mathbf{k})}{(\partial \mathcal{D} / \partial \omega)} \quad , \quad U(\mathbf{r}, t) = \frac{\delta \mathcal{D}(\mathbf{r}, t)}{(\partial \mathcal{D} / \partial \omega)} \,. \tag{11.119}$$

This result assumes that the plasma is unbounded, but a similar result can also be derived by bounded media, using appropriate mode decomposition and boundary conditions.

We can now follow the recipe of the previous photon-kinetic theory, and define a field autocorrelation function, as in eq. (11.59), and the corresponding Wigner function $W(\mathbf{r}, t, \omega, \mathbf{k})$, as the double Fourier transform of this autocorrelation function. Following the same procedure, we obtain an equation that is formally identical but physically distinct from the photon-kinetic equation

$$i \left( \frac{\partial}{\partial t} + \mathbf{v}_k \cdot \nabla \right) W = \int \frac{d\mathbf{q}}{(2\pi)^3} \int \frac{d\Omega}{2\pi} U(\Omega, \mathbf{q}) \tag{11.120}$$
$$[W_- - W_+] \exp(i\mathbf{q} \cdot \mathbf{r} - i\Omega t) \,,$$

where $W_\pm = W(\omega \pm \Omega/2, \mathbf{k} \pm \mathbf{q}/2)$, and $U(\Omega, \mathbf{q})$ are the Fourier components of the perturbation potential $U(\mathbf{r}, t)$. As in the case of photons, it is useful to assume a reduced Wigner function $W_\mathbf{k}$, as defined in eq. (11.64), now referring to unspecified quasi-particles (see a specific example below). The evolution equation of this reduced function will be formally identical to this equation, but where $W$ is replaced by $W_\mathbf{k}$. It is also useful to take the limit of geometric optics expansion, valid for $|\mathbf{k}| \gg |\mathbf{q}|$. This gives

$$W_\mathbf{k}^\pm \equiv W(\mathbf{k} \pm \mathbf{q}/2) = \sum_{n=0}^{\infty} \frac{1}{n!} \left( \pm \frac{\mathbf{q}}{2} \cdot \frac{\partial}{\partial \mathbf{k}} \right)^n W_\mathbf{k} \,, \tag{11.121}$$

and leads to

$$[W_- - W_+] = - \sum_{n=0}^{\infty} \frac{2}{(2n+1)!} \left( \frac{\mathbf{q}}{2} \cdot \frac{\partial}{\partial \mathbf{k}} \right)^{2n+1} W_\mathbf{k} \,. \tag{11.122}$$

Retaining only the first two terms in the expansion (11.121), we would simply have

$$W_\mathbf{k}^\pm \simeq W_\mathbf{k} \pm \frac{\mathbf{q}}{2} \cdot \frac{\partial}{\partial \mathbf{k}} W_\mathbf{k} + \frac{1}{2} \left( \frac{\mathbf{q}}{2} \cdot \frac{\partial}{\partial \mathbf{k}} \right)^2 W_\mathbf{k} \pm \dots \,. \tag{11.123}$$

Replacing this in eq. (11.121), we reduce this kinetic equation for quasi-particles, to

$$\left( \frac{\partial}{\partial t} + \mathbf{v}_\mathbf{k} \cdot \nabla - \nabla U(\mathbf{r}, t) \cdot \frac{\partial}{\partial \mathbf{k}} + \dots \right) W_\mathbf{k} = 0 \,. \tag{11.124}$$

This geometric optics limit neglects wave diffraction effects. The group velocity, $\mathbf{v_k}$, and the force $\mathbf{f_k}$ acting on the quasi-particles are

$$\frac{d\mathbf{r}}{dt} = \mathbf{v_k} \, , \quad \frac{d\mathbf{k}}{dt} = -\nabla U(\mathbf{r}, t) \equiv \mathbf{f_k} \, , \tag{11.125}$$

which reduces this equation to a Vlasov-type of equation, given by

$$\frac{dW_\mathbf{k}}{dt} = 0 \, , \quad \frac{d}{dt} \equiv \left( \frac{\partial}{\partial t} + \frac{d\mathbf{r}}{dt} \cdot \nabla + \frac{d\mathbf{k}}{dt} \cdot \frac{\partial}{\partial \mathbf{k}} \right) . \tag{11.126}$$

An additional equation should be added, usually ignored in ray-tracing codes, determines the quasi-particle frequency, as

$$\frac{d\omega}{dt} = \frac{\partial}{\partial t} U(\mathbf{r}, t) \, . \tag{11.127}$$

This generalizes eq. (11.37) for an arbitrary quasi-particle and can therefore be applied to the study of any form of wave turbulence. As an illustration, we consider the case of *lower-hybrid waves*, which are used to drive additional current in *tokamak devices* [214, 215]. They can be described by the dispersion relation

$$\mathcal{D}(\omega, \mathbf{k}) = 1 - k_\parallel^2 \epsilon_\parallel(\omega) - k_\perp^2 \epsilon_\perp(\omega) \, , \tag{11.128}$$

where the parallel and perpendicular dielectric functions are

$$\epsilon_\parallel(\omega) = -\frac{\omega_{pe}^2}{\omega^2} \, , \quad \epsilon_\perp(\omega) = 1 + \frac{\omega_{pe}^2}{\omega_{ce}^2} - \frac{\omega_{pi}^2}{\omega^2} \, . \tag{11.129}$$

Here, parallel and perpendicular refer to the direction of the static magnetic field $\mathbf{B}_0$. For slow-wave solutions, relevant to current drive, we usually have $\omega \ll \omega_{pe}$ [215]. Assuming large-scale density perturbations, $\delta n_e$, it can easily be shown that the group velocity is determined by

$$\mathbf{v_k} = -\frac{\omega^3}{k_\parallel^2 \omega_{pe}^2 + k_\perp^2 \omega_{pi}^2} \left[ \mathbf{k}_\parallel \epsilon_\parallel(\omega) + \mathbf{k}_\perp \epsilon_\perp(\omega) \right] \, . \tag{11.130}$$

The corresponding potential perturbation $U(\mathbf{r}, t)$ is given by

$$U = \frac{\omega^3}{2\omega_{pe}^2} \left( 1 - \frac{k_\parallel^2}{k_\perp^2} \frac{\omega_{ce}^2}{\omega^2} \right) \left( \frac{m_e}{m_i} + \frac{k_\parallel^2}{k_\perp^2} \right)^{-1} \frac{\delta n_e}{n_0} \, . \tag{11.131}$$

We can then use wave-kinetics to determine the dynamics of lower-hybrid waves (described as quasi-particles) in a perturbed plasma density background. For this purpose, we can consider an arbitrary large-scale density fluctuation spectrum, given by

$$\delta n_e(\mathbf{r}, t) = \int n(\mathbf{q}) \exp(i\mathbf{q} \cdot \mathbf{r} - i\Omega t) \frac{d\mathbf{q}}{(2\pi)^3} \, , \tag{11.132}$$

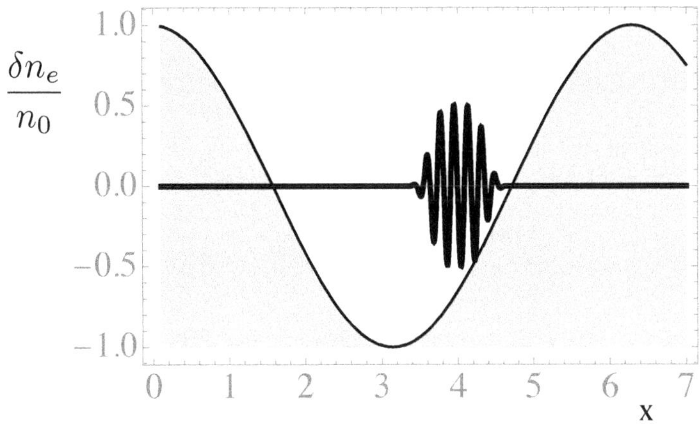


**FIGURE 11.6**
Quasi-particle (bold curve) moving in the presence of a large-scale density oscillation (thin curve), propagating along $Ox$.

This example is studied in detail elsewhere [216]. The case of single large-scale oscillation is illustrated in Figure 11.6. Phenomena as *quasi-particle trapping* and *quasi-particle Landau damping*, similar to those discussed in this chapter for photons, can then occur.

A more abstract, but nearly equivalent approach to wave-turbulence can be followed using more general equations to describe the fields. This shows the robustness of the concepts. We adopt a procedure used in [217]. We consider a generic turbulent mode in a plasma, such as a lower-hybrid wave as in the previous example, or another mode as a drift wave or an Alfvén wave. In many different situations, they can be described by an equation of the form [219]

$$\left(\frac{\partial}{\partial t} + i\omega_{\mathbf{k}}\right) a_{\mathbf{k}} + \int \mathcal{L}(\mathbf{p}, \mathbf{k} - \mathbf{p}) \Phi_{\mathbf{p}} a_{\mathbf{k}-\mathbf{p}} \frac{d\mathbf{p}}{(2\pi)^3} = 0\,, \tag{11.133}$$

where $\omega_{\mathbf{k}}$ is the linear mode frequency, $\mathbf{k}$ the corresponding wavevector, and $a_{\mathbf{k}}$ the wave amplitude. The wave propagates in a perturbed background, described by the low-frequency field $\Phi(\mathbf{p})$, and $\mathcal{L}(\mathbf{p}, \mathbf{k} - \mathbf{p})$ is the coupling parameter. In order to derive the field correlations, we multiply this equation by $a_{\mathbf{k}'}$, and add the symmetric equation. This leads to

$$\frac{\partial}{\partial t} K(\mathbf{k}', \mathbf{k}'') + i(\omega_{k'} + \omega_{k''}) K(\mathbf{k}', \mathbf{k}'') + [S_c(\mathbf{k}', \mathbf{k}'') + S_c(\mathbf{k}'', \mathbf{k}')] = 0\,, \tag{11.134}$$

where $K(\mathbf{k}', \mathbf{k}'') = a_{\mathbf{k}'} a_{\mathbf{k}''}$. Coupling with the perturbed medium is defined by

$$S_c(\mathbf{k}, \mathbf{k}') = a_{\mathbf{k}} \int \mathcal{L}(\mathbf{p}, \mathbf{k}' - \mathbf{p}) \Phi(\mathbf{p}) a_{\mathbf{k}'-\mathbf{p}} \frac{d\mathbf{p}}{(2\pi)^3}\,. \tag{11.135}$$

We then define the Wigner function of the quasi-particle field, in the usual way

$$W(\mathbf{r},\mathbf{k},t) = \int a_{\mathbf{k}'} a_{\mathbf{k}''}\, e^{i\mathbf{q}\cdot\mathbf{r}} \frac{d\mathbf{q}}{(2\pi)^3}\,, \tag{11.136}$$

where it is assumed that $\mathbf{k}' = \mathbf{k}+\mathbf{q}/2$, and $\mathbf{k}'' = \mathbf{q}/2-\mathbf{k}$. Multiplying equation (11.134) by $\exp(i\mathbf{q}\cdot\mathbf{r})$, and integrating over $\mathbf{q}$ space, it is then possible to obtain [217]

$$\frac{\partial W}{\partial t} + i\Delta_{\mathbf{k}} W + S_{\mathbf{k}} = 0\,, \tag{11.137}$$

where

$$S_{\mathbf{k}} = \int \left[S_c(\mathbf{k},\mathbf{k}'') + S_c(\mathbf{k}'',\mathbf{k}')\right] e^{i\mathbf{q}\cdot\mathbf{r}} \frac{d\mathbf{q}}{(2\pi)^3}\,, \tag{11.138}$$

and $\Delta_{\mathbf{k}}$ is the differential operator

$$\Delta_{\mathbf{k}} = \exp\left(-\frac{i}{2}\nabla\cdot\frac{\partial}{\partial\mathbf{k}}\right)\omega_k + c.c.\,. \tag{11.139}$$

Here, the operator $\partial/\partial\mathbf{k}$ only acts on the frequency, $\omega_{\mathbf{k}}$, not on the Wigner function $W \equiv W(\mathbf{r},\mathbf{k},t)$. In contrast, the operator $\nabla$ only acts on $W$, and not on $\omega_{\mathbf{k}}$. The source function $S_{\mathbf{k}}$, can also be transformed into

$$S_w(\mathbf{k}) = \int \Phi(\mathbf{p})\left[D_{\mathbf{k}} W_- + D_{-\mathbf{k}} W_+\right] e^{i\mathbf{p}\cdot\mathbf{r}} \frac{d\mathbf{p}}{(2\pi)^3}\,, \tag{11.140}$$

where we have defined

$$D_{\mathbf{k}} = \exp\left[\frac{i}{2}\nabla\cdot\frac{\partial}{\partial\mathbf{k}}\right]\mathcal{L}_{\mathbf{k}}\,, \tag{11.141}$$

with the additional notations $W_\pm \equiv W(\mathbf{k}\pm\mathbf{p}/2)$, and $\mathcal{L}_k \equiv \mathcal{L}(\mathbf{p},\mathbf{k}-\mathbf{p})$. The wave-kinetic equation (11.137) can finally be written as

$$\frac{\partial W}{\partial t} + i\Delta_k W + \int \phi(\mathbf{p})\left[D_k W_- + D_{-k} W_+\right] e^{i\mathbf{p}\cdot\mathbf{r}} \frac{d\mathbf{p}}{(2\pi)^3} = 0\,. \tag{11.142}$$

Using a series of simplifying assumptions, this can also be reduced to the Vlasov equation for the quasi-particle field [217], stressing the qualitative analogy, not only with the electromagnetic field, described as a photon gas, but also with the other plasma particles.

# 12

# *Quantum Plasmas*

Quantum plasma physics in the relativistic domain has been studied for long time, first in the context of Astrophysics [221], and later in intense laser-matter interactions [222, 223]. In contrast with the non-relativistic version, where the wave-kinetic theory and the magneto-hydrodynamic model is well established [224], the relativistic quantum plasma theory is not yet fully consolidated. The existing quantum kinetic models are, either too crude or too complicated (see the discussion of the end of this chapter). For this reason, we mainly focus on quantum plasma problems that can be studied with approximate and phenomenological approaches.

We start the chapter with the description a simple approach that is based on the Klein-Gordon equation. It leads to an approximate quantum kinetic equation that allows us to determine the dispersion relation of waves in a relativistic quantum plasma. This quantum kinetic equation can then be used to study other plasma effects, such as is the quantum Weibel instability, which has important implications on the excitation of intense magnetic fields in the astrophysical context.

Here, we also consider the free-electron laser (FEL) in the quantum regime. Our approach is based on the Dirac operator, and allows to demonstrate the existence of stimulated scattering in the quantum regime. This can be extended to waves with orbital angular momentum, leading to twisted FEL regimes.

This Chapter also includes a discussion of the Volkov solutions, describing the quantum state of a charged particle in the field of an electromagnetic wave. They were first established in 1935 [225], for particles in vacuum, and extended to plasmas in more recent years [226, 227]. The novelty in plasmas is the existence of solutions for particles in both electromagnetic and electrostatic waves, which are pertinent to wakefield acceleration.

Volkov states are also relevant to the discussion of Compton scattering processes. Recent experiments have shown the possible relevance of non-linear Compton scattering that involves the exchange of several quanta of electrostatic and electromagnetic wave energy. We also consider the inverse Bremsstrahlung mechanism of laser absorption in a plasma, where Volkov states play again an important role.

Finally, we conclude with a short review of quantum kinetic equations. For this purpose, we consider a formal approach to quantum kinetics, that starts from the Dirac equation and retains spin effects and the exclusion principle.

DOI: 10.1201/9781003514701-12

A particularly simple quantum kinetic equation is derived, its advantages and limitations when compared with other equations are then discussed.

## 12.1 Relativistic Quantum Kinetics

Here, we derive a wave-kinetic equation, valid in the relativistic quantum regime. For simplicity, we first neglect spin effects, which will be considered at the end of this chapter. These effects become more relevant in magnetized plasmas and ultra-intense fields.

### i) Wave-Kinetic Equation

We start with the Klein-Gordon equation, valid for a particle with charge $q$ and mass $m$, written in the following form

$$(i\partial_t - qV)^2 \psi = \left[m^2 + (i\nabla + q\mathbf{A})^2\right] \psi \,, \tag{12.1}$$

where $\partial_t \equiv \partial/\partial t$, $\psi$ is the particle wavefunction. The potentials $V$ and $\mathbf{A}$ are used to determine, not only the externally applied fields, but also the self-consistent field existing inside the medium. We also use $\hbar = 1$ and $c = 1$. We then define the auto-correlation function $K \equiv K(t, t'; \mathbf{r}, \mathbf{r}')$ as

$$K(t, t'; \mathbf{r}, \mathbf{r}') = \psi(t + t'/2, \mathbf{r} + \mathbf{r}'/2)\psi^*(t - t'/2, \mathbf{r} - \mathbf{r}'/2) \,. \tag{12.2}$$

In this definition we can identify, slow and fast time scales, $t$ and $t'$, as well as the long and short scale distances $\mathbf{r}$ and $\mathbf{r}'$. They play different roles in the physical description of the particle interaction with the fields. Using eq. (12.1), we can then derive an equation for the correlation function as

$$\begin{aligned}\left(\nabla \cdot \nabla' - \frac{\partial^2}{\partial t \partial t'}\right) K &= iq \left(V \frac{\partial}{\partial t} + \mathbf{A} \cdot \nabla\right) K + \\ +iq \left[\Delta^1 V \frac{\partial}{\partial t'} + \Delta^1 \mathbf{A} \cdot \nabla'\right] K &- \frac{q^2}{2}[\Delta^2 V - \Delta^2 A^2] K \,,\end{aligned} \tag{12.3}$$

where $\nabla \equiv \partial/\partial \mathbf{r}$ and $\nabla' \equiv \partial/\partial \mathbf{r}'$. We have also defined the operators

$$\Delta^j = 2 \sinh\left(\frac{j}{2}\mathbf{r}' \cdot \nabla + \frac{j}{2} t' \frac{\partial}{\partial t}\right) \,, \tag{12.4}$$

for $j = 1, 2$. Following the Wigner-Moyal procedure, we can define the quantum Wigner function as the Fourier transform of the correlation function, as for classical fields

$$W(\mathbf{r}, t; \Omega, \mathbf{p}) = \int dt' \int d\mathbf{r}' \, K(t, t'; \mathbf{r}, \mathbf{r}') \, \exp\left(-i\mathbf{p} \cdot \mathbf{r}' + i\Omega t'\right) \,. \tag{12.5}$$

And, using eq. (12.4), we arrive at the wave-kinetic equation [230]

$$i\left(\frac{\partial}{\partial\tau}+\mathbf{u}\cdot\nabla\right)W=\int g(\mathbf{k},\omega)\,[W_{-}-W_{+}]\,e^{i\mathbf{k}\cdot\mathbf{r}-i\omega t}\frac{d\omega d\mathbf{k}}{(2\pi)^4}, \tag{12.6}$$

with

$$W_{\pm}\equiv W(\mathbf{r},t;\Omega\pm\frac{1}{2}\omega,\mathbf{p}\pm\frac{1}{2}\mathbf{k}). \tag{12.7}$$

In this equation, the quantity $g(\mathbf{k},\omega)$ is the Fourier transformation of the potential function

$$g(\mathbf{r},t)=q(\gamma V-\mathbf{u}\cdot\mathbf{A}), \tag{12.8}$$

where $\gamma$ is the relativistic factor, $\mathbf{u}$ the covariant velocity, such that

$$m\gamma=(\Omega-qV)\,,\quad m\mathbf{u}=(\mathbf{p}-q\mathbf{A}), \tag{12.9}$$

and $\tau=t/\gamma$ is the proper time. Notice that the non-relativistic quantum kinetic equation is obtained in the limit $\tau\to t$ and $\gamma\to 1$, where the covariant velocity $\mathbf{u}$ reduces to the particle velocity $\mathbf{v}$. This means that the relativistic and non-relativistic quantum descriptions are formally identical at the kinetic level. Even more relevant is the similarity of eq. (12.6) with the photon-kinetic equation discussed in the previous chapter. What differs is the nature of the forces acting on photons and charged particles.

On the other hand, it is known that the particle energy and momentum are related by the so-called gap equation, which means that $\gamma^2=1+u^2$. Using eq. (12.9), this can be written in explicit form as

$$(\Omega-qV)^2=m^2(\mathbf{p}-q\mathbf{A})^2. \tag{12.10}$$

For a given momentum state $\mathbf{p}$, this equation determines the particle energy or frequency, as $\Omega_{\mathbf{p}}$. This allows us to define a reduced function $W(\mathbf{r},\mathbf{p},t)$, such that

$$W(\mathbf{r},t;\mathbf{p},\Omega)=2\pi\,W(\mathbf{r},\mathbf{p},t)\,\delta\left(\Omega-\Omega_{\mathbf{p}}\right). \tag{12.11}$$

Integration of eq. (12.6) over the particle energy spectrum leads to

$$i\left(\frac{\partial}{\partial\tau}+\mathbf{u}\cdot\nabla\right)W=\int g(\mathbf{k},t)\,[W_{-}-W_{+}]\,e^{i\mathbf{k}\cdot\mathbf{r}}\frac{d\mathbf{k}}{(2\pi)^3}, \tag{12.12}$$

where $W$ is now the reduced quasi-distribution.

### ii) Quasi-Classical Limit

It is now useful to consider the wave-kinetic equation in the quasi-classical limit. For fields with frequency $\omega$ and wavevector $\mathbf{k}$, this limit is determined by the inequalities $|\omega|\ll|\Omega|$ and $|\mathbf{k}|\ll|\mathbf{p}|$. In this case, we can use a Taylor expansion of the quantities $W_{\pm}$, as seen before. Keeping only the first terms of this expansion, we obtain

$$[W_- - W_+] \simeq -\mathbf{k} \cdot \frac{\partial}{\partial \mathbf{p}} \left[ 1 + \frac{1}{3 \times 2^3} \left( \mathbf{k} \cdot \frac{\partial}{\partial \mathbf{p}} \right)^2 \right] W \,. \tag{12.13}$$

Replacing this in eq. (12.11), we obtain

$$\left( \frac{\partial}{\partial \tau} + \mathbf{u} \cdot \nabla + \mathbf{F_p} \cdot \frac{\partial}{\partial \mathbf{p}} - \frac{1}{3 \times 2^3} \nabla^2 \mathbf{F_p} \cdot \frac{\partial^3}{\partial \mathbf{p}^3} \right) W = 0 \,, \tag{12.14}$$

where we have defined the force

$$\mathbf{F_p} = -\nabla g(\mathbf{r}, t) = -q \nabla \left( \gamma V - \mathbf{u} \cdot \mathbf{A} \right) . \tag{12.15}$$

Noting that, for two arbitrary vectors **a** and **b**, we have the vector identity

$$\nabla(\mathbf{a} \cdot \mathbf{b}) = (\mathbf{a} \cdot \nabla)\mathbf{b} + \mathbf{a} \times (\nabla \times \mathbf{b}) \,, \tag{12.16}$$

this force can also be written as

$$\mathbf{F_p} = -q \nabla \left[ \gamma \nabla V - (\mathbf{u} \cdot \nabla)\mathbf{A} - \mathbf{u} \times \mathbf{B}) \right] . \tag{12.17}$$

where $\mathbf{B} = \nabla \times \mathbf{A}$ is the magnetic field, Notice that, in the wave-kinetic equation (12.14), the last term contains the first-order quantum corrections. They describe the small quantum recoil effects that are associated with energy and momentum exchanges between the particles and the fields. If this term is neglected, we are reduced to a Vlasov equation, which states the total conservation of the particle quasi-probability $W$, as

$$\frac{d}{d\tau} W = 0 \,, \quad \frac{d}{d\tau} \equiv \frac{\partial}{\partial \tau} + \mathbf{u} \cdot \nabla + \frac{d\mathbf{p}}{d\tau} \cdot \frac{\partial}{\partial \mathbf{p}} \,. \tag{12.18}$$

If we want to understand the corrections introduced by both relativistic and quantum effects in the dispersion relation of waves in the medium, we therefore need to retain the recoil term. This is illustrated next.

### iii) Electron Plasma Waves

Let us assume a purely electrostatic field, such that $\mathbf{A} = 0$, and define the electric field as $\mathbf{E} = -\nabla V$. In this case, the force acting on the electrons is given by $\mathbf{F_p} = -e\gamma \mathbf{E}$, where we have used $q = -e$. Using $\mathbf{p} = m\mathbf{u}$, the quantum kinetic equation (12.14) is transformed into

$$\left( \frac{\partial}{\partial \tau} + \mathbf{u} \cdot \nabla - \frac{e\gamma}{m} \mathbf{E} \cdot \frac{\partial}{\partial \mathbf{u}} + \frac{e\gamma}{3 \times 2^3 m^3} \nabla^2 \mathbf{E} \cdot \frac{\partial^3}{\partial \mathbf{u}^3} \right) W = 0 \,. \tag{12.19}$$

On the other hand, assuming that the ions are immobile and provide a positive background, we can write Poisson's equation as

$$\nabla \cdot \mathbf{E} = \frac{en_0}{\epsilon_0}\left[1 - \int W(\mathbf{u})d\mathbf{u}\right], \tag{12.20}$$

where $n_0$ is the equilibrium plasma density. Let us now consider a wave perturbation $\tilde{W} = W - W_0$, where $W_0$ is the unperturbed quasi-distribution of the plasma electrons, with the normalization condition $\int W_0(\mathbf{u})d\mathbf{u} = 1$. Assuming that the perturbations evolve as $\exp(i\mathbf{k}\cdot\mathbf{r} - i\omega t)$, and linearizing eq. (12.19), we obtain

$$\tilde{W} = i\frac{e\gamma}{m(\gamma\omega - \mathbf{k}\cdot\mathbf{u})}\mathbf{E}\cdot\left[\frac{\partial}{\partial\mathbf{u}} + \left(\frac{\mathbf{k}}{3.2^3m^3}\cdot\frac{\partial}{\partial\mathbf{u}}\right)^2\frac{\partial}{\partial\mathbf{u}}\right]W_0\,. \tag{12.21}$$

From Poisson's equation, we then get

$$\mathbf{E} = i\frac{en_0}{\epsilon_0 k^2}\mathbf{k}\int \tilde{W}(\mathbf{u})d\mathbf{u}\,. \tag{12.22}$$

Assuming that the wave propagates along the $z$-direction, such that $\mathbf{k} = k\mathbf{e}_z$, we can derive a dispersion relation in the standard form

$$1 + \chi_e(\omega, \mathbf{k}) = 0\,, \tag{12.23}$$

where the electron susceptibility is defined as

$$\chi_e(\omega, \mathbf{k}) = -\frac{\omega_p^2}{k^2}\int \frac{\gamma d\mathbf{u}}{(u_z - \gamma\omega/k)}\frac{\partial W_{eff}}{\partial u_z}\,. \tag{12.24}$$

Here, quantum recoil effects are contained in the effective distribution $W_{eff}$ that incorporates the higher-order derivatives of the equilibrium distribution, and is defined as

$$W_{eff} = W_0 + \frac{k^2}{3\times 2^3m^2}\frac{\partial^2 W_0}{\partial u_z^2}\,. \tag{12.25}$$

As in classical plasmas, the integral that determines the electron susceptibility contains a pole that leads to Landau damping. Here, the pole is determined by the equality $u_z - u_0 = 0$, with $u_0 = \gamma\omega/k$. Expanding the denominator around this zero, we then get

$$\chi(\omega, \mathbf{k}) = -\frac{\omega_p^2}{(k^2 - \omega^2)}\int d\mathbf{u}_\perp \int \frac{\gamma du_z}{(u_z - u_0)}\frac{\partial W_{eff}}{\partial u_z}\,. \tag{12.26}$$

Notice that, in contrast with the non-relativistic situation, integration over the perpendicular velocity $\mathbf{u}_\perp$ and the use of a parallel quasi-distribution is forbidden, because the resonant parallel velocity $u_0$ depends on the relativistic factor, and therefore, also depends on the perpendicular velocity $\mathbf{u}_\perp$. We can nevertheless solve the integral in eq. (12.26) in the usual way, evaluating the principal part and the pole contribution separately, according to

$$\int \frac{du_z}{(u_z - u_0)} = \mathcal{P} \int \frac{du_z}{(u_z - u_0)} + i\pi\delta(u_z - u_0) , \tag{12.27}$$

where $\mathcal{P}$ represents the principal part of the integral, in the Cauchy sense. This means that we still have a complex value for the electron susceptibility, $\chi(\omega_r) = \chi_r(\omega_r) + i\chi_i(\omega_r)$, with $\omega = \omega_r + i\Gamma$ and $|\Gamma| \ll \omega_r$, as in the non-relativistic case. The real part of the susceptibility can then be calculated, using the approximate expression

$$\chi_r(\omega_r, \mathbf{k}) = -\frac{\omega_p^2}{(k^2 - \omega_r^2)} \int d\mathbf{u}_\perp \left[ \int \frac{\gamma W_{eff}}{u_0^2} \left( 1 + 3\frac{u_z^2}{u_0^2} \right) du_z \right] , \tag{12.28}$$

where $\omega_r$ represents the real part of the wave frequency.

At this point, it is useful to introduce the influence of the electron spin states, phenomenologically, because we know that the equilibrium state $W_0$ is that of a fermion gas. We therefore need to assume that $W_0$ is a relativistic Fermi-Dirac distribution, otherwise this description would not make sense. We therefore assume that

$$\langle \gamma \rangle \equiv \int \gamma W_0(\mathbf{u}) d\mathbf{u} = \gamma_F = \frac{E_F}{m} , \tag{12.29}$$

where $E_F$ is the Fermi energy for the degenerate electron gas. Along the same line, we also assume that

$$\langle \gamma/\gamma_\perp^2 \rangle \simeq 1/\gamma_F , \quad \langle \gamma u_z^2/\gamma_\perp^4 \rangle \simeq (1/3) u_F^2/\gamma_F^3 . \tag{12.30}$$

where we have used $\gamma_F = \sqrt{1 + u_F^2}$. This leads to the dispersion relation

$$1 - \frac{\omega_p^2}{\omega^2} \frac{1}{\gamma_F} \left[ 1 + \frac{k^2}{\omega^2} \left( 1 - \frac{\omega^2}{k^2} \right) v_F^2 + \frac{k^4}{4m^2} \frac{1}{\gamma_F^2} \right] = 0 , \tag{12.31}$$

where $v_F = u_F/\gamma_F$ is the Fermi velocity. Or, reintroducing $c$ and $\hbar$ in order to get explicit units

$$1 - \frac{\omega_p^2}{\omega^2} \frac{1}{\gamma_F} \left[ 1 + \frac{k^2}{\omega^2} \left( 1 - \frac{\omega^2}{k^2 c^2} \right) v_F^2 + \frac{\hbar^2 k^4}{4m^2} \frac{1}{\gamma_F^2} \right] = 0 . \tag{12.32}$$

It is now useful to discuss the classical and the quantum non-relativistic limits separately. We obtain the quantum non-relativistic limit if we take $c \to \infty$ and $\gamma_F \to 1$. This expression is reduced to

$$\omega^2 = \omega_p^2 + k^2 v_F^2 + \frac{\hbar^2 k^4}{4m^2} . \tag{12.33}$$

which is well-known [231]. The last term in this expression corresponds to

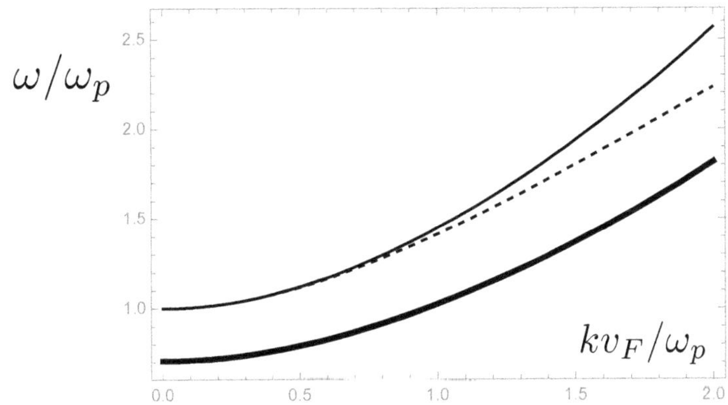


**FIGURE 12.1**
Dispersion relation of electron plasma waves in relativistic quantum plasmas: the normalized wave frequency $\omega/\omega_p$ is represented in terms of the normalized wavenumber $kv_F/\omega_p$ for (i) the relativistic quantum regime (in bold); (ii) the non-relativistic quantum regime (thin curve); and (iii) the non-relativistic classical regime (dotted), for comparison, where $v_F$ is replaced by the thermal velocity $v_{the}$.

the free streaming of electrons and becomes relevant in the short wavelength limit. A similar dispersion relation is also valid for Bogolioubov oscillations in dilute Bose-Einstein condensates of ultra-cold atoms.

In contrast, for a classical relativistic plasma, we have to use $\hbar \to 0$ and to replace $\gamma_F$ by the $\gamma_e = \sqrt{1 + 5T_e/2mc^2}$ as used previously, where $T_e$ is the electron temperature. The relation (12.31) is then transformed into

$$\omega^2 = \frac{\omega_p^2}{\gamma_e} \left[ 1 + \frac{k^2}{\omega^2} \left( 1 - \frac{\omega^2}{k^2 c^2} \right) \frac{5T_e}{2m} \right] . \tag{12.34}$$

Noting that $k^2c^2/\omega^2 \gg 1$ this is approximately equal to the dispersion relations discussed in Chapter 4. In particular, the wave cutoff is now determined by $\omega_p^2/\gamma_e$, as a result of the effective electron mass. The different dispersion regimes of relativistic quantum plasmas are illustrated in Figure 12.1.

In order to complete our discussion, let us go back to the electron susceptibility (12.26) and briefly discuss its imaginary part. As we have seen, the wave damping $\Gamma = \Im(\omega)$ is determined by

$$\Gamma = -\frac{\chi_i(\omega, \mathbf{k})}{(\partial \chi_r / \partial \omega)} \propto \left( \frac{\partial W_{eff}}{\partial u_z} \right)_{u_0} . \tag{12.35}$$

Notice that this quantity is now proportional do the derivative of effective distribution $W_{eff}$ at resonance, and not to the derivative of the equilibrium distribution as in the classical case. Taking the definition of $W_{eff}$ in eq.

(12.25), we realize that the $\Gamma$ is not only dependent on the first-order derivative, $\partial W_0/\partial\omega$ but it also receives a small contribution from the third-order derivative $\partial^3 W_0/\partial\omega^3$. This small correction is a result of quantum non-locality. A more exact calculation of the quantum effects would contain contributions from all odd derivatives of $W_0$.

## 12.2 Quantum Weibel

An interesting application of the wave-kinetic equation derived in the previous section is the generation of nearly static magnetic fields due to the occurrence of the Weibel instability. This problem was treated in a previous chapter in the classical limit. The quantum theory of the Weibel instability has been explored by several authors [228, 229], for non-relativistic plasmas. Here, we consider the case of an electron-ion plasma, where the ions are assumed immobile and the electrons are governed by the wave kinetic equation (12.12). But, in contrast with the previous example, we consider a purely electromagnetic problem, where the scalar potential can be taken equal to zero, $V = 0$, and we use the potential function

$$g(\mathbf{r}, t) = +e(\mathbf{u} \cdot \mathbf{A}) . \tag{12.36}$$

The vector potential $\mathbf{A}$ will have to satisfy the wave equation

$$\left(\nabla^2 - \frac{1}{c^2}\frac{\partial^2}{\partial t^2}\right)\mathbf{A} = -\mu_0 \mathbf{J}_e , \quad \mathbf{J}_e = -en_0 \int \mathbf{v} W(\mathbf{p}) d\mathbf{p} , \tag{12.37}$$

where the current is determined by the electron quasi-distribution $W(\mathbf{r}, \mathbf{p}, t)$, and $\mathbf{v} = \mathbf{p}/m\gamma$.

In order to study the Weibel instability, we assume a transverse electromagnetic wave propagating along the $z$-axis, characterized by the wavevector $\mathbf{k} = k\mathbf{e}_z$, and described by the vector potential

$$\mathbf{A} = \mathbf{A}_0 \exp(ikz - i\omega t) , \quad \mathbf{k} \cdot \mathbf{A}_0 = 0 . \tag{12.38}$$

We also assume that the quasi-distribution can be expanded as

$$W = W_0 + \tilde{W} \exp(ikz - i\omega t) , \quad |\tilde{W}| \ll |W_0| \tag{12.39}$$

where the equilibrium quasi-distribution $W_0$ satisfies normalization and symmetry relations of the form

$$\int W_0(\mathbf{p}) d\mathbf{p} = 1 , \quad \int \mathbf{p} W_0(\mathbf{p}) d\mathbf{p} = 0 . \tag{12.40}$$

Using eqs. (12.36)–(12.39), we can then derive a relation between the vector potential perturbation $\mathbf{A}_0$ and the quasi-distribution perturbation $\tilde{W}$ as

$$\left(\omega^2 - k^2c^2 - \frac{\omega_p^2}{\bar{\gamma}}\right)\mathbf{A}_0 = \omega_p^2 \int \frac{\mathbf{p}}{e\gamma}\tilde{W}(\mathbf{p})d\mathbf{p}\,. \tag{12.41}$$

In this expression, we have used the quantity $\bar{\gamma}^{-1} = \int \gamma^{-1} W_0(\mathbf{p})d\mathbf{p}$.

We now use eqs. (12.12) and (12.36) to determine the wave-kinetic perturbation $\tilde{W}$. This leads to

$$\tilde{W}(\mathbf{p}) = \frac{e}{\hbar}\frac{(\mathbf{u}\cdot\mathbf{A}_0)}{(\gamma\omega - ku_z)}\left[W_{0-} - W_{0+}\right], \tag{12.42}$$

where, obviously, the displaced distributions $W_{0\pm}$ are functions of the electron momentum $(\mathbf{p})$.

Replacing this in eq. (12.41), and noting that the mode is transverse, we obtain the dispersion relation

$$\omega^2 = k^2c^2 + \omega_p^2\left\{\frac{1}{\bar{\gamma}} + \int \frac{p_\perp^2\left[W_{0-} - W_{0+}\right]}{m\hbar\gamma(\gamma\omega - kp_z/m)}d\mathbf{p}\right\}. \tag{12.43}$$

In order to understand the physical meaning of this result, we can use the standard approach [93], and assume an equilibrium distribution of the form

$$W_0(\mathbf{p}) = W_0(p_\perp^2)\delta(p_z)\,, \quad W_0(p_\perp^2) = \delta(p_\perp^2 - p_b^2)\,, \tag{12.44}$$

where $p_b$ is a constant. The dispersion relation is then reduced to

$$\omega^2 = k^2c^2 + \omega_p^2\left[\frac{1}{\bar{\gamma}} + \frac{1}{m^2\gamma_k^3}\frac{p_b^2k^2}{(\omega^2 - \nu_k^2)}\right], \tag{12.45}$$

where the new quantities $\nu_k$ and $\gamma_k$ have been defined as

$$\nu_k = \frac{\hbar k^2}{2m\gamma_k}\,, \quad \gamma_k = \sqrt{\bar{\gamma}^2 + \left(\frac{\hbar k}{2mc}\right)^2}\,. \tag{12.46}$$

Given that we have assumed a cold distribution in the parallel direction $\mathbf{e}_z$, the average relativistic factor is simply given by $\bar{\gamma}^2 = 1 + (p_b/mc)^2$. The dispersion relation (12.45) predicts purely growing wave solutions that can easily be identified in the low frequency domain where we take $|\omega^2| \ll k^2c^2$. Using $\omega = i\Gamma$, we can then obtain an expression for the growth rates as

$$\Gamma \simeq \left[\frac{\omega_p^2}{m^2\gamma_k^3}\frac{p_b^2k^2}{(k^2c^2 + \omega_p^2/\bar{\gamma})} - \nu_k^2\right]^{1/2}. \tag{12.47}$$

Here, the influence of relativistic and quantum effects is clearly identified. Both effects tend to reduce the instability growth rate, due to the presence

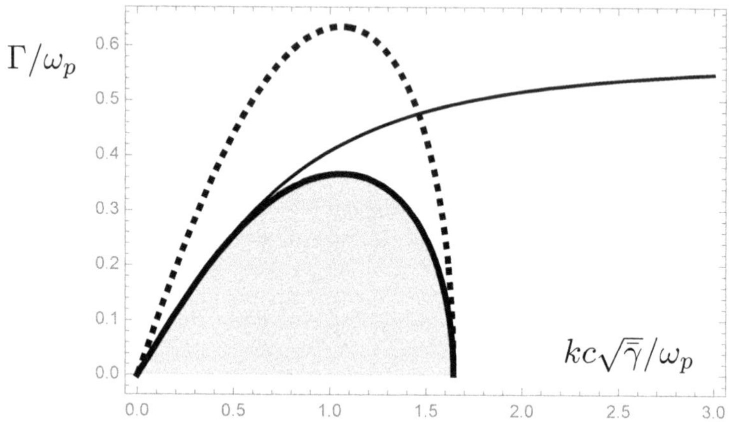


**FIGURE 12.2**
Weibel instability in relativistic quantum plasmas: Normalized growth rate $\Gamma/\omega_p$, as a function of the normalized wavenumber $kc\sqrt{\bar{\gamma}}/\omega_p$, for the (i) Relativistic quantum plasma, with $\bar{\gamma} = 3$, and quantum factor $H = 0.3$ (bold curve); (ii) Relativistic classical case, $\bar{\gamma} = 3$ and $H = 0$ (thin curve); (iii) Non-relativistic quantum case, $\bar{\gamma} = 1$ and $H = 0.3$ (dashed curve). The quantum factor is defined as $H = \hbar\omega_p/2mc^2\sqrt{\bar{\gamma}}$.

of $\gamma_k$, and $\nu_k$, as illustrated in Figure 12.2. The relativistic quantum regime is clearly shown. As we can see, the quantum effects tend to stabilize the instability for short wavelengths, which is due to quantum decoherence. In the classical limit, we take $\nu_k = 0$ and assuming that $k^2c^2 \simeq \omega_p^2/\bar{\gamma}$, we obtain

$$\Gamma \simeq \omega_p \frac{p_b}{mc} \gamma_k^{-3/2} . \tag{12.48}$$

This shows that the relativistic effects tend to reduce the growth rate, as $\Gamma \propto 1/\sqrt{\gamma_k}$, due to the increase of the electron effective mass.

As we have seen, the Weibel instability is associated with anisotropic electron distributions. Anisotropy can be generated by preferential heating along some direction $\mathbf{e}_z$, or by strong field interactions, or even by loss-cone effects. Thermal effects can be explored using the dispersion relation (12.43) with appropriate equilibrium distributions $W_0(\mathbf{p})$. This has been considered elsewhere [232].

## 12.3 Free Electron Laser

Another important application of quantum relativistic plasma is the *free-electron laser*, or simply FEL. This is a physical concept, first proposed by Madey in 1971 [233] to provide a source of coherent and tunable radiation, extending from the infrared to the XUV domain. In a FEL, radiation is coherently emitted by a relativistic electron beam moving in a periodic magnetic configuration, called the *wiggler*. The transverse electron oscillations produced by the magnetic wiggler are the source of radiation. In more modern concepts, the magnetic wiggler can eventually be replaced by an electron plasma wave or a laser beam.

Although this concept can be described by classical models, the quantum regimes have specific properties associated with the electron recoil during the emission process, which become important in the short wavelength domain [234]. Here, we mainly follow the approach proposed by Serbeto [235]. In this approach, we are able to derive a Schrödinger-type of equation, starting from the relativistic Dirac operator, $\mathcal{H}$. This was first shown by Preparata, following a quantum field theory method [236]. In order to define this operator we should keep in mind that, in a FEL, the electrons move in a slow ponderomotive potential $V \equiv V(\mathbf{r}, t)$, which results from the beating between the wiggler field and the radiation field produced by the electrons themselves. This self-interaction is an important ingredient of the coherent radiation process. The single electron Dirac operator can defined as

$$\mathcal{H} - V = \sqrt{p^2c^2 + m^2c^4}\,, \tag{12.49}$$

In a FEL, electron spin effects can be neglected, and coupling between positive and negative (or electron and positron) energy states can also be ignored. We can therefore assume that the electrons are in a well defined positive energy and spin state that can be described by a single wavefunction, $\psi \equiv \psi(\mathbf{r}, t)$. The Dirac operator can then be applied to the electron wave function $\psi$, leading to

$$i\hbar\frac{\partial}{\partial t}\psi = \mathcal{H}\psi\,. \tag{12.50}$$

In the absence of the wiggler, we have $V = 0$, and the electrons are in a state of constant energy $\mathcal{H} \to E = \hbar\omega_e$, with a constant momentum $\mathbf{p} = \hbar\mathbf{k}_e$. In the presence of a wiggler, $V \neq 0$ this energy state is modified. If the initial electron energy is much larger than the energy exchanges with radiation, we can assume that $V$ will only introduce small changes in space an time, and the deviations with respect to the initial state are small. This means that

$$\left|\frac{\partial\psi}{\partial t}\right| \ll \omega_e|\psi|\,, \quad |\nabla\psi| \ll |\mathbf{k}_e\psi|\,. \tag{12.51}$$

This allows us to make in eq. (12.50) the following replacements

$$\mathcal{H} \rightarrow \hbar \left( \omega_e - i\frac{\partial}{\partial t} \right) , \quad \mathbf{p}_e = \hbar(\mathbf{k}_e + i\nabla) . \tag{12.52}$$

This is equivalent to the *WKB approximation*, where the changes in energy and momentum occur on time and length-scales much longer than the typical period and wavelength of the electron wavefunction $\psi(\mathbf{r}, t)$. Using such replacements and developing the square-root operator, we get to the leading order, the following wave equation

$$i\hbar \left( \frac{\partial}{\partial t} + \mathbf{v}_e \cdot \nabla \right) \psi = V\psi - \frac{\hbar^2 c^2}{2\omega_e} \left[ \nabla^2 - \left( \frac{\mathbf{v}_e}{c} \cdot \nabla \right)^2 \right] \psi , \tag{12.53}$$

where we have defined the unperturbed electron velocity $\mathbf{v}_e$, as

$$\mathbf{v}_e = \frac{c^2 \mathbf{k}_e}{\omega_e} = c \left( 1 - \frac{m^2 c^4}{\hbar^2 \omega_e^2} \right)^{1/2} \mathbf{u} , \tag{12.54}$$

with the unit vector $\mathbf{u} = \mathbf{k}_e / |\mathbf{k}_e|$. This wave equation suggests a change of variables, from $(\mathbf{r}, t)$ to $(\mathbf{r}', t')$, such that

$$\mathbf{r}' = \mathbf{r} - \mathbf{v}_e t \quad t' = t , \tag{12.55}$$

Equation (12.53) is then transformed into

$$i\hbar \frac{\partial \psi}{\partial t'} = \left[ -\frac{\hbar^2 \nabla'^2}{2\gamma m} + V \right] \psi , \tag{12.56}$$

where the operator $\nabla'^2$ is defined as

$$\nabla'^2 = \left[ \frac{1}{\gamma^2} \frac{\partial^2}{\partial z'^2} + \frac{\partial^2}{\partial r'^2_\perp} \right] , \tag{12.57}$$

where we have defined the parallel and perpendicular coordinates according to $\mathbf{r}' = z'\mathbf{u} + \mathbf{r}'_\perp$. This is indeed a Schrödinger type of equation describing the evolution of relativistic quantum electrons. Notice that, in this equation, the relativistic gamma factor $\gamma$ corresponds to the unperturbed electron motion, and therefore we have $\gamma = \hbar\omega_e / mc^2$. Equation (12.56) will be the starting point of the quantum FEL model.

However, to simplify the description, we consider a purely one-dimensional problem, assuming that $\partial^2 / \partial r'^2_\perp = 0$. The above Schrödinger equation then reduces to

$$i\hbar \frac{\partial \psi}{\partial t'} = \left[ -\frac{\hbar^2}{2\bar{m}} \frac{\partial^2}{\partial z'^2} + V \right] \psi , \tag{12.58}$$

where $z' = z - v_z t$ and $\bar{m} = \gamma^3 m$ is a kind of effective electron mass, which incorporates the relativistic character of the motion. Although not strictly necessary, we can rewrite eq. (12.58) in a slightly different way, that includes the so-called *fundamental FEL parameter*, defined as [234]

$$\rho = \frac{1}{\gamma_r}\left(a_w \frac{\omega_p}{4k_w c}\right)^{2/3} , \tag{12.59}$$

This parameter depends on the wiggler field amplitude $a_w$, and on the wiggler wavenumber $k_w$, and well as on the resonant gamma factor $\gamma_r$, defined as

$$\gamma_r = \sqrt{(1+a_w)^2}\sqrt{\frac{k}{2k_w}} , \tag{12.60}$$

where $k$ is the wavenumber of the emitted radiation. Notice that, for $\gamma = \gamma_r$ the electron interacts resonantly with the ponderomotive potential created by the beating between the wiggler and the radiation field. We can now define the normalized time and space variables, $\tau$ and $\theta$, as

$$\tau = 2\rho k_w c t' \quad , \quad \theta = (k + k_w)z - \omega t' . \tag{12.61}$$

For short wavelength radiation, such that $k \gg k_w$, this space variable reduces to $\theta = kz'$. Written in terms of these new variables, the Schrödinger equation (12.58), takes the form

$$i\hbar \frac{\partial \psi}{\partial \tau} = \left[ -\frac{1}{2\mu}\frac{\partial^2 \psi}{\partial \theta^2} + \bar{V} \right] \psi , \tag{12.62}$$

where the quantities $\mu$ and $\bar{V}$ are defined as

$$\mu = 2\bar{m}\frac{k_w c}{\hbar k^2}\rho , \quad \bar{V} = \frac{V}{2\hbar \rho k_w c} . \tag{12.63}$$

At this point, it is useful to specify the ponderomotive potential, using

$$\bar{V} = -i\frac{\bar{m}}{2k^2}\left[A(\tau)e^{i\theta} - c.c.\right] , \tag{12.64}$$

In order to solve the wave equation (12.62), we need to provide a self-consistent description of the ponderomotive potential. For that purpose, we only need to include an evolution equation for the field $A(\tau)$, which results from the interference between the wiggler field and the radiation field. It can be written in dimensionless form as

$$\left(\frac{\partial}{\partial \tau} - i\bar{\delta}\right) A = \int_0^{2\pi} |\psi(\tau,\theta)|^2 e^{-i\theta}\frac{d\theta}{2\pi} \tag{12.65}$$

where $\bar{\delta}$ is a detuning factor. This completes the set of free-electron laser equations. In order to understand their meaning, it is now useful to write in terms of the electron quasi-distribution. First, we write the auto-correlation function $K$ in terms of the normalized space and time variables. It can be written as

$$K \equiv K(\tau, \theta, s) = \psi^*(\tau, \theta - s/2)\psi(\tau, \theta + s/2)\,, \tag{12.66}$$

We then define the corresponding Wigner function $W(\tau, \theta, p)$, in the usual way

$$W(\tau, \theta, p) = \int_{-\infty}^{\infty} K(\tau, \theta, s) e^{-ips} ds\,, \tag{12.67}$$

We then follow the Wigner-Moyal procedure to derive the evolution equation for $W$, which is the wave-kinetic equation associated with the Shrödinger equation (12.62). It takes the form

$$i\hbar \left( \frac{\partial}{\partial \tau} + \frac{\hbar p}{\mu} \frac{\partial}{\partial \theta} \right) W = \int \bar{V}(\tau, q)[W_- - W_+] e^{iq\theta} \frac{dq}{2\pi} \tag{12.68}$$

where we have $W_\pm = W(p \pm q/2)$. Here, the quantity $\bar{V}(\tau, q)$ is the Fourier transform of the ponderomotive potential

$$\tilde{V}(\tau, q) = \int \tilde{V}(\tau, \theta) e^{-iq\theta} d\theta\,. \tag{12.69}$$

Now, using the potential defined by eq. (12.64), we can write the wave-kinetic equation in explicit form, as

$$\left( \frac{\partial}{\partial \tau} + \frac{\hbar p}{\mu} \frac{\partial}{\partial \theta} \right) W = \mu (A e^{i\theta} + c.c.)(W_- - W_+) \tag{12.70}$$

Let us now consider a perturbation of the radiation field such that

$$A(\tau) = \tilde{a} \exp(-i\Omega\tau)\,, \quad W(\tau, \theta, p) = W_0(\theta, p) + \tilde{W} \exp(iq\theta - i\Omega\tau)\,, \tag{12.71}$$

where $W_0$ is the unperturbed value for the electron quasi-distribution, and $\Omega$ is a detuning frequency. Linearizing eq. (12.70) with respect to the perturbations $\tilde{a}$ and $\tilde{W}$, we obtain

$$\tilde{W} = -i \frac{\bar{\rho}\tilde{a}}{(\Omega - q\bar{v})} [W_- - W_+]\,. \tag{12.72}$$

Similarly, if we linearize the field equation (12.65), we get

$$\tilde{a} = \frac{i}{(\Omega + \bar{\delta})} \int \frac{dp}{2\pi} \int d\theta\, \tilde{W}(\theta, p)\,. \tag{12.73}$$

Using these two equations, we can now derive a dispersion relation for the field perturbations occurring in a FEL. It takes the form

$$1 - \frac{\bar{\rho}}{(\Omega + \bar{\delta})} \int \frac{(G_{0-} - G_{0+})}{(\Omega - q\bar{v})} dp = 0\,, \tag{12.74}$$

where we have used the auxiliary function

$$G_0(p) = \frac{1}{2\pi} \int_0^{2\pi} W_0(\theta, p) d\theta \,, \tag{12.75}$$

and the obvious notation $G_{0\pm} = G_0(p \pm q/2)$. As we can see, this new quantity represents the equilibrium electron distribution, averaged over the angular variable $\theta$. We consider the case of a mono-kinetic electron beam, determined by $G_0(p) = G_0\delta(p - p_0)$, where $p_0$ is a constant. In this case, integration of the above dispersion relation is trivial and leads to

$$1 + \frac{\bar{\rho}}{(\Omega + \bar{\delta})} \left[ \frac{1}{(\Omega_+ - q\bar{v}_0)} - \frac{1}{(\Omega_- - q\bar{v}_0)} \right] = 0 \,, \tag{12.76}$$

where we have defined

$$\bar{v}_0 = \frac{\hbar p_0}{m} \,, \quad \Omega_\pm = \Omega \pm \frac{\hbar q^2}{2m} \,. \tag{12.77}$$

This new dispersion relation can be rewritten as

$$(\Omega + \bar{\delta}) \left[ (\Omega - q\bar{v}_0)^2 - \left( \frac{\hbar q^2}{2m} \right)^2 \right] = \hbar q^2 \frac{\mu}{m} \,. \tag{12.78}$$

This clearly shows the existence of two different excitation modes, coupled by the quantity on the right-hand side. They are defined by the simple dispersion relations

$$(\Omega + \delta) = 0 \,, \quad (\Omega - qv_0) = \frac{\hbar q^2}{2m} \,. \tag{12.79}$$

The first mode is trivial. It is a mode that has zero frequency in the absence of detuning, $\Omega = 0$. The second mode corresponds to a Doppler shifted frequency with a photon recoil term. These two modes can be described as a field and a beam mode, coupled by the FEL parameter $\mu$. The simplest version of the coupled mode dispersion relation (12.78), corresponds to $\bar{v}_0 = 0$, and $\hbar q^2(\mu/m) = 1$. It takes the form

$$(\Omega + \bar{\delta}) \left( \Omega^2 - \frac{1}{4\mu} \right) = 1 \tag{12.80}$$

The case of quantum FEL instability, with a growth rate $\Gamma = \Im(\Omega) > 0$, was first discussed by [234]. It corresponds to a mono-kinetic electron beam distribution and is illustrated in Figure 12.3. As we can see, the unstable range and the instability growth rate decrease in the presence of detuning. Furthermore, the growth rate depends on the quantum parameter $\mu$ which contains, not only on the physical properties of the wiggle field, but also on the relativistic properties of the electron beam. More general quantum instability regimes can be found when the energy dispersion of the electron beam distribution is included, as discussed in [235].

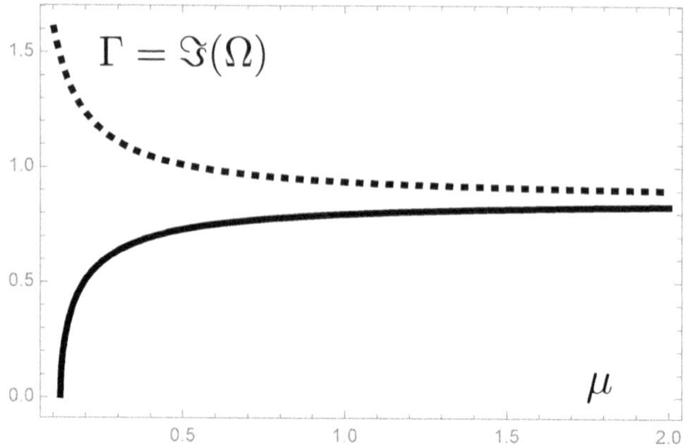


**FIGURE 12.3**
Quantum FEL instability: growth rate $\Gamma = \Im(\Omega)$ as determined by eq. (12.80), valid for a mono-kinetic electron beam distribution, with (i) $\bar{\delta} = 0.1$ (in bold), and (ii) $\bar{\delta} = 0$ (dashed).

## 12.4 Volkov States

We now focus on the quantum states of an electron in the field of a nearly monochromatic wave. These are called the *Volkov states*, which are exact solutions of the relativistic quantum equations in vacuum. When the electrons are immersed in a plasma, the wave dispersion changes and the solutions are non longer exact. Another novelty introduced by the plasma is that we can also consider electron states in electrostatic waves, thus extending the original concept. A further extension of the Volkov states concerns the existence of structured waves with orbital angular momentum, both in vacuum and in plasmas.

### 12.4.1 Electromagnetic Wave

We start with the study of electron states in the field of an electromagnetic wave. Neglecting spin, we can use the Klein-Gordon equation (12.1). For the description of transverse electromagnetic waves, we can take the scalar potential equal to zero, and write this equation as

$$\left[ \left( \nabla^2 - \frac{1}{c^2} \frac{\partial^2}{\partial t^2} \right) - \left( 1 + \mathbf{a}^2 \right) k_C^2 \right] \psi = 0 \,, \tag{12.81}$$

where we have used the normalized vector potential $\mathbf{a}$ and the Compton wavenumber $k_C$, defined by

$$\mathbf{a} = \frac{e\mathbf{A}}{mc}, \quad k_C = \frac{mc}{\hbar}. \tag{12.82}$$

We now consider a quasi-monochromatic wavepacket with frequency $\omega$ and wavevector $\mathbf{k}$, described as

$$\mathbf{a}(\mathbf{r}, t) = \mathbf{a}_0 f(\tau), \quad \tau = t - \frac{\mathbf{k} \cdot \mathbf{r}}{\omega}, \tag{12.83}$$

where $\mathbf{a}_0$ is the wave amplitude, and the function $f(\tau)$ describes the shape of the wavepacket, assuming that it depends only on the variable $\tau$ and stays invariant along propagation. Assuming the traditional Volkov approach, we assume a solution of eq. (12.81) of the form

$$\psi(\mathbf{r}, t) = \Psi(\tau) \exp(i\theta), \quad \theta = \frac{1}{\hbar}(\epsilon_e t - \mathbf{p}_e \cdot \mathbf{r}), \tag{12.84}$$

where $\epsilon_e$ and $\mathbf{p}_e$ are the electron energy and momentum. According to the mass-gap equation, they have to satisfy the relation

$$\epsilon_e = c\sqrt{\mathbf{p}_e^2 + m^2c^2(1 + \bar{a}^2)}, \tag{12.85}$$

where $\bar{a}^2$ is defined as the time average of $\mathbf{a}^2$ and can be taken as $a_0^2$. This can also be written as

$$\left(\frac{\omega_e^2}{c^2} - \frac{p_e^2}{\hbar^2}\right) = k_C^2 \left(1 + \bar{a}^2\right). \tag{12.86}$$

In our analysis we also need to assume that the fast time-scales of the electron wave wavefunction (12.84) are contained in the phase function $\theta$. This is valid for waves with frequencies such that $\omega \ll mc^2/\hbar$, which correspond to photon energies much less than the electron energy. Keeping that in mind, we can then derive from eq. (12.81) an evolution equation for the slowly varying amplitude $\Psi$, of the form

$$\left(\frac{k^2c^2}{\omega^2} - 1\right)\frac{\partial^2 \Psi}{\partial \tau^2} + 2ig(\tau)\frac{\partial \Psi}{\partial \tau} + F(\tau)\Psi = 0, \tag{12.87}$$

where we have used the auxiliar functions

$$g(\tau) = \omega_e - \frac{k_C}{\omega}(\mathbf{a} \cdot \mathbf{k})c^2 - \frac{(\mathbf{p}_e \cdot \mathbf{k})}{\hbar\omega}c^2, \tag{12.88}$$

and

$$F(\tau) = -2c^2\frac{k_C}{\hbar}(\mathbf{a} \cdot \mathbf{p_e}) - c^2k_C^2\left[\gamma - (1 + \mathbf{a}^2\right]. \tag{12.89}$$

As can be seen from eq. (12.87), the first term is identically equal to zero because $k^2c^2 - \omega^2$. The same is not true for waves in a medium. In the case of a cold plasma, we can use the dispersion relation

$$\omega^2 = k^2c^2 + \frac{\omega_p^2}{\gamma}, \quad \gamma = \sqrt{1 + a_0^2} \tag{12.90}$$

An approximate solution can still be obtained for wave frequencies well above cutoff, such that $(\omega_p/\omega)^2 \ll \gamma$. In both cases, we can neglect the first term in eq. (12.87). This leads to

$$\Psi(\mathbf{r}, t) = \Psi_0 \exp[iS(\tau)], \quad S(\tau) = -\frac{1}{2} \int \frac{F(\tau')}{g(\tau')} d\tau' . \tag{12.91}$$

For a purely sinusoidal wave, such that $f(\tau) = \cos(\omega\tau)$, the phase function becomes

$$S(\tau) = s_1 \sin(\omega\tau) + s_2 \sin(2\omega\tau) . \tag{12.92}$$

with the coefficients

$$s_1 = c^2 k_C \frac{k_C}{\hbar} (\mathbf{a}_0 \cdot \mathbf{p_e}), \quad s_2 = -\left[\gamma - (1 + \mathbf{a}^2)\right] \frac{c^2 k_C^2}{2\omega\omega_e} a_0^2 . \tag{12.93}$$

These expressions are valid if we one retains the dominant term in $g(\tau) \simeq \omega_e$. We notice that Bessel function expansions can be used for the phase factor, leading to

$$\exp[iS(\tau)] = \sum_{n_1, n_2} J_{n_1}(s_1) J_{n_2}(s_2/2) \exp[-i(n_1 + 2n_2)\omega\tau] . \tag{12.94}$$

This completes the derivation of the Volkov solutions. As we can see, they are exact solutions in vacuum, and only approximate in a plasma. The validity of the approximation increases for $\omega^2 \gg \omega_p^2/\gamma_e$.

### 12.4.2 Electrostatic Wave

A similar derivation for the quantum state of an electron in a nearly monochromatic electron plasma wave can be made. In this case, the wave can is described by a scalar potential, and the Klein-Gordon equation becomes

$$\left[\left(\nabla^2 - \frac{1}{c^2}\frac{\partial^2}{\partial t^2}\right) + \frac{2i}{c} k_C \mathcal{U} \frac{\partial}{\partial t^2} \left(1 + \mathcal{U}^2\right) k_C^2\right] \psi = 0 , \tag{12.95}$$

where we have used the normalized potential

$$\mathcal{U} = \frac{eV}{mc^2} , \quad \mathcal{U}(\mathbf{r}, t) = \mathcal{U}_0 f(\tau) . \tag{12.96}$$

The same procedure then leads to a solution of the form (12.84), and to an equation formally identical to eq. (12.87), but where now the auxiliary functions (12.88) and (12.89) are replaced by

$$g(\tau) = \omega_e + \mathcal{U} \frac{k_C}{\omega} (\mathbf{a} \cdot \mathbf{k}) c^2 - \frac{(\mathbf{p}_e \cdot \mathbf{k})}{\hbar\omega} c^2 , \tag{12.97}$$

and

$$F(\tau) = 2ck_C\omega_e\mathcal{U} + c^2 k_C^2 \mathcal{U}^2 . \tag{12.98}$$

At this point, it should be noticed that the wave dispersion relation is quite different from the electromagnetic case. The first term in eq. (12.87) now has to be evaluated as

$$\left(\frac{k^2c^2}{\omega^2} - 1\right) \simeq \frac{c^2}{S_e^2}\left(\gamma_e - \frac{\omega_p^2}{\omega^2}\right), \tag{12.99}$$

where $S_e$ is the electron thermal velocity. We notice that the frequency of these waves is very close to the plasma frequency, $\omega^2 \simeq \omega_p^2$. This means that the contribution of this term can usually be neglected. We are then led to a Volkov solution similar to eq. (12.91), with a different value for the phase function, in accordance with the new expressions for the functions $g(\tau)$ and $F(\tau)$. For a sinusoidal wave potential $\mathcal{U}(\tau) = \mathcal{U}_0 \cos(\omega\tau)$, we obtain

$$S(\tau) = s_0 \sin(\omega\tau) . \quad s_0 = 2\frac{ck_C}{\omega}\mathcal{U}_0 , \tag{12.100}$$

In this case, the Bessel expansion of the phase function then leads to

$$\exp[iS(\tau)] = \sum_n J_n(s_0)\exp(-in\omega\tau) . \tag{12.101}$$

The above analysis can be expanded to new physical geometries, using nearly the same methods. First of all, Volkov states of electrons in the simultaneous presence of two different waves can be derived. This is particularly useful to describe nonlinear Compton scattering, to be considered later. Second, Volkov solutions for one wave can be extended to waves with orbital angular momentum. In this case, twisted electron wavefunctions can be obtain, as shown next.

### 12.4.3 Twisted Wave

Let us consider a spinless electron in the field of a twisted electromagnetic wave in a plasma. The problem is now intrinsically three-dimensional and, in the Lorentz gauge, eq. (12.81) is replaced by

$$\left[\left(\nabla^2 - \frac{1}{c^2}\frac{\partial^2}{\partial t^2}\right) - 2ik_C\,(\mathbf{a}\cdot\nabla) - (1+\mathbf{a}^2)k_C^2\right]\psi = 0 . \tag{12.102}$$

If the electromagnetic wave described by the vector potential $\mathbf{a}$ is a twisted wave, with a finite amount of angular momentum, propagating along the $z$-axis, we can use

$$\mathbf{a}(\mathbf{r}, t) = \mathbf{a}_0 F_0(\mathbf{r}_\perp, z) f(\tau) . \tag{12.103}$$

The radial function $F_0(\mathbf{r}_\perp, z)$ describes the transverse field properties of the wave in the transverse coordinates, $\mathbf{r}_\perp = (r, \theta)$ and is assumed to satisfy the paraxial equation

$$\left(\nabla_\perp^2 + 2ik\frac{\partial}{\partial z}\right) F_0 = 0 . \tag{12.104}$$

These are usually called the *Laguerre-Gauss modes*, as we have seen previously. Let us take a particular solution, corresponding to the angular and radial quantum numbers $(\ell_0, \nu_0)$, as defined by

$$F_0(\mathbf{r}_\perp, z) = C_{\ell_0\nu_0} X^{|\ell_0|} L_{\nu_0}^{|\ell_0|} \exp\left(i\ell_0\theta - \frac{X}{2}\right), \tag{12.105}$$

where $X = r^2/w(z)^2$, $w(z)$ is the beam waist, and $C_{\ell_0\nu_0}$ the normalization constant.

In order to solve eq. (12.102) for this twisted vector potential, we assume an electron wavefunction $\psi$ similar to eq. (12.84), which now also depends on the radial coordinates, as

$$\psi(\mathbf{r}, t) = \Phi(\mathbf{r}, \tau) \exp(i\theta), \tag{12.106}$$

where a slow dependence of $\Phi$ with the variable $z$ can be retained. Let us also assume that the wave is transverse, such that $\mathbf{a} \cdot \mathbf{e}_z = 0$. The evolution equation for $\Phi$, can now be written as

$$-\frac{\nu_p^2}{\omega^2}\frac{\partial^2\Phi}{\partial\tau^2} + 2ig(\tau)\frac{\partial\Phi}{\partial\tau} - 2ik_C\mathbf{a}\cdot\nabla_\perp\Phi - \mathbf{a}^2k_C^2\Phi = 0, \tag{12.107}$$

where $g(\tau)$ is given by eq. (12.88). To simplify the calculations we also consider circular polarization, such that

$$\mathbf{a} \cdot \nabla_\perp = \pm\frac{|\mathbf{a}|}{r}\frac{\partial}{\partial\theta}, \tag{12.108}$$

where $\pm 1$ defined the two possible polarization states. The amplitude wavefunction $\Phi$ can then be defined as a superposition of LG modes, as defined by

$$\Phi(\mathbf{r}, \tau) = \sum_{\ell p} F_{\ell p}(\mathbf{r}_\perp, z) G_{\ell p}(\tau). \tag{12.109}$$

This kind of solution shows that twisted electron states can be generated in the presence of twisted electromagnetic waves. But, more interestingly, several twisted states described by different LG quantum numbers can be excited by a single LG mode of the laser field [238].

In order to understand this property, we take all radial numbers identical, $p = \nu_0$. Replacing this in eq. (12.107), integrating in the radial coordinates, and using the orthogonality of the LG functions, we obtain an equation for the coefficients $G_\ell$, as

$$-\frac{\nu_p^2}{\omega^2}\frac{\partial^2 G_\ell}{\partial\tau^2} + 2ig(\tau)\frac{\partial G_\ell}{\partial\tau} + \sum_{\ell'}\left(2iB_{\ell,\ell_0} - C_{\ell,\ell_0}\right) G_{\ell'} = 0, \tag{12.110}$$

where $\ell_0$ represents the LG state of the twisted radiation, and new coefficients were introduced, given by

$$B_{\ell,\ell_0} = \sigma\ell_0 a_0 k_C \left\langle F_{\ell-\ell_0}\left|\frac{1}{r}F_{\ell_0}F_\ell\right.\right\rangle, \quad C_{\ell,\ell_0} = a_0^2k_C^2\left\langle F_{\ell_0}\left|F_{\ell_0}^2F_\ell\right.\right\rangle. \tag{12.111}$$

The notation $\langle . \, | . \rangle$ represents integration over the perpendicular plan. We can use a perturbative approach and first assume that mode coupling is negligible, reducing the above equation to

$$-\frac{\nu_p^2}{\omega^2}\frac{\partial^2 G_\ell}{\partial \tau^2} + 2ig(\tau)\frac{\partial G_\ell}{\partial \tau} - a_0^2 k_C^2 G_{\ell'} = 0 \,. \tag{12.112}$$

We can solve this equation using a solution of the form

$$G_\ell(\tau) = U_\ell \exp\left[i \int^\tau \Omega(\tau')d\tau'\right] , \tag{12.113}$$

where $U_\ell$ is an amplitude and $\Omega(\tau)$ a frequency. Replacing this in eq. (12.112), we get

$$\Omega(\tau) = \frac{g(\tau)}{\hbar^2 \nu_p^2}\left[1 - \sqrt{1 + a_0^2 k_C^2 \frac{\hbar^2 \omega}{g(\tau)}}\right] . \tag{12.114}$$

This formally solves the unperturbed problem. Next, we consider coupling between different modes $G_{\ell'}$, as predicted by eq. (12.110). This leads to the following mode coupling relation

$$\frac{\partial U_\ell}{\partial \tau} = iWU_{\ell-2} \,, \quad W = \frac{\omega^2}{2\Omega \nu_p^2} C_{\ell,\ell-2} \,. \tag{12.115}$$

This shows that different twisted electron modes can be excited, on a time-scale of order $1/|W|$. If we choose initial conditions where only the Gaussian mode $\ell = 0$ is present, then the other modes $\ell > 0$ emerge, one after the other, on this time-scale. This interesting mode coupling process is illustrated in Figure 12.4.

## 12.5 Compton Scattering

Compton scattering was discovered in 1923 [173], and corresponds to scattering of an quantum of light, a photon, by a free electron. Being one of the elementary processes of light-matter interaction, it has been described in many textbooks (see, for instance, [240]). The traditional explanation is based on quantum theory and corresponds to what is now called the linear Compton scattering, when the process takes place in vacuum. With the development of very intense radiation sources, and the need for interpretation of laser-plasma experiments, it became clear that a nonlinear description would be necessary, and the contributions of the background medium would have to be taken into account.

Here, we give a description of nonlinear Compton scattering in plasmas. This description includes the possible occurrence of Compton scattering of

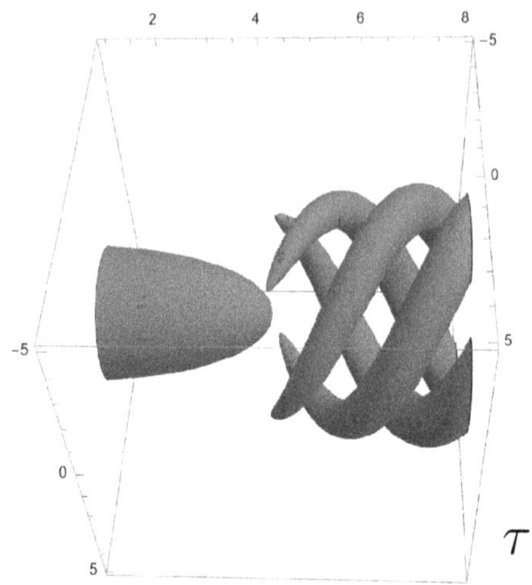


**FIGURE 12.4**
Emergence of twisted electron state $\ell = 2$, in the presence of a twisted laser field $\ell_0 = 1$, when we start from a purely Gaussian electron state.

plasmons and is based on a straightforward generalization of the electron Volkov states to the case of two waves.

### i) Volkov States in Two Waves

Let us first study the Volkov states of an electron in the presence of two waves. We start with the case of two transverse electromagnetic waves with frequencies $\omega$ and $\omega'$, and wavevectors $\mathbf{k}$ and $\mathbf{k}'$ respectively. We then go back to the Klein-Gordon equation (12.81), but now assume that the normalized vector potential (12.83) is replaced by

$$\mathbf{a}_{tot}(\mathbf{r}, t) = \mathbf{a}_0 f(\tau) + \mathbf{a}'_0 f'(\tau') , \quad \tau = t - \frac{\mathbf{k} \cdot \mathbf{r}}{\omega} , \tag{12.116}$$

where $\mathbf{a}_0$ and $\mathbf{a}'_0$ are the two wave amplitudes, and the form functions $f(\tau)$ and $f'(\tau')$, describing the two wavepackets, depend on two different time variables $\tau$ and $\tau'$, defined as

$$\tau = t - \frac{\mathbf{k} \cdot \mathbf{r}}{\omega} , \quad \tau' = t - \frac{\mathbf{k}' \cdot \mathbf{r}}{\omega'} . \tag{12.117}$$

The Klein-Gordon equation can now be written as

$$\left[ \left( \nabla^2 - \frac{1}{c^2} \frac{\partial^2}{\partial t^2} \right) - \left( 1 + \mathbf{a}_{tot}^2 \right) k_C^2 \right] \psi = 0 , \tag{12.118}$$

where we have assumed that the waves are transverse, and therefore $\mathbf{a} \cdot \mathbf{k} = \mathbf{a}' \cdot \mathbf{k}' = 0$, and defined the quantity

$$\mathbf{a}_{tot}^2 = \mathbf{a}^2 + \mathbf{a}'^2 + 2\left(\mathbf{a} \cdot \mathbf{a}'\right) . \tag{12.119}$$

We now follow a similar procedure, and try a solution of eq. (12.118) of the form

$$\psi(\mathbf{r}, t) = e^{i\theta_e} \Phi(\tau, \tau') , \tag{12.120}$$

with

$$\theta_e = -\frac{1}{\hbar} p^\mu x_\mu = \frac{1}{\hbar} \left(\mathbf{p}_e \cdot \mathbf{r} - \epsilon_e t\right) . \tag{12.121}$$

Here, we have the 4-momentum, $p^\mu = (\epsilon_e / c, \mathbf{p}_e)$, and the electron energy and momentum, $\epsilon_e$ and $\mathbf{p}_e$ . We now replace this in eq. (12.118) and derive an evolution equation for the function $\Psi(\tau, \tau')$. This equation is simplified if we assume that the plasma dispersion relations and mass-gap equation are valid. This means that we assume a relation between the frequency and wavenumber of the two waves given by

$$\left(\frac{k^2}{\omega^2} - \frac{1}{c^2}\right) = \frac{\omega_p^2}{\gamma_a} , \quad \left(\frac{k'^2}{\omega'^2} - \frac{1}{c^2}\right) = \frac{\omega_p^2}{\gamma_a} , \tag{12.122}$$

where the effective gamma-factor is given by $\gamma_a \simeq \sqrt{1 + a_{tot}^2}$. Of particular interest is the case where one of the waves is dominant, $a_0 \gg a_0'$. In this case, we have $a_{tot}^2 \simeq a_0^2$. On the other hand, the electron mass-gap equation can be written as

$$\left(\frac{\epsilon_e^2}{c^2} - p_e^2\right) = m^2 \gamma_a c^2 . \tag{12.123}$$

This allows us to derive an evolution equation for the reduced wavefunction $\Phi(\tau, \tau')$, of the form

$$2ig \frac{\partial \Phi}{\partial \tau} + 2ig' \frac{\partial \Phi}{\partial \tau'} + F\Phi = 0 . \tag{12.124}$$

In this equation we have used the new quantities,

$$g \equiv g(\tau, \tau') = \frac{\epsilon_e}{\hbar c^2} - \frac{(\mathbf{p}_e \cdot \mathbf{k})}{\hbar \omega} - k_C \frac{(\mathbf{a}' \cdot \mathbf{k})}{\omega} , \tag{12.125}$$

and

$$g' \equiv g'(\tau, \tau') = \frac{\epsilon_e}{\hbar c^2} - \frac{(\mathbf{p}_e \cdot \mathbf{k}')}{\hbar \omega'} - k_C \frac{(\mathbf{a} \cdot \mathbf{k}')}{\omega'} . \tag{12.126}$$

Notice that these two functions depend on both waves, because of the last term. We also have defined

$$F \equiv F(\tau, \tau') = -k_C^2 \left(\gamma_a - 1 + a_{tot}^2\right) - 2k_C \frac{\mathbf{p}_e}{\hbar} \cdot (\mathbf{a} + \mathbf{a}') . \tag{12.127}$$

As in the previous case of Volkov states for a single wave, this new evolution equation is valid under the assumption that the fast time-scales of the electron wavefunction are concentrated in the phase function $\theta_e$. It is the case of wave photon energies $\hbar\omega$ and $\hbar\omega'$ much smaller than the electron rest energy, $mc^2$. The second derivative terms, $\partial^2\Phi/\partial\tau^2$, $\partial^2\Phi/\partial\tau'^2$ and $\partial^2\Phi/\partial\tau\partial\tau'$, were neglected, following the arguments of the previous section. We now look for solutions of the form

$$\Phi(\tau, \tau') = T(\tau)T'(\tau'). \tag{12.128}$$

This leads to

$$2ig\frac{1}{T}\frac{\partial T}{\partial \tau} + 2ig'\frac{1}{T'}\frac{\partial T'}{\partial \tau'} + F\Phi = 0. \tag{12.129}$$

We also note that

$$F(\tau, \tau') \simeq G(\tau) + G'(\tau'), \tag{12.130}$$

with

$$G(\tau) = -2k_C\frac{\mathbf{p}_e}{\hbar}\cdot\mathbf{a}, \quad G'(\tau') = -2k_C\frac{\mathbf{p}_e}{\hbar}\cdot\mathbf{a}'. \tag{12.131}$$

This means that we can replace eq. (12.129) by two (nearly) decoupled equations, of the form

$$2ig\frac{dT}{d\tau} = G(\tau)T, \quad 2ig'(\tau, \tau')\frac{dT'}{d\tau'} = G'(\tau')T'. \tag{12.132}$$

They can solved as

$$T(\tau) = T_0 \exp\left[-\frac{i}{2}\int_0^\tau \frac{G(z)}{g(z, \tau')}dz\right] \tag{12.133}$$

and

$$T'(\tau') = T_0' \exp\left[-\frac{i}{2}\int_0^{\tau'} \frac{G'(z')}{g'(\tau, z')}dz'\right], \tag{12.134}$$

where $T_0$ and $T_0'$ are constants. Finally, using these solutions, we are able to reconstruct the expected solution for the Klein-Gordon equation, which takes the form

$$\psi(\mathbf{r}, t) = \psi_0 e^{i\theta_e} e^{iS(\tau, \tau')}, \tag{12.135}$$

with $\psi_0 = T_0 T_0'$, and

$$S(\tau, \tau') = -\int_0^\tau \frac{G(z)}{2g(z, \tau')}dz - \int_0^{\tau'} \frac{G'(z')}{2g'(\tau, z')}dz'. \tag{12.136}$$

They can be seen as the Volkov solutions for relativistic electron state in the presence of two waves. Similar solutions could also be found for an electron in two waves, where one is transverse electromagnetic and the other is an electron plasma wave. This result is now used to define the nonlinear Compton scattering states and to determine the corresponding recoil formula, as discussed next.

### ii) Compton Recoil Formula

In order to understand the physical meaning of these Volkov solutions, we assume that the waves as monochromatic, as described by the following expressions

$$a(\tau) = a_0 \cos(\omega\tau) , \quad a'(\tau') = a_0' \cos(\omega'\tau') . \tag{12.137}$$

When $a_0 \gg a_0'$, they can describe an incident wave with frequency $\omega$, and a scattered wave with frequency $\omega'$. Noting that the functions appearing in the denominator of $S(\tau, \tau')$ defined above are of order $g \simeq \epsilon_e/\hbar c^2$ and $g \simeq \epsilon_e/\hbar c^2$, we can reduce eq. (12.136) to

$$S(\tau, \tau') = -\beta \sin(\omega\tau) - \beta' \sin(\omega'\tau') , \tag{12.138}$$

with the two normalized amplitude factors

$$\beta = c^2 k_C \frac{(\mathbf{p}_e \cdot \mathbf{a}_0)}{\hbar\omega\epsilon_e} , \quad \beta' = c^2 k_C \frac{(\mathbf{p}_e \cdot \mathbf{a}'_0)}{\hbar\omega'\epsilon_e} . \tag{12.139}$$

Replacing this in the wavefunction solution (12.135), and developing the exponential factors in Bessel functions, we obtain

$$\psi(\mathbf{r}, t) = \psi_0 e^{i\theta_e} \sum_{n,n'} J_n(\beta) J_{n'}(\beta') e^{-in\omega\tau + n'\omega'\tau'} . \tag{12.140}$$

This solution contains a very complicated phase structure, from where we can isolate those that correspond to Compton scattering. These processes involve incident photons with frequency $\omega$, that are scattered by electrons with initial energy $\epsilon_e$ and momentum $\mathbf{p}_e$. It means that, associated with the scattered photons with frequency $\omega'$, we should expect to observe electrons with a final energy state $\epsilon_e'$ and final momentum $\mathbf{p}'_e$.

We can easily realize that these new energy and momentum states of the electron are only possible if one of the terms contained in the sum of eq. (12.140) is equal to the new electron phase function

$$\theta_e' = -\frac{1}{\hbar} \left( \epsilon_e' t - \mathbf{p}'_e \cdot \mathbf{r} \right) . \tag{12.141}$$

This means that the following identity should be verified

$$\theta' = \theta + \theta_s(n, n') . \tag{12.142}$$

Splitting the space and time terms, we can replace this phase identity in two separate expressions, as

$$\begin{aligned} \epsilon_e' &= \epsilon_e + n\hbar\omega - n'\hbar\omega' , \\ \mathbf{p}'_e &= \mathbf{p}_e + n\hbar\mathbf{k} - n'\hbar\mathbf{k}' . \end{aligned} \tag{12.143}$$

They explicitly state the energy and momentum conservation relations that

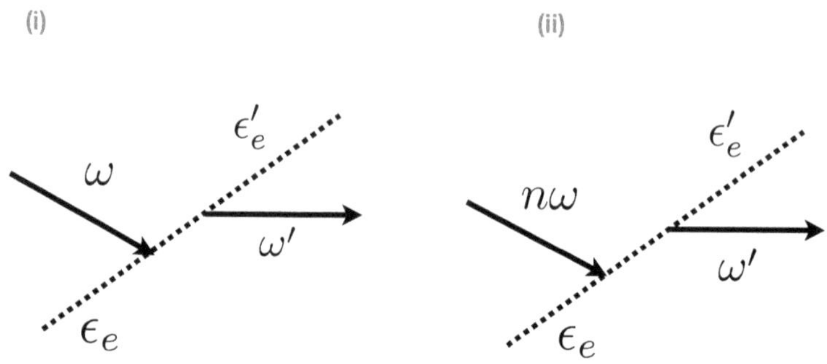


**FIGURE 12.5**
Linear and nonlinear Compton scattering: (i) single scattering process, corresponding to $n = n' = 1$ and (ii) nonlinear scattering process, with emission of a single high energy photon $n' = 1$, by a large number of $n \gg 1$ incident photons.

are associated with nonlinear Compton scattering processes in a plasma. They involve an arbitrary number of $n$ incident photons and $n'$ scattered photons, and include plasma dispersion effects, as shown next. The relevant case of the emission of a single high energy photon $n' = 1$, due to scattering of $n \gg 1$ incident photons is illustrated in Figure 12.5.

From this analysis, we can now retrieve the Compton recoil formula, that incorporates the exchange of momentum. In the non-relativistic limit, known as *Thomson scattering*, this exchange of momentum would vanish. For such purpose, we assume that the electron is initially at rest, with initial momentum, $p_e \ll mc$, such that it can be ignored. In this case, we use $\epsilon_e \simeq m_a c^2$, with $m_a = m\gamma_a$, in the first eq. (12.143), and obtain

$$p_e'^2 c^2 = \left[m_a c^2 + \hbar(\omega' - n\omega)\right]^2 - m_a^2 c^4 . \tag{12.144}$$

We now turn to the second eq. (12.143), assuming $\mathbf{p}_e \simeq 0$. We get

$$p_e'^2 c^2 = c^2 \left[n^2 \hbar^2 k^2 + \hbar^2 k'^2 - 2n\hbar^2 k k' \cos\varphi\right] , \tag{12.145}$$

where $\varphi$ is the angle between the incident and scattered wavevectors, $\mathbf{k}$ and $\mathbf{k}'$. We now use the photon dispersion relations (12.122). Equating the two previous expressions, we obtain

$$\begin{aligned} &(n\omega + n'\omega')^2 - 2m_a c^2 (n\omega + n'\omega')/\hbar = \\ &\left[n^2\omega^2 + n'^2\omega'^2 - \left(n^2 + n'^2\right)\omega_p^2 + 2nn' f(\varphi)\right] , \end{aligned} \tag{12.146}$$

with

$$f(\varphi) = \sqrt{(\omega^2 - \omega_p^2)(\omega'^2 - \omega_p^2)} \cos\varphi . \tag{12.147}$$

At this point, we simplify this expression by taking the high-frequency limit, where $\omega'^2 \gg \omega^2 \gg \omega_p^2$. We can then use the approximation

$$\sqrt{(\omega^2 - \omega_p^2)(\omega'^2 - \omega_p^2)} \simeq \omega\omega' \left(1 - \frac{1}{2}\frac{\omega_p^2}{\omega^2}\right) . \tag{12.148}$$

We also focus on a relevant situation for intense lasers, where a large number of incident photons $n \gg 1$ coalesce with a single scattered photon, described by $n' = -1$. Noticing that $m_a = \gamma_a \hbar k_C / c$, we obtain

$$\left(\frac{c}{n\omega} - \frac{c}{\omega'}\right) \gamma k_C = 1 - (n^2 + 1)\frac{\omega_p^2}{2n\omega\omega'} - \left(1 - \frac{\omega_p^2}{2\omega^2}\right) \cos\varphi . \tag{12.149}$$

This is our final expression for the electron recoil formula associated with the nonlinear Compton scattering in a plasma. In order to reduce it to a more familiar formula, we take the limit where plasma dispersion effects can be ignored, and $\omega_p^2 \to 0$. Writing $\gamma_a$ explicitly, we get

$$\left(\frac{c}{n\omega} - \frac{c}{\omega'}\right) = \frac{1}{k_C}\frac{(1 - \cos\varphi)}{\sqrt{1 + a_0^2}} . \tag{12.150}$$

This coincides with the well-known formula for nonlinear Compton scattering of photons in vacuum. An illustration of this recoil formula is shown in Figure 12.6.

To conclude this section, we now briefly discuss the transition probability $P(n, n')$, associated with the nonlinear Compton scattering processes. It can be defined as

$$P(n, n') \equiv |\psi(n, n')|^2 , \tag{12.151}$$

where $\psi(n, n')$ are given by the Volkov solutions (12.140), for processes that satisfy the energy and momentum conservation relations (12.143). We can then easily get

$$P(n, n') = |\psi_0|^2 \, |J_n(\beta)|^2 \, |J_{n'}(\beta')|^2 . \tag{12.152}$$

It is useful to derive an expression for the probability, valid when the scattered wave amplitude is very small, $a_0' \ll 1$. Using the asymptotic expansion of the Bessel functions for small arguments,

$$J_n(z) = \frac{1}{2^n}\frac{z^n}{n!} , \quad (z \ll 1, \quad n \geq 0) , \tag{12.153}$$

we write

$$P(n, 1) = \frac{1}{4}\beta'^2 \, |\psi_0|^2 \, |J_n(\beta)|^2 . \tag{12.154}$$

The minimum value of $\beta'$ corresponds to a single photon, for which we have $a_0'^2 = \hbar/\epsilon_0\omega'$. This corresponds to

$$\beta' = \frac{1}{\epsilon_0}\frac{mc^2}{\omega'^2} . \tag{12.155}$$

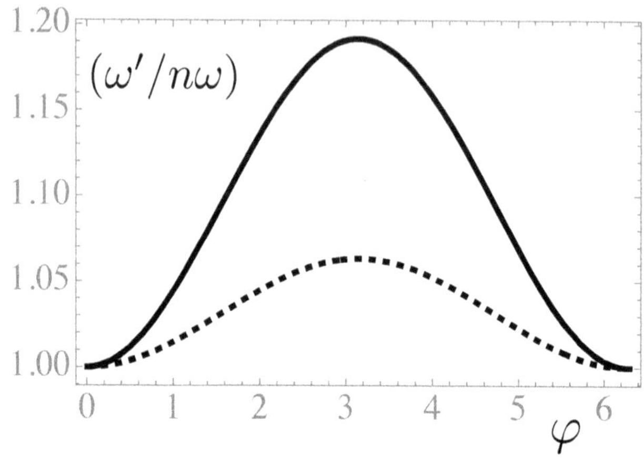


**FIGURE 12.6**
Nonlinear Compton scattering: Relative frequency $(\omega'/n\omega)$ of a single scattered photon, as a function of he scattering angle $\varphi$, with $\omega'/ck_C = 0.1$, for low intensity $a_0^2 = 0.1$ (black curve) and high intensity incident lasers $a_0^2 = 10$ (dashed curve),as given by eq. (12.150).

We should notice that the transition probability $P(n, 1)$ becomes negligible for a large number of photons $n \gg 1$. It also strongly depends on the incident laser amplitude $a_0$, and on the initial electron momentum $p_e$, who are included in the parameter $\beta$. Typically, it increases with the laser amplitude, at low intensities, $a_0 \ll 1$, as it should be expected, but it will eventually increase for large values $a_0 \gg 1$, due to electron inertia. This completes our discussion on the nonlinear Compton scattering.

## 12.6 Inverse Bremsstrahlung

The inverse Bremsstrahlung is the main processes of absorption of the intense laser beam energy when it interacts with a plasma, and plays an important role in laser fusion. This basic process can be described in the following way. First, the electrons are forced to oscillate in the field of the incident laser beam. If we neglect scattering, the energy going into these oscillations is completely reversible, and cannot lead to a net laser absorption. But, if during their oscillation the electrons collide with an ion, then the reversible motion is interrupted and dissipation can occur. This dissipation takes the form of thermal energy flowing into the plasma, and is characterized by an *effective collision frequency* $\nu_{eff}$, moderated by the laser fields.

An early account of the process was given by Silin in 1965 [241], and more recent discussions can be found in [242, 243] and others. Here we derive the effective collision frequency, using a relativistic quantum approach to the inverse Bremsstrahlung. Our starting point will be the electron Volkov states in the incident laser field, as derived in Section 4.

Let us consider the electron quantum states, in the presence of an electromagnetic wave, as determined by the Klein-Gordon (KG) equation (we use $\hbar = 1, c = 1$ in this section), describing the space-time evolution of the electron wave function $\psi$, as

$$(i\partial_t + eV)^2\psi = \left[(i\nabla - e\mathbf{A})^2 + m^2\right]\psi\,, \tag{12.156}$$

where we first use $V = 0$, and

$$\mathbf{A} \equiv \mathbf{A}(\tau) = \mathbf{A}_0 f(\tau)\,, \tag{12.157}$$

where $\mathbf{A}_0$ is the wave amplitude and $f(\tau)$ and $\tau = t - (\mathbf{k}\cdot\mathbf{r})/\omega$, as before. Let us then write the solutions of the wave equation in the form

$$\psi(\mathbf{r}, t) = \varphi(\tau)\exp(i\mathbf{q}\cdot\mathbf{r} - i\omega_q t)\,, \tag{12.158}$$

where $\mathbf{q}$ and $\omega_q$ are constants, representing the electron momentum and energy states, as we have seen. Replacing this in the wave equation (12.1), we obtain

$$-\Omega_p^2\varphi'' + 2ig_q\varphi' - \left[2e(\mathbf{A}\cdot\mathbf{q}) + e^2A^2\right]\varphi = 0\,, \tag{12.159}$$

where we have introduced the auxiliary quantities

$$\Omega_p^2 = \left(\frac{k^2}{\omega^2} - 1\right)\,, \quad g_q = \left(\omega_q - \frac{\mathbf{q}\cdot\mathbf{k}}{\omega}\right)\,. \tag{12.160}$$

We notice that, in vacuum, the first term in (12.159) is absent, because we have $\omega = k$ and therefore, $\Omega_p^2 = 0$. Again, the plasma dispersion relation is given by $\omega^2 = k^2 + \omega_p^2/\gamma_a$ and $\gamma_a = \sqrt{1 + a_0^2}$, with the wave amplitude $a_0 = e|A_0|/m$. This electron wavefunction can now be used to study the electron-ion collisions and to calculate the transition probabilities for electron momentum states.

We have seen that, in the absence of collisions, the electron quantum states in the field of an intense laser pulse with frequency $\omega$ can be written as

$$\psi_q(\mathbf{r}, t) = \tilde{\psi}_q e^{i\theta_q} e^{iS_q(\tau)}, \tag{12.161}$$

where $\tilde{\psi}_q$ is a normalization constant, and $\theta_q = (\mathbf{q}\cdot\mathbf{r} - \omega_q t)$. For circular polarization, this can be written as

$$\psi_q(\mathbf{r}, t) = \tilde{\psi}_q e^{i(\theta_q + 2\eta_q\theta)} \sum_n J_n(\zeta_q) e^{in\theta}\,, \tag{12.162}$$

with

$$\eta_q = \frac{e}{\Omega_q}\frac{A_0}{\omega}(\mathbf{q}\cdot\mathbf{e})\,, \quad \nu_q = \frac{e^2}{4\Omega_q}\frac{A_0^2}{\omega} - \frac{\Omega_p^2}{4\Omega_0}\zeta_q^2\omega\,. \tag{12.163}$$

These wave function solutions satisfy the orthogonality relation

$$\langle \psi_q | \psi_{q'} \rangle \equiv \int \psi_q^*(\mathbf{r}, t) \psi_{q'}(\mathbf{r}, t) d\mathbf{r} = \tilde{\psi}_q^* \tilde{\psi}_{q'} \sum_{n,n'} J_n(\zeta_q) J_{n'}(\zeta_{q'}^*) I_{n,n'}(\mathbf{q}, \mathbf{q}'), \tag{12.164}$$

where we have defined the integral

$$I_{n,n'}(\mathbf{q}, \mathbf{q}') = \int e^{-i(\theta_q - \theta_{q'})} e^{-2i(\eta_q - \eta_{q'})\theta} e^{-i(n-n')\theta} d\mathbf{r}. \tag{12.165}$$

In the dipole approximation, such that $\mathbf{k} \cdot \mathbf{r} \to 0$, this would take a much simpler form, as

$$I_{n,n'}(\mathbf{q}, \mathbf{q}') = 2\pi e^{i(n-n')\omega t} \delta(\mathbf{q} - \mathbf{q}'). \tag{12.166}$$

If we now take an average over a large time interval $\Delta t \gg 1/\omega$, and use the *addition theorem* of Bessel function, and an appropriate normalization constant, we obtain

$$|\langle \psi_q | \psi_{q'} \rangle| = \delta(\mathbf{q} - \mathbf{q}'). \tag{12.167}$$

This normalization condition is valid in general, not just in the dipole approximation where $\mathbf{k} \cdot \mathbf{r} \neq 0$. It shows that the electron wave equation is a linear superposition of orthonormal quantum states. Let us now consider that the electrons collide with an ion located at some position $\mathbf{r} = 0$. The ion can be described by the Debye-screened potential

$$V(\mathbf{r}) = \frac{Ze^2}{4\pi\epsilon_0 r} \exp\left( -\frac{r}{\lambda_D} \right), \tag{12.168}$$

where $Ze$ is the charge of the ion, $\lambda_D = v_{the}/\omega_p$ is the electron Debye length, and $v_{the}$ the electron thermal velocity. Different electron states are then coupled by this collision potential, and the resulting electron wave function can be represented by a superposition of quantum states, given by

$$\psi(\mathbf{r}, t) = \int C_q(t) \tilde{\psi}_q e^{i\theta_q} e^{iS_q(\tau)} \frac{d\mathbf{q}}{(2\pi)^3}. \tag{12.169}$$

We now need to determine the coupling coefficients $C_q(t)$. For that purpose, we replace this expression in the KG equation (12.156), where now the electrostatic potential $V$ is not zero, but is determined by eq. (12.168). For intense laser fields, the scalar potential $V$ can be seen as a small perturbation. Using a perturbative approach, we get

$$i\frac{dC_q}{dt} = -e \int d\mathbf{r} \int \frac{d\mathbf{q}'}{(2\pi)^3} V(\mathbf{r}) C_{q'}(t) e^{-i(\theta_q - \theta_{q'}) - i(S_q - S_{q'})}. \tag{12.170}$$

After integration, we obtain

$$C_q(t) = i\pi \sum_n \delta(\omega_{q'} - \omega"_n) \int C_{q'}(t) H_n(\mathbf{q}, \mathbf{q}') \frac{d\mathbf{q}'}{(2\pi)^3}. \tag{12.171}$$

with the interaction matrix elements

$$H_n(\mathbf{q}, \mathbf{q}') = eV(\mathbf{q}"_n) J_n(\zeta) , \tag{12.172}$$

and $\zeta = \zeta_q + \zeta^*_{q'}$. In this expression, $V(\mathbf{q}"_n)$ is the Fourier transform of the interaction potential

$$V(\mathbf{q}) = \int V(\mathbf{r}) e^{-i\mathbf{q}\cdot\mathbf{r}} d\mathbf{r} , \tag{12.173}$$

calculated for the particular values $\mathbf{q} = \mathbf{q}"_n$. We have also used the quantities

$$\omega"_n = \omega_q + n\omega + 2(\eta_q - \eta_{q'})\omega , \quad \mathbf{q}"_n = (\mathbf{q} - \mathbf{q}') + n\mathbf{k} + 2(\eta_q - \eta_{q'})\mathbf{k} . \tag{12.174}$$

For initial conditions such that, at $t = 0$, the electron momentum state is well defined, we have $C_{q'} = \delta(\mathbf{q}' - \mathbf{q}_i)$. The transition probability between this initial state $\mathbf{q}_i$ and an arbitrary final state $\mathbf{q}$ can then be defined as

$$P(\mathbf{q}' \to \mathbf{q}) \equiv |C_q(t)|^2 . \tag{12.175}$$

Or, in explicity form

$$P(\mathbf{q}' \to \mathbf{q}) = \sum_n P(n, \mathbf{q}' \to \mathbf{q}) , \tag{12.176}$$

with

$$P(n, \mathbf{q}' \to \mathbf{q}) = \pi^2 |H_n(\mathbf{q}, \mathbf{q}')|^2 \delta(\omega_{q'} - \omega"_n) , \tag{12.177}$$

This expression shows that $P(n, \mathbf{q}' \to \mathbf{q})$ are transition probabilities between two different electron momentum states, due to absorption (or emission) of $n$ laser photons. This expression includes the recoil due to photon absorption.

In the dipole approximation, the dependence of $H_n$ on the photon momentum would vanish, and the quantity $\mathbf{q}"_n$ would reduce to $\mathbf{q}" = (\mathbf{q} - \mathbf{q}')$. As a result, the relevant Fourier component of the Yukawa potential would reduce to $V(\mathbf{q}")$, and would be independent of the number of absorbed or emitted photons. The recoil effects vanish in this limit, and the transition probability would simply be given by

$$P(n, \mathbf{q}' \to \mathbf{q}) = \pi^2 |V(\mathbf{q}")|^2 J_n^2(\zeta) \delta(\omega_{q'} - \omega"_n), \tag{12.178}$$

This dipole approximation generalizes those of [242], for Debye screened potentials. We are now in conditions to determine the *effective collision frequency*, which characterizes the laser energy dissipation due to laser-assisted electron-ion collisions.

In order to derive such a quantity, we have to consider the electron kinetic equation with the collision terms associated with the laser induced electron-ion collisions. A kinetic equation for the electron distribution, $f_e(\mathbf{q})$ that incorporates collisions can be derived using a detailed balance between the different electron states, as in the *Boltzmann integral*. It takes the form

$$\frac{\partial}{\partial t} f_e(\mathbf{q}) = \sum_{n \neq 0} \int \frac{d\mathbf{q}'}{(2\pi)^3} \left[ P(n, \mathbf{q}' \to \mathbf{q}) f_e(\mathbf{q}') - P(n, \mathbf{q} \to \mathbf{q}') f_e(\mathbf{q}) \right] . \tag{12.179}$$

It is important to notice that, in general, the transition probabilities are not symmetric, $P(n, \mathbf{q}' \to \mathbf{q}) \neq P(n, \mathbf{q} \to \mathbf{q}')$, because the Fourier transform of the Yukawa potential, $V(\mathbf{q}")$, is not symmetric with respect to the exchange of $\mathbf{q}$ with $\mathbf{q}'$. However, this symmetry is recovered in the dipole approximation.

If the plasma is in equilibrium at a given temperature $T$, we can use the relativistic Maxwell distribution

$$f_e(\mathbf{q}) = C \exp\left[-(\omega_q - m)/T\right]. \tag{12.180}$$

For simplicity we take the constant $C = 1$. Due to the existence of delta functions $\delta(\omega_{q'} - \omega"_n)$ in the expression of the transition probabilities (12.178), we can write

$$f_e(\mathbf{q}') = f_e(\mathbf{q})\delta_n(\mathbf{q}, \mathbf{q}'), \tag{12.181}$$

where

$$\delta_n(\mathbf{q}, \mathbf{q}') = \exp\left\{-\frac{\omega}{T}\left[n + 2(\eta_q - \eta_{q'})\right]\right\}. \tag{12.182}$$

Replacing this in the kinetic equation (12.179, we obtain

$$\frac{\partial}{\partial t} f_e(\mathbf{q}) = f_e(\mathbf{q}) \int \frac{d\mathbf{q}'}{(2\pi)^3} \nu_{coll}(\mathbf{q}, \mathbf{q}'), \tag{12.183}$$

where we have defined the effective collision frequency $\nu_{coll}$ as

$$\nu_{coll}(\mathbf{q}, \mathbf{q}') = \sum_n \left[\delta_n(\mathbf{q}, \mathbf{q}')\, P(n, \mathbf{q}' \to \mathbf{q}) - P(n, \mathbf{q} \to \mathbf{q}')\right]. \tag{12.184}$$

An average collision frequency can then be obtained if we integrate this over the different electron states. This leads to

$$\langle \nu_{coll} \rangle = \int \frac{d\mathbf{q}}{(2\pi)^3} \int \frac{d\mathbf{q}'}{(2\pi)^3} \nu_{coll}(\mathbf{q}, \mathbf{q}') f_e(\mathbf{q}). \tag{12.185}$$

In order to get a more explicit expression for the collision frequency $\nu_{eff}$, we use the transition probabilities $P(n, \mathbf{q}' \to \mathbf{q})$, which are not symmetrical with respect to the exchange between $\mathbf{q}$ and $\mathbf{q}'$, and obtain

$$\nu_{coll}(\mathbf{q}, \mathbf{q}') = \pi^2 e^2 \sum_n \left\{\delta_n(\mathbf{q}, \mathbf{q}') - R_n\right\} \left|V(q"_n)\right|^2 J_n^2(\eta) \delta(\omega_{q'} - \omega"_n), \tag{12.186}$$

where the Fourier transform of the Yukawa potential $V(q"_n)$ is

$$\left|V(q"_n)\right|^2 = \left(\frac{Ze}{\epsilon_0}\right)^2 \frac{\lambda_D^4}{(\lambda_D^2 q"^2_n + 1)^2}, \tag{12.187}$$

and the asymmetry factor $R_n$ is determined by

$$R_n = \frac{(\mathbf{q} - \mathbf{q}') + n\mathbf{k} + 2(\eta_q - \eta_{q'})\mathbf{k}}{(\mathbf{q}' - \mathbf{q}) + n\mathbf{k} - 2(\eta_{q'} - \eta_q)\mathbf{k}}. \tag{12.188}$$

It is now important to write this result in the non-relativistic limit, which corresponds to weak laser fields, such that $a_0 < 1$. In this case, we can use $\omega_q \simeq m + q^2/2m$, assume that $\zeta \ll 1$, and retain only the first term $n = 1$ in the above summation. The quantities (12.174) can also be simplified to

$$\omega_n^{"} \simeq \omega_q + n\omega \,, \quad \mathbf{q}_n^{"} \simeq (\mathbf{q} - \mathbf{q}') + n\mathbf{k}. \tag{12.189}$$

This leads an expression of the effective collision frequency (12.186), in the low intensity limit, of the form

$$\nu_{coll}(\mathbf{q}, \mathbf{q}') \simeq \nu_0 \left\{ \left( e^{-\omega/T} - R_1 \right) \delta(\Delta + \omega) - \left( e^{\omega/T} - \frac{1}{R_1} \right) \delta(\Delta - \omega) \right\} , \tag{12.190}$$

where $\Delta \equiv (\omega_q - \omega_{q'})$, and the amplitude $\nu_0$ is

$$\nu_0 = \pi^2 e^2 \left( \frac{\lambda}{2\omega} \right)^2 \left| V(q_1^{"}) \right|^2 . \tag{12.191}$$

If we take this result in the limit of $\omega \ll \Delta$ and $\lambda_D^2 q_1^{''2} \ll 1$ , we recover the famous result first derived by Silin [241]. To complete our discussion, let us now take the ultra-relativistic limit, where $a_0 \gg 1$. In this case, the parameter $\zeta$, is now given by

$$\zeta = \frac{\lambda'}{\omega} \,, \quad \lambda' = a\frac{m}{q} . \tag{12.192}$$

It is also useful to assume that $\lambda' \gg \omega$. For intense laser fields, the multiphoton processes corresponding to $n \gg 1$ will dominate. In this limit, we can take $R_n = 1$ and use

$$\sum_n J_n^2 \left( \frac{\lambda'}{\omega} \right) \delta(\Delta - n\omega) \simeq J_x^2 \left( \frac{\lambda'}{\omega} \right) \sum_n \delta(\Delta - n\omega) \simeq \frac{1}{2} \left[ \delta(\Delta - \lambda') - \delta(\Delta + \lambda') \right] \tag{12.193}$$

where $x = \Delta/\omega$, and $J_x^2$ has a maximum for $\Delta = \pm\lambda'$. This is essentially the result obtained by [242] in the ultra-intensity limit.

## 12.7 Plasma QED

Laser systems able to attain ultra-high intensities above the PetaWatt level are now in operation [245]. This stimulates interest on vacuum QED effects that can be explored with such devices, including the creation of matter from vacuum by intense electric fields. The theoretical description of this process is well established, as described in a number of reviews [246, 244]. In this last section, we intend to discuss the relativistic electron processes associated with spin, which has been neglected so far. For this purpose, we need to replace the Klein-Gordon equation by the Dirac equation, and look for the possible consequences for particle acceleration in plasmas. We also discuss the derivation of a quantum kinetic equation with spin.

### i) Dirac Equation

We start with the Dirac equation, which describes the electron states in a in a generic 4-vector potential $A^\mu = (V, \mathbf{A})$. It can be written as

$$[\gamma^\mu (i\partial_\mu - eA_\mu) - m] \psi = 0 , \tag{12.194}$$

where the four matrix components of $\gamma^\mu$ can be written in terms of the $(2 \times 2)$ unit matrix, and the Pauli matrices. This is written in $\hbar = c = 1$ units. It is well-known that, in a constant potential, positive and negative energy solutions $\psi^\pm(x)$, are given by

$$\psi^+(x) = u(\mathbf{p}) \exp(ip_\mu x^\mu) , \quad \psi^-(x) = v(-\mathbf{p}) \exp(-ip_\mu x^\mu) , \tag{12.195}$$

with the 4-momentum $p^\mu = (p^0, \mathbf{p})$. As seen above, the energy $E \equiv p^0$ is related to $\mathbf{p}$ by the mass-gap (or dispersion) equation

$$E = eV + \left[ (\vec{\sigma} \cdot \mathbf{P})^2 + m^2 \right]^{1/2} , \tag{12.196}$$

where the components of $\vec{\sigma}$ are the Pauli matrices, and $\mathbf{P} = \mathbf{p} - e\mathbf{A}$. In order to compare the Dirac equation with the Klein-Gordon equation (12.1), it is also useful to write it in the equivalent form

$$\left( i\frac{\partial}{\partial t} + eV \right) \psi = \left[ \bar{\bar{\alpha}} \cdot (i\nabla - e\mathbf{A}) + \bar{\bar{\beta}} m \right] \psi , \tag{12.197}$$

where $\gamma^0 = \bar{\bar{\beta}}$ and $\gamma^j = \bar{\bar{\beta}}\bar{\bar{\alpha}}^j$ for $j = 1, 2, 3$. In terms of the normalized potentials $\mathcal{U}$ and $\mathbf{a}$, this can also be written as

$$\left( i\frac{\partial}{\partial t} + k_C \mathcal{U} \right) \psi = \left[ \bar{\bar{\alpha}} \cdot (i\nabla - k_C \mathbf{a}) + \bar{\bar{\beta}} m k_C \right] \psi , \tag{12.198}$$

Multiplying this by the operator on $\psi$, we can then derive the *quadratic Dirac equation*, given by

$$(i\partial + k_C a)^2 \psi = - \left[ k_C \left( \bar{\bar{\alpha}} \cdot \mathbf{e} + \bar{\bar{\Sigma}} \cdot \mathbf{b} \right) - k_C^2 \right] \psi , \tag{12.199}$$

where we have used the compact notation $\partial = (\partial_t, \nabla)$, and $a = (\mathcal{U}, \mathbf{a})$. We have also used the normalized electric and magnetic fields $\mathbf{e}$ and $\mathbf{a}$, and the operator $\bar{\bar{\Sigma}}$, are defined by

$$\mathbf{e} = -\partial_t \mathbf{a} - \nabla \mathcal{U} , \quad \mathbf{b} = \nabla \times \mathbf{a} , \quad \bar{\bar{\Sigma}} = -\frac{i}{2} \bar{\bar{\alpha}} \times \bar{\bar{\alpha}} . \tag{12.200}$$

Apart from the new term associated with $\bar{\bar{\Sigma}}$ that describes spin effects, this quadratic Dirac equation strongly resembles Klein-Gordon equation. This completes our short presentation of the Dirac equation. More detailed discussions can be found elsewhere [240, 247].

### ii) Quantum Acceleration

Let us now discuss particle acceleration as a mean-field process that, in the frame of the Dirac equation necessarily includes spin. This naturally extends the particle acceleration concept to the relativistic quantum domain and contributes to a broader understanding of the acceleration phenomenology. The relativistic quantum approach, not only gives a more fundamental description already well understood at the classical level, but also identifies purely quantum effects of particle (electron) acceleration in plasmas.

Let us then assume an electron interacting with a space and time dependent potential $V(\vec{r}, t)$. The relativistic Hamiltonian operator can be written as

$$H = \bar{\bar{\beta}} m + \bar{\bar{\alpha}} \cdot \mathbf{p} + V(\mathbf{r}, t) , \tag{12.201}$$

where $\mathbf{r}$ and $\mathbf{p}$ are the position and momentum operators. In the Hilbert space associated with $H$, the Heisenberg equations for $\mathbf{r}$ and $\mathbf{p}$ are determined by

$$\frac{d\mathbf{r}}{dt} = i[H, \mathbf{r}] = \bar{\bar{\alpha}} , \quad \frac{d\mathbf{p}}{dt} = i[H, \vec{p}] = -\frac{\partial V}{\partial \mathbf{r}} . \tag{12.202}$$

We now assume that the scalar potential moves with constant velocity $\mathbf{u}$, such that $V(\mathbf{r}, t) = V(\mathbf{r} - \mathbf{u}t)$. This form of potential can be used to describe the electrostatic wakefield perturbations produced by an intense laser pulse in a plasma. In this case, the value of the velocity $u$ is nearly equal to the laser group velocity. If we use a new position operator, $\mathbf{r}' = \mathbf{r} - \mathbf{u}t$, this leads to a new Hamiltonian operator, that is not explicitly dependent on time, and can be written as

$$H' \equiv H(\mathbf{r}') - \mathbf{u} \cdot \mathbf{p} = \bar{\bar{\beta}} m + (\bar{\bar{\alpha}} - \mathbf{u}) \cdot \mathbf{p} + V(\mathbf{r}') . \tag{12.203}$$

Because it is independent of time, the expectation value of $H'$, is an invariant

$$\langle H' \rangle = E - \mathbf{u} \cdot \langle \mathbf{p} \rangle \tag{12.204}$$

where $E = \left\langle \bar{\bar{\beta}} m + \bar{\bar{\alpha}} \cdot \mathbf{p} + V \right\rangle$ is the energy expectation value. This shows that, in general, both the particle energy $E$ and momentum $\langle \mathbf{p} \rangle$ vary in time, but $\langle H' \rangle$ stays constant.

A more explicit description of the electron quantum states in the moving potential can now be given by the Dirac equation which, for $\mathbf{a} = 0$, takes the form

$$i \frac{\partial}{\partial t} \psi = H \psi , \quad H \equiv \bar{\bar{\beta}} m + V - i \bar{\bar{\alpha}} \cdot \nabla . \tag{12.205}$$

We can now write the particle wavefunction as $\psi = (\varphi, \chi)$ and the spinors $\varphi$ and $\chi$ are determined by the coupled equations

$$i\frac{\partial \varphi}{\partial t} = (V+m)\varphi - i\vec{\sigma}\cdot\frac{\partial \chi}{\partial \mathbf{r}}\,, \quad i\frac{\partial \chi}{\partial t} = (V-m)\chi - i\vec{\sigma}\cdot\frac{\partial \varphi}{\partial \mathbf{r}}\,. \tag{12.206}$$

Here, the components of the vector operator $\vec{\sigma}$ are the Pauli matrices. Let us now focus the one-dimensional problem, where the fermion particles move along the $x$-axis. If we write $\varphi = (\varphi^1, \varphi^2)$ and $\chi = (\chi^1, \chi^2)$, we can easily identify the two pairs of coupled functions $(\varphi^1, \chi^2)$ and $(\varphi^2, \chi^1)$ with two spinors that propagate independently from each other. To simplify, we can use the notation $(\varphi, \chi)$ to refer to any of these two pairs. The potential perturbations described in this one-dimensional problem by $V(x,t) = V(x-ut)$ suggest the use of *D'Alembert operators*, $(\xi = x - ut, \eta = x + ut)$. The coupled equations (12.206) can be written in terms of these new operators as

$$iu\left(\frac{\partial}{\partial \eta} - \frac{\partial}{\partial \xi}\right)\varphi = [V(\xi)+m]\varphi - i\left(\frac{\partial}{\partial \xi} + \frac{\partial}{\partial \eta}\right)\chi\,, \tag{12.207}$$

and

$$iu\left(\frac{\partial}{\partial \eta} - \frac{\partial}{\partial \xi}\right)\chi = [V(\xi)-m]\chi - i\left(\frac{\partial}{\partial \xi} + \frac{\partial}{\partial \eta}\right)\varphi\,. \tag{12.208}$$

Noting that the moving potential is independent of $\eta$, we can make a Fourier transformation in $\eta$, such that

$$\varphi(\xi,\eta) = \int \varphi_q(\xi) e^{iq\eta}\frac{dq}{2\pi}\,, \tag{12.209}$$

with an identical expression for $\chi(\xi,\eta)$. The coupled equations for the Fourier components can then be written as

$$-iu\frac{\partial}{\partial \xi}\varphi_q = [V(\xi)+uq+m]\varphi_q + \left(q - i\frac{\partial}{\partial \xi}\right)\chi_q\,, \tag{12.210}$$

and

$$-iu\frac{\partial}{\partial \xi}\chi_q = [V(\xi)+uq-m]\chi_q + \left(q - i\frac{\partial}{\partial \xi}\right)\varphi_q\,. \tag{12.211}$$

Let us consider the case where the potential $V(\xi)$ is a slowly varying function of $\xi$ (when compared with the particle length-scale). This is usually valid for laser wakefield acceleration, because the electron Compton wavelength is much smaller than the length-scale of the potential perturbations. In this case, we can use solutions of these equations of the form

$$\varphi_q(\xi) = \varphi_{q0}\exp\left[i\int^{\xi} k(\xi')d\xi'\right]\,, \quad \chi_q(\xi) = \chi_{q0}\exp\left[i\int^{\xi} k(\xi')d\xi'\right]\,. \tag{12.212}$$

Replacing this in eqs. (12.210) and (12.211), we obtain

$$\{V(\xi) + m + u[q - k(\xi)]\}\phi_q = -[q + k(\xi)]\psi_q\,, \tag{12.213}$$

and

$$\{V(\xi) - m + u[q - k(\xi)]\}\chi_q = -[q + k(\xi)]\varphi_q \tag{12.214}$$

In can easily be seen that these equations satisfy the dispersion relation

$$\{V(\xi) + m + u[q - k(\xi)]\}\{V(\xi) - m + u[q - k(\xi)]\} = [q + k(\xi)]^2\,, \tag{12.215}$$

which is locally valid. These approximate solutions imply that the two equalities

$$p(\xi) = q + k(\xi)\,, \quad E(\xi) = u[k(\xi) - q]\,, \tag{12.216}$$

are also satisfied. This reduces the above dispersion relation to a much simpler expression, as

$$[E(\xi) - V(\xi)]^2 = p^2(\xi) + m^2\,, \tag{12.217}$$

where the energy $E(\xi)$ and momentum $p(\xi)$ are slowly varying functions of the spacetime variable $\xi$. In the special case of a constant potential $V(\xi) \equiv V_0$ this expression would simply state the invariance of the particle energy and momentum. But, in the general case of a moving potential the only constant of motion that survives is

$$E(\xi) - up(\xi) = -2uq\,. \tag{12.218}$$

It is obvious that this quantity is an invariant for each quantum state $q$, because the velocity $u$ is constant. One should also notice that the above solutions completely determine the values of $E(\xi)$ and $p(\xi)$ as functions of $\xi$, therefore, as functions of space and time.

A bridge can be established between the present relativistic quantum description and the classical models of acceleration. The classical description of acceleration by a moving potential (a wakefield) is given here in terms of the quantum invariant (12.218). The energy increase $dE$ of the particle over a distance $d\xi$ is determined by

$$dE = \frac{1}{2}\left[E\left(1 - \frac{1}{u^2}\right) - V - \frac{2q}{u}\right]^{-1} (\nabla_\xi V)\, d\xi\,. \tag{12.219}$$

This equation shows that any potential gradient $\nabla_\xi V$ is able to increase or decrease of the particle energy, as expected. This is common to both classical and quantum models. The spinor solution of a fermion moving in a variable potential $V(\xi)$ can then be written as

$$\psi_q(\xi) = [\varphi_{q0}, \chi_q(\xi)] \exp\left[\pm i \int^{\xi} k(\xi')d\xi'\right]\,, \tag{12.220}$$

where

$$\chi_q(\xi) = \varphi_{q0} \, \frac{p(\xi)}{E(\xi) + m - V(\xi)} \,. \tag{12.221}$$

Finally we note that the present theoretical model can also be easily applied to other physical problems, such as the neutrino deceleration by shock fronts of particular interest to *super-novae explosions* [248], hadron acceleration by shocks and waves in a *quark-gluon plasma*, which can be relevant to *heavy-ion collisions* [249], or the search for *dark matter* [250].

### iii) Quantum Kinetic Equation

Let us now discuss the derivation of quantum kinetic equations. For that purpose, we should use the second quantized Dirac field. It is known that the quantized Dirac field satisfying eq. (12.194) can be written as

$$\psi(x) = \sum_{s,\mathbf{p}} \left[ b_{s\mathbf{p}} u_{s\mathbf{p}} e^{-i\omega_{\mathbf{p}} t} + d^\dagger_{s\mathbf{p}} v_{s\mathbf{p}} e^{+i\omega_{\mathbf{p}} t} \right] e^{i\mathbf{p}\cdot\mathbf{r}/\hbar} \,, \tag{12.222}$$

where $s$ specifies the spin states. The pair of creation and annihilation operators, $(b^\dagger, d^\dagger)$ and $(b, d)$, correspond to positive and negative energy states, and $u_{s\mathbf{p}}$ and $v_{s\mathbf{p}}$ are the respective spinors.

In contrast with our previous discussion, we now assume a purely electromagnetic problem, with $V = 0$, and $\mathbf{A} \neq 0$. Consider an infinitesimal and abrupt change of the vector potential at a given time $t = 0$. This can be described as

$$\mathbf{A}(t) = A^\mu(t \leq 0) + \delta\mathbf{A} \, H(t) \,, \tag{12.223}$$

where $H(t)$ is the Heaviside function. Due to this sudden change, the energy $E$ and momentum $p = |\mathbf{p} - e\mathbf{A}|$ of a particle will also verify infinitesimal increments, as given by

$$\delta E = -\frac{e}{E} (\mathbf{p} - e\mathbf{A}) \cdot \delta\mathbf{A} \,, \quad \delta p = \frac{E}{p} \delta E \,, \tag{12.224}$$

Noting that the Dirac field solutions (12.222) are well-defined at all times, and in particular at $t = 0$, we have to assume that there is a continuity relation between the initial operators, valid for $t < 0$, and the final operators valid after the perturbation, for $t > 0$. This implies that the operators $b_p$ and $d_p^\dagger$ have to suffer an increment dictated by

$$\tilde{b}_p = A_p b_p + \delta B_p \tilde{d}_p^\dagger \,, \quad \tilde{d}_p^\dagger = -\delta B_p b_p + A_p d_p^\dagger \,. \tag{12.225}$$

Furthermore, the two coefficients, $A_p$ and $\delta B_p$, have to satisfy the relation $|A_p|^2 + |\delta B_p|^2 = 1$. To simplify these expressions, we have used the notation $p \equiv (s, \mathbf{p})$, such that for a given spin state $s$, corresponding operators are

$$\tilde{b}_p \equiv \tilde{b}_{s,\mathbf{p}} \,, \quad \tilde{d}_p \equiv \tilde{d}_{s' \neq s, -\mathbf{p}} \,. \tag{12.226}$$

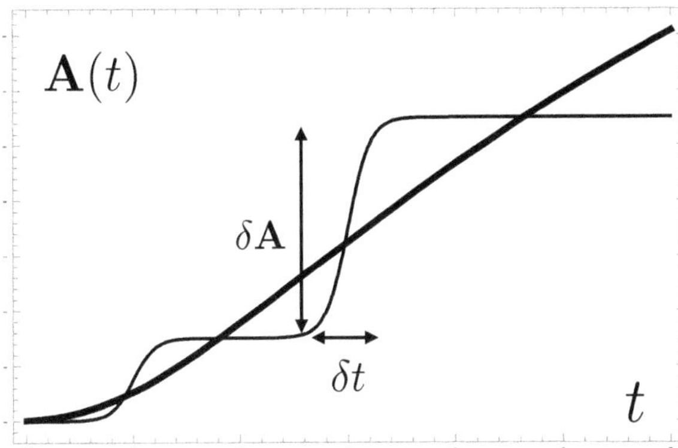


**FIGURE 12.7**
An arbitrary vector potential $\mathbf{A}(t)$, can be seen as an infinite series of infinitesimal time steps, associated with infinitesimal variations $\delta\mathbf{A}$ at consecutive elementary time intervals $\delta t$.

This necessarily implies that

$$A_p = 1 - \frac{1}{2}(\delta B_p)^2 \,, \quad \delta B_p = \frac{m}{2E^2}\delta p \,. \tag{12.227}$$

Notice that the momentum variation is determined by $\delta p = -e\mathcal{E}(t)\delta t$, where $\mathcal{E} = -\partial_t \mathbf{A}$ is the electric field.

Of course, this is only valid for infinitesimal perturbations at $t = 0$. But, an arbitrary time variation of the vector potential $A(t)$ can be defined as a series of infinitesimal temporal steps, as illustrated in Figure 12.7. This means that, after a finite time $t$, the accumulated change of the operators is given by an integral over these infinitesimal steps, and can be written as

$$\begin{aligned} b_p(t) &= \mathcal{A}(t)b_p(0) + \mathcal{B}(t)\tilde{d}_p^\dagger(0) \,, \\ \tilde{d}_p^\dagger(t) &= -\mathcal{B}(t)b_p(0) + \mathcal{A}(t)d_p^\dagger(0) \,, \end{aligned} \tag{12.228}$$

where the new coefficients are determined by

$$\mathcal{A}(t) = \cos[\theta_p(t)] \,, \quad \mathcal{B}(t) = \sin[\theta_p(t)] \,, \tag{12.229}$$

and the angle $\theta_p(t)$, is defined as

$$\theta_p(t) = \int_0^t B_p(t') \exp[2i\varphi_p(t')] dt' \,. \tag{12.230}$$

Notice that the amplitude $B_p(t)$ and the phase $\varphi_p(t)$ depend on the variation of the particle energy $E(t)$ over the time interval, as

$$B_p(t) = -(em/2E^2)\mathcal{E}(t) \,, \quad \varphi_p(t) = \int^t E(t')dt' \,, \tag{12.231}$$

and

$$E^2(t) = m^2 + p_\perp^2 + (p_\parallel - eA)^2 \,, \qquad (12.232)$$

At this point, we should also notice that the perpendicular momentum, $\mathbf{p}_\perp$ is absent from the above expression, because it remains constant and can be always included in the mass $m$. We ignore it, and use $p$ to refer to the parallel momentum. For particles nearly at rest, we can use $p = eA$, and obtain

$$E^2(t) = m^2 \left[1 + a^2(t)\right] \,, \quad \mathbf{a(t)} = e\mathbf{A}(t)/m \,. \qquad (12.233)$$

The probability for particle-pair creation by an applied electric field $\mathcal{E}(t)$ can be established from the above analysis, as

$$P_p(t > t_0) = |\mathcal{B}(t)|^2 = \sin^2[\theta_p(t)] \,. \qquad (12.234)$$

This quantity is usually very small, and we can use the approximation $\sin\theta = \theta$ [251].

The same approach can be used to derive an evolution equation for the distribution functions $f_\pm(t)$, where the minor sign refers to electron states and the plus sign to positrons. For this purpose, we can take the derivative of eqs. (12.228), and obtain

$$\frac{d}{dt} b_p = B_p \, \tilde{d}_p^\dagger \exp[2i\varphi_p(t)] \,, \quad \frac{d}{dt} d_p^\dagger = -B_p \, b_p \exp[-2i\varphi_p(t)] \,. \qquad (12.235)$$

Let us now define the particle distribution functions $f_\pm(p)$ as the occupation numbers

$$f_-(p) = \langle 0 | \, b_p^\dagger b_p \, | 0 \rangle \,, \quad f_+(p) = \langle 0 | \, d_p^\dagger d_p \, | 0 \rangle \,. \qquad (12.236)$$

Taking the time derivative of the first of these distribution functions, and then use eqs. (12.235), we arrive at

$$\frac{df_-}{dt} = 2B_p \Re \left\{ \Phi(p) \right\} \,, \qquad (12.237)$$

where we have used the auxiliary quantity

$$\Phi(p) = \langle 0 | \, d_p b_p \, | 0 \rangle \exp[-2i\varphi_p(t)] \,. \qquad (12.238)$$

Taking the time derivative of this new expression, we can then easily get

$$\frac{d\Phi(p)}{dt} = -B_p \left[1 - (f_- + f_+)\right] - 2iE(t)\Phi(p) \,. \qquad (12.239)$$

where $[1 - (f_- + f_+)] \equiv [1 - 2f_-]$ is the well-known Pauli blocking factor. This is easily integrated, leading to

$$\Phi(p) = -\int^t B_p(t') \left[1 - 2f_-(t')\right] \exp[2i\varphi_p(t') - 2i\varphi_p(t)] dt' \,. \qquad (12.240)$$

Replacing this in eq. (12.237), we finally get the evolution equation for the electron distribution function in closed form, as

$$\frac{df_-}{dt} = -2B_p(t) \int^t B_p(t') \, [1 - 2f_-(t')] \cos[2\varphi_p(t') - 2\varphi_p(t)] dt' \, . \qquad (12.241)$$

This is equation was first derived in [251], and is usually called the quantum kinetic, or simply *QK equation*. A similar equation can be obtained for the positron distribution $f_+$. Notice the appearance of the Pauli factor $(1 - 2f_-)$ in the integrand, which results from the *Exclusion Principle*.

A different and eventually more rigorous form of quantum kinetic equation can be derived, when we apply the Wigner-Moyal procedure to the Dirac equation. This new formalism is sometimes called the Dirac-Heisenberg-Wigner, or simply the *DHW equation* [252]. However the use of this approach is not very practical because it involves the use of a Wigner matrix that contains five different terms describing separately the mass density, the charge density, the current density and the spin density. For this reason, it has only be applied to simple cases, where the difference between the QK and the QHW results is marginal [253]. Relation between these two formalisms is discussed in [254].

# Bibliography


[1] H.M. Mott-Smith, "History of 'Plasmas'," *Nature* **233**, 219 (1971).

[2] J-L. Delcroix and A. Bers, *Physique des Plasmas*, vols. 1 and 2, InterEditions/CNRS Editions, Paris (1994).

[3] N.A. Krall and A.W. Trivelpiece, *Principles of Plasma Physics*, McGraw-Hill, New York (1973).

[4] F.F. Chen, *Introduction to Plasma Physics and Controlled Fusion*, Springer, New York, 3rd ed. (2016).

[5] J.A. Bittencourt, *Fundamentals of Plasma Physics*, Springer-Verlag, New York (2004).

[6] D.R. Nicholson, *Introduction to Plasma Theory*, John Wiley & Sons, New York (1983).

[7] J.T. Mendonça and H. Terças, *Physics of Ultra-Cold Matter (Atomic Clouds, Bose Einstein Condensates and Rydberg Plasmas)*, Springer, New York, (2012).

[8] C.S. Liu, V.K. Tripathi and B. Eliasson, *High-Power Laser-Plasma Interaction*, Cambridge University Press, Cambridge, U.K. (2019).

[9] P. Michel, *Introduction to Laser-Plasma Interactions*, Springer Cham, New York (2023).

[10] M. Lyon and S.L. Rolston, "Ultracold neutral plasmas", *Rep. Prog. Phys.* **80**, 017001 (2017).

[11] L.V. Keldish, "Ionization in the Field of a Strong Electromagnetic Wave", *Sov. Phys. JETP* **20**, 1307 (1965).

[12] H.R. Reiss, "Theoretical Methods in Quantum Optics: S-Matrix and Keldish Techniques for Strong-Field Processes", *Prog. Quant. Electr.* **16**, 1 (1992).

[13] L. Landau and J. Lifshitz, *Mechanics*, 3rd. edition, Butterworth-Heinemann, Oxford (1976).

[14] H. Goldstein, C.P. Poole and J.L Safko, *Classical Mechanics*, 3rd. edition, Pearson, New York (2002).

[15] J.D. Jackson, *Classical Electrodynamics*, 3rd. edition, John Wiley & Sons, New York (1999).

[16] L.D. Landau and E.M. Lifshitz, *The Classical Theory of Fields*, Butterworth-Heinemann, Oxford (1975).

[17] R.B. White, *Theory of Toroidally Confined Plasmas*, World Scientific, Singapore (2013).

[18] A.A. Chernikov, T. Tél, G. Vattay and G.M. Zaslavsky, "Chaos in the Relativistic Generalization of the Standard Map", *Phys. Rev. A* **40**, 4072 (1989).

[19] Y. Nomura, Y.H. Ichikawa and W. Horton, "Nonlinear Dynamics of the Relativistic Standard Map", *Phys.Rev. A* **45**, 1103 (1992).

[20] B.V. Chirikov, "A Universal Instability of Many-Dimensional Oscillators", *Phys. Rep.* **52**, 263 (1979).

[21] E. Ott, *Chaos in Dynamical Systems*, 2nd edition, Cambridge University Press, Cambridge (2014).

[22] A. J. Lichtenberg and M. A. Lieberman, *Regular and Chaotic Dynamics*, Springer-Verlag, New York (1992)

[23] M. N. Rosenbluth, R. Z. Sagdeev, J. B. Taylor, and G. M. Zaslavskii, "Destruction of Magnetic Surfaces by Magnetic Field Irregularities" *Nucl. Fusion* **6**, 297 (1966).

[24] J.T. Mendonça, "Diffusion of magnetic field fines in a toroidal geometry", *Phys. Fluids B*, **3**, 87 (1991).

[25] E. Fermi, "On the Origin of the Cosmic Radiation", *Phys. Rev.* **75**, 1169 (1949).

[26] L.O'C. Drury, "An Introduction to the Theory of Diffusive Shock Acceleration of Energetic Particles in Tenuous Plasmas", *Rep. Prog. Phys.* **46**, 973 (1983).

[27] R. Blandford and D. Eichler, "Particle Acceleration at Astrophysical Shocks: A Theory of Cosmic Ray Origin", *Phys. Rep.* **154**, 1 (1987).

[28] S.M. Ulam, "On Some Statistical Properties of Dynamical Systems," *Proc. Fourth Berkeley Symp. on Math. Stat. and Prob.* **3**, 315 (1960).

[29] Y. Kominis, A. K. Ram, and K. Hizanidis, "Interaction of Charged Particles with Localized Electrostatic Waves in a Magnetized Plasma", *Phys. Rev. E* **85**, 016404 (2012).

[30] N.N. Bogoliubov, *Problems of a Dynamical Theory in Statistical Physics*, State Technical Press, Moscow (1946).

[31] M. Born and H. S. Green, *A General Kinetic Theory of Liquids*, Cambridge University Press, Cambridge, U.K. (1949).

[32] J. G. Kirkwood, "The Statistical Mechanical Theory of Transport Processes", *J. Chem. Phys.* **14**, 180 (1946); **15**, 72 (1947).

[33] J. Yvon, *La Theorie des Fluides et l'Equation d'Etat*, Hermann et Cie, Paris, France (1935)

[34] D.C. Montgomery and D.A. Tidman, *Plasma Kinetic Theory*, McGraw-Hill, New York (1964).

[35] Yu. L. Klimontovich, *The Statistical Theory of Non-equilibrium Processes in a Plasma*, M.I.T. Press, Cambridge, Mass. (1967).

[36] G.B. Arfken and H.J. Weber, *Mathematical Methods for Physicists*, Elsevier, Amsterdam, 6th ed. (2005).

[37] F. Jüttner, "Das Maxwellsche Gesetz der Geschwindigkeitsverteilung in der Relativtheorie", *Annalen der Physik*, **339**, 856 (1911).

[38] H. Alfvén, "Existence of Electromagnetic-Hydrodynamic Waves", *Nature* **150**, 405 (1942).

[39] L.D. Landau and E.M. Lifshitz, *Fluid Mechanics*, Butterworth-Heinemann, Oxford (1975).

[40] K. Hattori, M. Hongo and X.-G. Huang, "New Developments in Relativistic Magnetohydrodynamics", *Symmetry* **14**, 1851 (2022).

[41] R.D. Hazeltine and S.M. Mahajan, "Fluid Description of Relativistic Magnetized Plasma", *Astrophys. J.* **567**, 1271 (2002)

[42] P.A. Andreev, "Microscopic Model for Relativistic Hydrodynamics of Ideal Plasmas", *Eur. Phys. J. D* **77**, 145 (2023).

[43] P.L. Bhatnagar, E.P. Gross, M. Krook, "A Model for Collision Processes in Gases" *Phys. Rev.* **94**, 511 (1954).

[44] L. Spitzer and R. Härm, "Transport Phenomena in a Completely Ionized Gas, *Phys. Rev* **89**, 977 (1953).

[45] T.H. Stix, *The Theory of Plasma Waves*, McGraw-Hill, New York (1962).

[46] W.P. Allis, S.J. Buchsbaum and A. Bers, *Waves in Plasmas*, Wiley & Sons, New York (1962).

[47] A.I. Akhiezer, I.A. Akhiezer, R.V. Polovin, A.G. Sitenko and K.N. Stepanov, *Plasma Electrodynamics*, Pergamon Press, Oxford (1972)

[48] D.G. Swanson, *Plasma Waves*, Institute of Physics Publ., Bristol (2003),

[49] J. Bergmann and B. Eliasson, "Linear Wave Dispersion Laws in Unmagnetized Relativistic Plasma: Analytical and Numerical Results", *Phys. Plasmas* **8**, 1482 (2001); Erratum: *Phys. Plasmas* **16**, 129902 (2009).

[50] F. Pegoraro and F. Porcelli, "Equation of State for Relativistic Plasma Waves", *Phys. Fluids* **27**, 1665 (1984).

[51] R. Hakim, *Introduction to Relativistic Statistical Mechanics*, World Scientific, Singapore (2011).

[52] B.D. Fried and S.D. Conte, *The Plasma Dispersion Function*, Academic Press, New York (1961).

[53] L. Landau, "On the vibrations of the electronic plasma", *Zh. Eksp. Teor. Fiz.* **16**, 574 (1946). [reprinted in *Collected Papers of Landau*, ed. D. ter Haar, vol. 2. Pergamon Press, Oxford (1965)].

[54] H. Alfvén, "Existence of Electromagnetic–Hydrodynamic Waves", *Nature*, 150, 405 (1942).

[55] M. Gedalin, "Linear Waves in Relativistic Anisotropic Magnetohydrodynamics". *Phys. Rev. E* **47**, 4354 (1993).

[56] B.A. Trubnikov, in *Plasma Physics and the Problem of Controlled Thermonuclear Reactions*, vol. 3,ed M.A. Leontovich, Pergamon Press, Oxford (1959), p. 122.

[57] I.P. Shkarofsky, "New Representations of Dielectric Tensor Elements in Magnetized Plasma", *J. Plasma Phys.* **35**, 319 (1986).

[58] M. Brambilla, *Kinetic Theory of Plasma Waves*, Clarendon Press, Oxford (1998).

[59] A.I. Akhiezer and R.V. Polovin, "Theory of Wave Motion of an Electron Plasma", *Sov. Phys. JETP* **3**, 696 (1956).

[60] P. Kaw and J. Dawson, "Relativistic Nonlinear Propagation of Laser Beams in a Cold Overdense Plasmas", *Phys. Fluids* **13**, 472 (1970).

[61] P.K. Kaw and A. Sen, "Nonlinear Propagation of Relativistically Intense Electromagnetic Waves in a Collisionless Plasma", *Pramana - J. Phys.* **48**, 675 (1997).

[62] G. Lehmann, E.W. Laedke and K.H. Spatschek, "Existence and Stability of Nonlinearly Coupled Electromagnetic and Electrostatic Relativistic Wave Solutions during Laser-Plasma-Interaction", *37th EPS Conference on Plasma Physics*, P4.209 (2010).

[63] I.H. Hutchinson, "Electron Holes in Phase Space: What They Are and Why They Matter", *Phys. Plasmas* **24**, 055601 (2017).

[64] I.B. Bernstein, J.M. Greene and M.D. Kruskal, "Exact Nonlinear Plasma Oscillations", *Phys. Rev.* **108**, 546 (1957).

[65] H. Shammel, "Electron Holes, Ion Holes and Double Layers", *Phys. Rep.* **140**, 161 (1986).

[66] B. Eliasson and P.K. Shukla, "Theory of Relativistic Electron Holes in Hot Plasmas", *Phys. Lett. A* **340**, 237 (2005).

[67] H. Shammel and S. Bujarbarua, "Solitary Plasma Hole via Ion-Vortex Distribution", *Phys. Fluids* **23**, 2498 (1980); S. Bujarbarua and H. Shammel, "Theory of finite-amplitude electron and ion holes", *J. Plasma Phys.* **25**, 515 (1981).

[68] R.K. Dodd, J.C. Eilbeck, J.D. Gibbon and H.C. Morris, *Solitons and Nonlinear Wave Equations*, Academic Press, London (1982).

[69] T.Zh. Esirkepov, F.F. Kamenets, S.V. Bulanov and N.M. Naumova, "Low-frequency Relativistic Electromagnetic Solitons in Collisionless Plasmas", *JETP Lett.* **68**, 36 (1998).

[70] D. Farina and S.V. Bulanov, "Relativistic Electromagnetic Solitons in the Electron-Ion Plasma", *Phys. Rev. Lett.* **86**, 5289 (2001).

[71] N. Bloembergen, *Nonlinear Optics*, Benjamin, New Youk (1964).

[72] R.W. Boyd, *Nonlinear Optics*, Academic Press, San Diego (1992).

[73] R.Z. Sagdeev and A.A. Galeev, *Nonlincar Plasma Theory*, W.A. Benjamin, New York (1969).

[74] V.N. Tsytovich, *Theory of Turbulent Plasma*, Springer, New York (2012).

[75] B.B. Kadomtsev, *Plasma Turbulence*, Academic Press, London (1965).

[76] J. Weiland and H. Wilhelmsson, *Coherent Non-Linear Interaction of Waves in Plasmas*, Pergamon Press, Oxford (1977).

[77] J.-M. Wersinger, J.M. Finn and E. Ott, "Bifurcation and "Strange" Behavior in Instability Saturation by Nonlinear Three-Wave Mode Coupling", *Phys. Fluids* **23**, 1142 (1980).

[78] R. Weigand, J.T. Mendonça and H. Crespo, "Cascaded Nondegenerate Four-Wave-Mixing Technique for High-Power Single-Cycle Pulse Synthesis in the Visible and Ultraviolet Ranges", *Phys. Rev. A* **79**, 063838 (2009).

[79] P. B. Corkum, "Plasma Perspective on Strong Field Multiphoton Ionization", *Phys. Rev. Lett.* **71**, 1994 (1993).

[80] S. Bulanov, N. Naumova, and F. Pegoraro, "Interaction of an Ultrashort, Relativistically Strong Laser Pulse with an Overdense Plasma", *Phys. Plasmas* **1**, 745 (1994).

[81] S. Gordienko, A. Pukhov, O. Shorokhov, and T. Baeva, "Relativistic Doppler Effect: Universal Spectra and Zeptosecond Pulses", *Phys. Rev. Lett.* **93**, 115002 (2004).

[82] T. Baeva, S. Gordienko and A. Pukhov, "Interaction of an Ultrashort, Relativistically Strong Laser Pulse with an Overdense Plasma" *Phys. Rev. E* **74**, 046404 (2006).

[83] E. Esarey, Ph. Sprangle, J. Krall and A. Ting, "Self-Focusing and Guiding of Short Laser Pulses in Ionizing Gases and Plasmas", *IEEE J. Quantum Electron.* **33**, 1879 (1997).

[84] A. Alfvén, "On the Motion of Cosmic Rays in Interstellar Space", *Phys. Rev.* **55**, 425 (1939).

[85] D.A. Hammer and N. Rostoker, "Propagation of High Current Relativistic Electron Beams", *Phys. Fluids* **13**, 1831 (1970).

[86] J.D. Lawson, "On the Classification of Electron Streams", J. Nucl. Energy, Part C Plasma Phys, **1**, 31 (1959).

[87] M. Abramovitz and I.A. Stegun, "Handbook of Mathematical Functions", National Bureau of Standards, Washington, 10th ed. (1972).

[88] R.C. Davidson, "Nonlinear Stability Theorem for High-Intensity Charged Particle Beams", *Phys. Rev. Lett.* **81**, 991 (1998).

[89] R. C. Davidson, *Physics of Nonneutral Plasmas*, Addison- Wesley Publishing Co., Reading, MA (1990).

[90] K. Molvig, "Filamentation Instability of a Relativistic Electron Beam", *Phys. Rev. Lett.* **35**, 1504 (1975).

[91] A. Bret, M.C. Firpo and C. Deutsch, "Collective electromagnetic modes for beam-plasma interaction in the whole k space", *Phys. Rev. E* **70**, 046401 (2004).

[92] E.S. Weibel, "Spontaneously Growing Transverse Waves in a Plasma Due to an Anisotropic Velocity Distribution", *Phys. Rev. Lett. 2, 83 (1959*

[93] B.D. Fried, "Mechanism for Instability of Transverse Plasma Waves", *Phys. Fluids* **2**, 237 (1959).

[94] T. Tajima, "Plasma Astrophysics", CRC Press, Boca Raton (2002).

[95] J.T. Mendonça, "Modulational Instabilities in Relativistic Pair Plasmas", *Plasma Phys. Rep.* **42**, 537 (2016).

[96] D. Strickland and G. Mourou, "Compression of amplified chirped optical pulses", *Opt. Commun.* **55**, 447 (1985).

[97] M. Litos et al., "High-Efficiency Acceleration of an Electron Beam in a Plasma Wakefield Accelerator", *Nature* **515**, 92 (2014).

[98] J.B. Rosenzweig, "Trapping, Thermal Effects, and Wave Breaking in the Nonlinear Plasma Wakefield Accelerator", *Phys. Rev. A*, **38**, 3634 (1988).

[99] J.T. Mendonça and A. Serbeto, "Effective charge of intense laser pulses in plasmas", *EuroPhys. J. D* **71**, 221 (2017).

[100] N.A. Krall and A.W. Trivelpiece, *Principles of Plasma Physics*, McGraw-Hill, New York (1973), p. 110.

[101] J.M. Dawson, "Nonlinear Electron Oscillations in a Cold Plasma", *Phys. Rev.* **113**, 383 (1959).

[102] A.I. Akhiezer and P.V. Polovin, "Theory of Wav Motion of an Electron Plasma", *Sov. Phys. JETP* **3**, 696 (1956).

[103] T.P. Coffey, "Breaking of Large Amplitude Plasma Oscillations", *Phys. Fluids* **14**, 1402 (1971).

[104] T. Katsouleas and W. B. Mori, "Wave-Breaking Amplitude of Relativistic Oscillations in a Thermal Plasma", *Phys. Rev. Lett.* **61**, 90 (1988).

[105] Z.M. Sheng and J. Meyer-ter-Wehn, "Relativistic Wave breaking in Warm Plasmas", *Phys. Plasmas* **4**, 493 (1997).

[106] W. Lu et al., "A Nonlinear Theory for Multidimensional Relativistic Plasma Wave Wakefields", *Phys. Plasmas* **13**, 056709 (2006).

[107] A. Golovanov, I. Yu. Kostyukov, A. Pukhov and V. Malka, "Energy-Conserving Theory of the Blowout Regime of Plasma Wakefield", *Phys. Rev. Lett.* **130**, 105001 (2023).

[108] D.L. Andrews and M. Babiker, *The angular momentum of light*, Cambridge University Press, Cambridge (2013).

[109] A.E. Siegman, *Lasers*, University Science Books, Sausalito, CA (1986).

[110] L. Allen, M.W. Beijersbergen, R.J.C. Spreeuw, and J.P. Woerdman, "Orbital Angular Momentum of Light and the Transformation of Laguerre-Gaussian Laser Modes", *Phys. Rev. A* **45**, 8185 (1992).

[111] J.T. Mendonça, B. Thidé, and H. Then, "Stimulated Raman and Brillouin Backscattering of Collimated Beams Carrying Orbital Angular Momentum", *Phys. Rev. Lett.* **102**, 185005 (2009).

[112] L. Rego et al., "Generation of Extreme-Ultraviolet Beams with Time-Varying Orbital Angular Momentum", *Science* **364**, eaaw9486 (2019).

[113] G. Pariente and F. Quéré, "Spatio-temporal light springs: Extended encoding of orbital angular momentum in ultrashort pulses", *Opt. Lett.* **40**, 2037 (2015).

[114] M.N. Rosenblush and C.S. Liu, "Excitation of Plasma Waves by Two Laser Beams", *Phys Rev Lett* **29**, 701 (1972).

[115] L. P. Pitaevskii, "Electric Forces in a Transparent Dispersive Medium", *Sov. Phys. JETP* **12**, 1008 (1961).

[116] A.D. Steiger and C.H. Woods, "Intensity-Dependent Propagation Characteristics of Circularly Polarized High-Power Laser Radiation in a Dense Electron Plasma", *Phys. Rev. A* **5**, 1467 (1972).

[117] J. Deschamps, M. Fitaire, and M. Lagoutte, "Inverse Faraday Effect in a Plasma", *Phys. Rev. Lett.* **25**, 1330 (1970).

[118] Z. Najmudin et al., "Measurements of the Inverse Faraday Effect from Relativistic Laser Interactions with an Underdense Plasma", *Phys. Rev. Lett.* **87**, 215004 (2001).

[119] S. Ali, J.R. Davies and J.T. Mendonça, "Inverse Faraday Effect with Linearly Polarized Laser Pulses", *Phys. Rev. Lett.* **105**, 035001 (2010).

[120] J.T. Mendonça, "Kinetic Description of Electron Plasma Waves with Orbital Angular Momentum", *Phys. Plasmas* **19**, 112113 (2012).

[121] J. Vieira and J.T. Mendonça, "Nonlinear Laser Driven Donut Wakefields for Positron and Electron Acceleration", *Phys. Rev. Lett.* **112**, 215001 (2014).

[122] J. Vieira, J.T. Mendonça and F. Quéré, "Optical Control of the Topology of Laser-Plasma Accelerators", *Phys. Rev. Lett.* **121**, 054801 (2018).

[123] J. Faure et al., "Multi-GeV Electron Beams from Capillary-Discharge-Guided Subpetawatt Laser Pulses in the Self-Trapping Regime", *Nature*, **431**, 541 (2004).

[124] S.P.D. Mangles et al., "Monoenergetic beams of relativistic electrons from intense laser–plasma interactions", *Nature*, **431**, 535 (2004).

[125] C. Geddes et al., "High-quality electron beams from a laser wakefield accelerator using plasma-channel guiding", *Nature*, **431**, 538 (2004).

[126] T. Tajima and J.M. Dawson, "Laser Electron Accelerator", *Phys. Rev.Lett.*, **43**, 267 (1979).

[127] R. Bingham, J.T. Mendonça and P.K. Shukla, "Plasma Based Charged-Particle Acelerators", *Plasma Phys. Control. Fusion*, **64**, R-1 (2004).

[128] G.A. Mourou, T. Tajima and S.V. Bulanov, "Optics in the Relativistic Regime", *Rev. Mod. Phys.*, **78**, 309 (2006).

[129] S.P.D. Mangles et al., "Electron Acceleration in Cavitated Channels Formed by a Petawatt Laser in Low-Density Plasma", *Phys. Rev. Lett.*, **94**, 245001 (2005).

[130] J. T. Mendonca and F. Doveil, "Stochasticity in Plasmas with Electromagnetic Waves", *J. Plasma Phys.* **28**, 485 (1982).

[131] J.T. Mendonça, "Threshold for Electron Heating by Two Eletromagnetic Waves", *Phys. Rev. A*, **28**, 3592 (1983).

[132] Z.-M. Sheng et al., "Stochastic Heating and Acceleration of Electrons in Colliding Laser Fields in Plasma", *Phys. Rev. Lett.*, **88**, 055004 (2002).

[133] L. Chopineau et al., "Identification of Coupling Mechanisms between Ultraintense Laser Light and Dense Plasmas", *Phys. Rev. X*, **9**, 011050 (2019).

[134] J.T. Mendonça, "Laser Wakefield Acceleration in the Petawatt Regime", *Plasma Phys. Control. Fusion* **51**, 024007 (2009).

[135] W. Lu, M. Tzoufras, C. Joshi, et al., "Generating Multi-GeV Electron bunches using Single Stage Laser Wakefield Acceleration in a 3D Nonlinear Regime", *Phys. Rev. ST Accel. Beams* **10**, 061301 (2007).

[136] A. Rousse et al., "Production of a keV X-Ray Beam from Synchrotron Radiation in Relativistic Laser-Plasma Interaction", *Phys. Rev. Lett.* **93**, 135005 (2004).

[137] A.H. Nayfeh and D.T. Mook, *Nonlinear Oscillations*, Wiley and Sons, New York (1979).

[138] A. Pukhov, Z.-M. Sheng and J. Meyer-ter-Vehn. "Particle Acceleration in Relativistic Laser Channels", *Phys. Plasmas* **6**, 2847 (1999).

[139] . A.V. Arefiev, A. P. L. Robinson and V. N. Khudik, "Novel Aspects of Direct Laser Acceleration of Relativistic Electrons", *J. Plasma Phys.* **81**, 475810404 (2015).

[140] C.J. McKinstrie and E.A. Startsev, "Electron Acceleration by a Laser Pulse in a Plasma", *Phys. Rev. E*, **54**, R1070 (1996).

[141] J.T. Mendonça, L.O. Silva and R. Bingham, "Reflection of an Electron Beam by a Photon Mirror", *J. Plasma Phys.*, **73**, 627 (2007).

[142] T. Katsouleas and J.M. Dawson, "Unlimited Electron Acceleration in Laser-Driven Plasma Waves", *Phys. Rec. Lett.* **51**, 846 (1983).

[143] M Moisan and Z. Zakrzewski, "Plasma Sources Based on the Propagation of Electromagnetic Surface Waves", *J. Phys. D: Appl. Phys.* **24**, 1025 (1991).

[144] R. Bingham, R.A. Cairns and J.T. Mendonça, "Particle Acceleration in Plasmas by Perpendicularly Propagating Waves", *J. Plasma Phys.* **64**, 481 (2000).

[145] L. O'C. Drury, "An Introduction to the Theory of Diffusive Shock Acceleration of Energetic Particles in Tenuous Plasmas", *Rep. Prog. Phys.* **46**, 973 (1983).

[146] W.K.H. Panofsky and W.A. Wenzel, "Some considerations concerning the transverse deflection of charged particles in radio-frequency fields", *Rev Sci Instrum* **27**, 967 (1956).

[147] X.Q. Yan et al., "Generating High-Current Monoenergetic Proton Beams by a CircularlyPolarized Laser Pulse in the Phase-StableAcceleration Regime", *Phys. Rev. Lett.* **100**, 135003 (2008).

[148] H. Schwoerer et al., "Laser-Plasma Acceleration of Quasi-Monoenergetic Protons from Microstructured Targets", *Nature*, **439**, 445 (2006).

[149] A.P.L. Robinson et al., "Relativistically Correct Hole-Boring and Ion Acceleration by Circularly Polarized Laser Pulses", *Plasma Phys. Control. Fusion*, **51**, 024004 (2009).

[150] N. Naumova et al., "Hole Boring in a DT Pellet and Fast-Ion Ignition with Ultraintense Laser Pulses", *Phys. Rev. Lett.*, **102**, 025002 (2009).

[151] V.K. Tripati, C.S. Liu, X. Shao, B. Eliasson and R.Z. Sagdeev, "Laser Acceleration of Monoenergetic Protons in a Self-Organized Double Layer from Thin Foil", *Plasma Phys. Control. Fusion*, **51**, 024014 (2009).

[152] T. Ditmire, "Simulation of Exploding Clusters Ionized by High-Intensity Femtosecond Laser Pulses", *Phys. Rev. A* **57**, R4094 (1998).

[153] M. Eloy, R. Azambuja, J.T. Mendonça and R. Bingham, "Interaction of Ultrashort High-Intensity Laser Pulses with Atomic Clusters", *Phys. Plasmas* **8**, 1084 (2001).

[154] G.J. Cao, A. Lindstrom, E. Adli, S. Corde and S. Gessner, "Positron Acceleration in Plasma Wakefields", *Phys. Rev. Accel. Beams* **27**, 034801 (2024).

[155] G. Bekefi, *Radiation Processes in Plasmas*, John Wiley and Sons, New York (1966).

[156] R. Bingham and R.A. Cairns, "Generation of auroral kilometric radiation by electron horseshoe distributions", *Phys. Plasmas* **7**, 3089 (2000).

[157] R. Bingham, R.A. Cairns and B.J. Kellet, "Coherent cyclotron maser radiation from UV Ceti", *A & A* **370**, 1000 (2001).

[158] V.L. Ginzburg and I .M. Frank, "Radiation of a Uniformly Moving Electron due to its Transition from one Medium to Another", *Sov. Phys. JETP* **16**, 15 (1946).

[159] G.M. Garibian, "Contribution to the Theory of Transition Radiation", *Sov. Phys. JETP* **6**, 1079 (1958).

[160] P. Goldsmith and J.V. Jelley, "Optical Transition Radiation from Protons Entering Metal Surfaces", *Philos. Mag.* **4**, 836 (1959).

[161] B. Pardo and J.M. André, "Transition Radiation from Periodic Stratified Structures", *Phys. Rev. A* **40**, 1918 (1989).

[162] B. Dolgoshein, "Transition Radiation Detectors", *Nucl. Instr. and Meth. Phys. Res. A* **326**, 436 (1993).

[163] J. J. Santos, F. Amiranoff, S. D. Baton, et al., "Fast Electron Transport in Ultraintense Laser Pulse Interaction with Solid Targets by Rear-Side Self-Radiation Diagnostics", *Phys. Rev. Lett.* **89**, 025001 (2002).

[164] Jian Zheng, K.A. Tanaka, T. Miyakoshi, Y. Kitagawa, R. Kodama, T. Kurahashi, and T. Yamanaka, "Theoretical Study of Transition Radiation from Hot Electrons Generated in the Laser–Solid Interaction", *Phys. Plasmas* **10**, 2994 (2003).

[165] C. B. Schroeder, E. Esarey, J. van Tilborg and W. P. Leemans, "Theory of Coherent Transition Radiation Generated at a Plasma-Vacuum Interface", *Phys. Rev. E* **69**, 016501 (2004).

[166] K.T. McDonald, *On the History of the Radiation Reaction*, *Princeton University* Princeton NJ (March 18, 2020).

[167] T.G. Blackburn, "Radiation Reaction in Electron–Beam Interactions with High-Intensity Lasers", *Rev. Mod. Plasma Phys.* **4**, 5 (2020).

[168] H.A. Lorentz, "Weiterbildung der Maxwellschen Theorie. Elektronentheorie". *Encykl. Mathe. Wiss.* **14**, 145 (1904).

[169] M. Abraham, *Theorie der Elektrizitaet*, vol. 2 (Teubner, Leipzig, 1905).

[170] P.A.M. Dirac, "Classical Theory of Radiating Electrons", *Proc. Roy. Soc. A* **167**, 148 (1938).

[171] F. Niel, C. Riconda, F. Amiranoff, R. Duclous and M. Grech, "From Quantum to Classical Modelling of Radiation Reaction: A Focus on Stochasticity Effects", *Phys. Rev. E* **97**, 043209 (2018).

[172] J.J. Thomson, "Conduction of Electricity Through Gases", Cambridge University Press, Cambridge, UK (1906).

[173] A.H. Compton, "A Quantum Theory of the Scattering of X-Rays by Light Elements", *Phys. Rev.* **21**, 483 (1923).

[174] O. Klein and Y. Nishina (1929). "Über die Streuung von Strahlung durch freie Elektronen nach der neuen relativistischen Quantendynamik von Dirac". *Z. Phys.* **52**, 853 (1929).

[175] S.Y. Chen, A. Maksimchuk and D. Umstadter, "Experimental Observation of Nonlinear Thomson Scattering", *Nature (London)* **396**, 653 (1998).

[176] E.S. Sarachik and G.T. Schappert, "Classical Theory of the Scattering of Intense Laser Radiation by Free Electrons", *Phys. Rev. D* **1**, 2738 (1970).

[177] E. Esarey, S.K. Ride and P. Sprangle, "Nonlinear Thomson Scattering of Intense Laser Pulses from Beams and Plasmas", *Phys. Rev. E* **48**, 3003 (1993).

[178] Y.Y. Lau, Fei He, D. Umstadter and R. Kowalczyk, "Nonlinear Thomson Scattering: a Tutorial", *Phys. Plasmas* **10**, 2155 (2003).

[179] C.V. Raman, and K.S. Krishnan, "A New Type of Secondary Radiation", *Nature* **121**, 501 (1928).

[180] G. Eckhardt, R.W. Hellwarth, F.J. McClung, S.E. Schwartz, D. Weiner and E.J. Woodbury, "Stimulated Raman Scattering from Organic Liquids", *Phys. Rev. Lett.* **9**, 455 (1962).

[181] R.C. Prince, R.R. Frontiera and E.O Potma, "Stimulated Raman Scattering: From Bulk to Nano", *Chem. Rev.* **117**, 5070 (2017).

[182] K. Nishikawa, "Parametric Excitation of Coupled Waves I. General Formulation", *J. Phys. Soc. Japan* **24**, 916, 1152 (1968).

[183] G. Bekefi, *Radiation Processes in Plasmas*, John Wiley and Sons, New York (1966).

[184] P.H. Yoon, R. Schlickeiser and U. Kolberg, "Thermal fluctuation levels of magnetic and electric fields in unmagnetized plasma: The rigorous relativistic kinetic theory", *Phys. Plasmas* **21**, 032109 (2014).

[185] W.G. Unruh, "Notes on Black Hole Evaporation", *Phys. Rev.* **14**, 870 (1976).

[186] N.D. Birrel and P.C.W. Davies, *Quantum Fields in Curved Space*, Cambridge Univ. Press, Cambridge, UK (1982).

[187] P. Chen and T. Tajima, "Testing Unruh Radiation with Ultraintense Lasers", *Phys. Rev. Lett.* **83**, 256 (1999).

[188] R. Schützhold, G. Schaller and D. Habs, "Signatures of the Unruh Effect from Electrons Accelerated by Ultrastrong Laser Fields" *Phys. Rev. Lett.* **97**, 121302 (2006).

[189] C.W. Misner, K.P. Thorne and J.A. Wheeler, *Gravitation*, Princeton Univ. Press. (2017).

[190] W. Rindler, "Kruskal space and the Uniformly Accelerated Frame, *Am. J. Phys.* **34**, 1174 (1966).

[191] P.M. Alsing and P.W. Milonni, "Simplified Derivation of the Hawking-Unruh Temperature for the Accelerated Observer in Vacuum", *Am. J. Phys.* **72**, 1524 (2004).

[192] G.B. Arfken and H.J. Weber, *Mathematical Methods for Physicists*, 6th ed., Elsevier, Amsterdam (2005).

[193] J.T. Mendonça and A. Serbeto, "Gamma-ray Sources using Imperfect Relativistic Mirrors", *Phys. Plasmas* **15**, 113105 (2008).

[194] J.T. Mendonça, *Theory of Photon Acceleration*, Institute of Physics Publishing, Bristol (2001).

[195] R. Bingham, J. T. Mendonça and J. M. Dawson, "Photon Landau Damping", *Phys. Rev. Lett.* **78**, 247 (1997).

[196] J.T. Mendonça and A. Serbeto, "Effective Charge of Intense Laser Pulses in Plasmas", *Eur. Phys. J. D* **71**, 221 (2017).

[197] J.T. Mendonça and A. Serbeto, "Photon Landau Damping of Electron Plasma Waves with Photon Recoil", *Phys. Plasmas*, **13**, 102109 (2006).

[198] Y.B. Zel'dovich and E.V. Levich, "Bose Condensation and Shock Waves in Photon Spectra", *Sov. Phys. JETP* **28**, 1287 (1969).

[199] J.T. Mendonça and H. Terças, "Bose-Einstein Condensation of Photons in a Plasma", *Phys. Rev. A* **95**, 063611 (2017).

[200] J.L. Figueiredo, J.T. Mendonça and H. Terças, "Bose-Einstein Condensation of Photons in Microcavity Plasmas", *Phys. Rev. E* **108**, L013201 (2023).

[201] M. Born and E. Wolf, *Principles of Optics*, Cambridge University Press, Cambridge, 7th edition (1999).

[202] M. Kline and I.W. Kay, *Electromagnetic Theory and Geometric Optics*, John Wiley, New York (1965).

[203] P.W. Anderson, "Plasmons, Gauge Invariance, and Mass", *Phys. Rev.* **130**, 439 (1963).

[204] C. Ren, R.G. Hemker, R.A. Fonseca, B.J. Duba and W.B. Mori, "Mutual Attraction of Laser Beams in Plasmas: Braided Light", *Phys. Rev. Lett.* **85**, 2124 (2000).

[205] E. Wigner, "On the Quantum Correction For Thermodynamic Equilibrium", *Phys. Rev.*, **40**, 749 (1932).

[206] J.E. Moyal, "Quantum mechanics as a statistical theory", *Proc. Cambridge Phil. Soc.*, **45**, 99 (1949).

[207] W.E. Brittin and W.R. Chappell, "The Wigner Distribution Function and Second Quantization in Phase Space", *Rev. Mod. Phys.*, **34**, 620 (1962).

[208] M. Hillary, R. F. O' Connel, M. O. Scully, and E. P. Wigner, "Distribution Functions in Physics: Fundamentals", *Phys. Rep.*, **106**, 121 (1984).

[209] N. L. Tsintsadze and J. T. Mendonça, "Kinetic Theory of Photons in a Plasma", *Phys. Plasmas*, **5**, 3609 (1998).

[210] J.T. Mendonça and N.L. Tsintsadze, "Analog of the Wigner-Moyal Equation for the Electromagnetic Field", *Phys. Rev. E*, **62**, 4276 (2000).

[211] A.S. Kompaneets, "The Establishment of Thermal Equilibrium between Quanta and Electrons", *Soviet Physics JETP* **4**, 730 (1957).

[212] Landau and Lifshitz, *Electrodynamics of Continuous media*, Pergamon Press, Oxford (1963).

[213] J.T. Mendonça and F. Haas, "Compton Scattering of Plasmons", *Phys. Scripta* **98**, 065603 (2023).

[214] C.F.F. Karney and N.J. Fisch, "Current in Wave Driven Plasmas", *Phys. Fluids* **29**, 180 (1986).

[215] W. Horton, M. Goniche, Y. Peysson, J. Decker, A. Ekedahl and X. Litaudon, "Penetration of lower hybrid current drive waves in tokamaks", *Phys. Plasmas* **20**, 112 508 (2013).

[216] J.T. Mendonça, W. Horton, R.M.O. Galvão and Y. Elskens, "Transport Equations for Lower Hybrid Waves in a Turbulent Plasma", *J. Plasma Phys.* **81** 905810206 (2015).

[217] J.T. Mendonça and K. Hizanidis, "Improved model of quasi-particle turbulence (with applications to Alfvén and drift wave turbulence)", *Phys. Plasmas* **18**,112306 (2011).

[218] L. Pitaevskii and S. Stringari, *Bose-Einstein Condensation*, Clarendon Press, Oxford (2003).

[219] P.H. Diamond, S.-I. Itoh, K. Itoh, T.S. Hahm, "Zonal flows in plasma - A review", Plasma Phys. Control. Fusion 47, R35 (2005)

[220] R. Trines, R. Bingham, L.O. Silva, J.T. Mendonça, P.K. Shukla and W.B. Mori, "Quasiparticle Approach to the Modulational Instability of Drift Waves Coupling to Zonal Flows", *Phys. Rev. Lett.* **94**, 165002 (2005).

[221] G.G. Raffelt, *Stars as Laboratories for Fundamental Physics*, University of Chicago Press, Chicago (1996).

[222] M. Marklund and P.K. Shukla, "Nonlinear Collective Effects in Photon–Photon and Photon–Plasma Interactions", *Rev. Mod. Phys.* **78**, 591 (2006).

[223] A. Di Piazza, C. Müller, K.Z. Hatsagortsyan and C.H. Keitel, "Extremely High-Intensity Laser Interactions with Fundamental Quantum Systems", *Rev. Mod. Phys.* **84**, 1177 (2012).

[224] F. Haas, *Quantum Plasmas: An Hydrodynamic Approach*, Springer, New York (2011).

[225] D.M. Wolkow, "Über eine Klasse von Lösungen der Diracschen Gleichung", slZ. Phys. **94**, 250 (1935).

[226] J.T. Mendonça and A. Serbeto, "Volkov solutions for relativistic quantum plasmas", *Phys. Rev. E* **83**, 026406 (2011).

[227] E. Raicher and S. Eliezer, "Analytical solutions of the Dirac and the Klein-Gordon equations in plasma induced by high-intensity laser", *Phys.Rev. A* **88**, 022113 (2015).

[228] F. Haas, "Quantum Weibel Instability", *Phys. Plasmas* **15**, 022104 (2008).

[229] L.N. Tsintsadze and P.K. Shukla, "Weibel instabilities in dense quantum plasmas", *J. Plasma Phys.* **74**, 431 (2008).

[230] J.T. Mendonça, "Wave kinetics of relativistic quantum plasmas", *Phys. Plasmas* **18**, 062101 (2011).

[231] F. Haas, G. Manfredi, and M. R. Feix, "Multistream Model for Quantum Plasmas", *Phys. Rev. E* **62**, 2763 (2000).

[232] J.T. Mendonça and G. Brodin, "Weibel Instability in Relativistic Quantum Plasmas", *Phys. Scripta* **90**, 088003 (2015).

[233] J.M.J. Madey, "Stimulated Emission of Bremsstrahlung in a Periodic Magnetic Field", *J. Appl. Phys.* **42**, 1906 (1971).

[234] R. Bonifacio, N. Piovella, and G. R. M. Robb, Nucl. Instrum. Methods Phys. Res. A 543, 645 2005.

[235] A. Serbeto, J.T. Mendonça, K.H. Tsui and R. Bonifacio, "Quantum wave kinetics of high-gain free-electron lasers", *Phys. Plasmas* **15**, 013110 (2008).

[236] G. Preparata, "Quantum Field Theory of the Free-Electron Laser", *Phys. Rev. A* **38**, 233 (1988).

[237] M. Hillary, R.F. O'Connell, M.O. Scully, E.P. Wigner, "Distribution Functions in Physics: Fundamentals", *Phys. Rep.* **106**, 121 (1984).

[238] J.T. Mendonça and A. Serbeto, "Twisted Volkov States in a Plasma", *EuroPhys. Lett.* **125**, 65002 (2019).

[239] A.H. Compton, "A Quantum Theory of the Scattering of X-Rays by Light Elements", *Phys. Rev* **21**, 483 (1923).

[240] C. Itzykson and J.-B. Zuber, *Quantum Field Theory*, McGraw-Hill, New York (1980).

[241] V. P. Silin, "Nonlinear High-Frequency Plasma Conductivity", *Sov. Phys. JETP* **20**, 1510 (1965).

[242] J.F. Seely and E.G. Harris, "Heating of a Plasma by Multiphoton Inverse Bremsstrahlung", *Phys. Rev. A* **7**, 1064 (1973).

[243] P. Mora, "Theoretical Model of Absorption of Laser Light by a Plasma", *Phys. Fluids* **25**, 1051 (1982).

[244] A. Gonoskov, T.G. Blackburn, M. Marklund, and S.S Bulanov, "Charged Particle Motion and Radiation in Strong Electromagnetic Fields", *Rev. Mod. Phys.* **94**, 045001 (2022).

[245] J.W. Yoon, Y.G. Kim, I.W. Choi, J.H. Sung, H.W. Lee, S.K. Lee and C.H. Nam, "Realization of Laser Intensity over $10^{23}\, W/cm^2$", *Optica* **8**, 630 (2021).

[246] A. Fedotov, A. Ilderton, F. Karbstein, B. King, D. Seipt, H. Taya and G. Torgrimsson, "Advances in QED with Intense Background Fields", *Phys. Rep.* **1010**, 1 (2023).

[247] J.T. Mendonça, *The Quantum Nature of Light: From Photon States to Quantum Fluids of Light*, Institute of Physics Publishing, Bristol (2022).

[248] H. Suzuki, "Neutrinos from Core-Collapse Supernova Explosions", *Prog. Theor. Exp. Phys.*, 05B101 (2024).

[249] W. Busza, K. Rajagopal and W. van der Schee, "Heavy Ion Collisions: The Big Picture and the Big Questions", *Annu. Rev. Nucl. Part. Sci.* **68**, 339 (2018).

[250] H. Terças, J.T. Mendonça and R. Bingham, "Resonant Axion-Plasmon Conversion in Neutron Star Magnetospheres", *Phys. Rev. Lett.* **135**, 111001 (2025).

[251] S. Schmidt, D. Blaschke, G. Röpke, S.A. Smolyansky, A.V. Prozorkevich and V.D. Toneev, "A Quantum Kinetic Equation for Particle Production in the Schwinger Mechanism", *Int. J. Mod. Phys. E*, **7**, 709 (1998).

[252] Z. Bialynicka-Birula and I. Bialynicki-Birula, "Nonlinear Effects in Quantum Electrodynamics. Photon Propagation and Photon Splitting in an External Field", *Phys. Rev. D* **2**, 2341 (1970).

[253] H. Al-Naseri, J. Zamanian and G. Brodin, "Plasma Dynamics and Vacuum Pair Creation using the Dirac-Heisenberg-Wigner Formalism", *Phys. Rev. E* **104**, 015207 (2021).

[254] F. Hebenstreit, R. Alkofer and H. Gies, "Schwinger Pair Production in Space- and Time-dependent Electric Fields:Relating the Wigner Formalism to Quantum Kinetic Theory", *Phys, Rev. D* **82**, 105026 (2010).

# Index

For Product Safety Concerns and Information please contact our EU
representative GPSR@taylorandfrancis.com
Taylor & Francis Verlag GmbH, Kaufingerstraße 24, 80331 München, Germany

www.ingramcontent.com/pod-product-compliance
Ingram Content Group UK Ltd.
Pitfield, Milton Keynes, MK11 3LW, UK
UKHW022215190926
475702UK00003B/375
* 9 7 8 1 0 3 2 8 4 3 6 7 4 *